Also in the Variorum Collected Studies Series:

DAVID A. KING
Astronomy in the Service of Islam

CHARLES BURNETT
Numerals and Arithmetic in the Middle Ages

JULIO SAMSÓ
Astronomy and Astrology in al-Andalus and the Maghrib

J.L. MANCHA
Studies in Medieval Astronomy and Optics

MENSO FOLKERTS
The Development of Mathematics in Medieval Europe
The Arabs, Euclid, Regiomontanus

PAUL KUNITZSCH
Stars and Numbers
Astronomy and Mathematics in the Medieval Arab and Western Worlds

RAYMOND MERCIER
Studies on the Transmission of Medieval Mathematical Astronomy

GERARD L'E. TURNER
Renaissance Astrolabes and their Makers

BRUCE S. EASTWOOD
The Revival of Planetary Astronomy in Carolingian and Post-Carolingian Europe

DONALD R. HILL
Studies in Medieval Islamic Technology
From Philo to al-Jazari – from Alexandria to Diyar Bakr

EDWARD S. KENNEDY
Astronomy and Astrology in the Medieval Islamic World

EMMANUEL POULLE
Astronomie planétaire au Moyen Âge latin

RICHARD LORCH
Arabic Mathematical Sciences
Instruments, Texts and Transmission

VARIORUM COLLECTED STUDIES SERIES

Astrolabes from Medieval Europe

For Adrian

David A. King

Astrolabes from Medieval Europe

Routledge
Taylor & Francis Group
LONDON AND NEW YORK

First published 2011 by Ashgate Publishing

Published 2016 by Routledge
2 Park Square, Milton Park, Abingdon, Oxon OX14 4RN
711 Third Avenue, New York, NY 10017, USA

Routledge is an imprint of the Taylor & Francis Group, an informa business

ISBN 9781409425939 (hbk)

British Library Cataloguing in Publication Data
King, David A.
Astrolabes from medieval Europe.
– (Variorum collected studies series)
1. Astrolabes – Europe – History – To 1500. 2. Astronomy, Medieval – Europe.
I. Title II. Series
522.4'094'0902–dc22

Library of Congress Control Number: 2010940591

VARIORUM COLLECTED STUDIES SERIES CS977

CONTENTS

This volume contains xvi + 404 pages

Publisher's Note

The articles in this volume, as in all others in the Variorum Collected Studies Series, have not been given a new, continuous pagination. In order to avoid confusion, and to facilitate their use where these same studies have been referred to elsewhere, the original pagination has been maintained wherever possible.

Each article has been given a Roman number in order of appearance, as listed in the Contents. This number is repeated on each page and is quoted in the index entries.

PREFACE*

Medieval astronomical instruments constitute a substantial corpus of mainly untapped historical sources. They are at once objects of importance for the history of science and of technology as well as for the history of art and of material objects. Whilst the astrolabe is the most common instrument, the celestial globes, quadrants, sundials and horary devices should not be overlooked.

Most medieval European instruments are neither signed nor dated. This is one reason why they have not received the attention due to them: signed and dated instruments have more value in the auction world and are easier to associate with a geographical and temporal milieu. As far as astrolabes are concerned, there are usually several factors – such as rete design, star positions, geographical information on the plates, saints' names on calendrical scales, month-names in regional varieties of medieval Latin or even in vernaculars – that can be used to assign a given piece to a particular milieu. Sometimes, even these indicators are insufficient to assign a provenance, though it must be stressed that the study of these instruments is still in its initial stages. In this volume we shall mainly encounter astrolabes that are unsigned and undated, but this in no way lessens their historical interest.

The 1990s saw a great deal of activity at the Institute for the History of Science at Frankfurt University relating to medieval astronomical instruments, generously funded by the German Research Council (DFG). I prepared basic descriptions of a substantial number of instruments, Islamic and European, predating *ca.* 1500. However, whereas with Islamic instruments one can usually assume that the star-names are correct and the star-positions reasonable, this is by no means the case with European instruments: each piece presents problems of its own that go beyond the limits of a descriptive catalogue. During the research, I was constantly confronted with the uncertain future of the Institute (eventually closed when I retired in 2007); the lack of continuous large-scale funding; the lack of continuity of auxiliary personnel; and not least the difficulty of obtaining adequate photographic material (in those days, inevitably, always in the form of black and white prints). More to the point, it was the discovery or, in some cases, new understanding, of a series of very

* The following abbreviations refer to three recent books of mine and the leading journal in the history of Islamic science:

The Ciphers of the Monks ... , Stuttgart, 2001

In Synchrony with the Heavens ... , 2 vols., Leiden, 2004-05

Astrolabes and Angels, Epigrams and Enigmas ... , Stuttgart, 2007

Suhayl – Journal for the History of the Exact and Natural Sciences in Islamic Civilisation (Barcelona), available on the Internet (usd.proves.ub.edu/suhayl/)

special 'landmark' instruments that demanded, so I thought at the time, immediate attention.[1] It was the last of these factors that led to most of the papers collected in this volume.

I 'Astronomical instruments between East and West'. This is an overview of the field and the attendant surprises awaiting investigators. It was prepared for a conference held at the Institut für Realienkunde des Mittelalters und der frühen Neuzeit at Krems, Austria, for a group of Medievalists, Byzantinists and Islamicists. These colleagues were already converted to the realization that material objects (*Realien*) merit their rightful place as historical sources. (Most of our colleagues tend to treat only textual sources or artistic objects as valid historical sources.) The paper concludes with the first overviews of all relevant instruments in Austrian collections and all instruments known from 15th-century Vienna, then the leading school of astronomy in Europe. On some instruments from this school see also XI below.

Most of the remaining studies deal with individual instruments. I would have liked to include here a detailed description of a very remarkable astrolabe from 14th-century Spain with inscriptions in Hebrew, Latin and Arabic. Alas, this would have filled half of a standard Variorum volume and so it could not be reprinted here. In any case, the study is available on the Internet.[2] The star-names on that piece are analyzed in VII.

II 'The earliest known European astrolabe in the light of other early astrolabes'. This is an attempt to still some of the controversy around the 'Destombes astrolabe', otherwise known as the 'Carolingian astrolabe'. At the 29th International Congress of the History of Science held at Saragossa in 1993, a whole session was devoted to this remarkable piece.[3] The resulting publication contains a basic description of it. There were still some colleagues – albeit none with any experience with medieval instruments – tending to think that it was a fake. In my paper, I consider all of the earliest known Islamic astrolabes as well as all of the earliest known European astrolabes. I argue that the available evidence points to an origin for this piece in 10th-century Catalonia. A study presented by Anscari M. Mundó demonstrated that the distinctive forms of the letters in the engraving were attested in, and only in,

---

[1] These 'landmark' astrolabes are identified and discussed in *Astrolabes and Angels, Epigrams and Enigmas*, pp. 202–19.

[2] See 'An astrolabe from 14th-century Christian Spain with inscriptions in Latin, Hebrew and Arabic', *Suhayl* 3 (2002/03), pp. 9–156, also reprinted in *In Synchrony with the Heavens*, XV.

[3] See Wesley Stevens, Guy Beaujouan & Anthony J. Turner, eds, *The Oldest Latin Astrolabe*, a special issue of *Physis – Rivista di storia della scienza* (Rome), published as Nuova serie 32:2–3 (1995).

10th-century Catalonia.[4] I also outline the findings of Kurt Maier in this regard, namely, that even one of the additional markings – the name of one zodiacal sign added in the 13th or 14th century – is in the Catalan form of medieval Latin. It is a sad fact of astrolabic life that once an instrument has been deemed suspicious by people who do not understand it, the piece is essentially doomed forever and it is impossible to reinstate it to its rightful place in history.

III 'Rewriting history through instruments: the secrets of a medieval astrolabe from Picardy'. This astrolabe, from 14th-century Picardy, bears a later inscription dated 1522 indicating that it was presented by the Benedictine monk Paschasius Berselius of Liège to his teacher of Greek in Louvain, Hadrianus Amerotius, both men friends of Erasmus. Otherwise, all of the numbers engraved on the 'Berselius astrolabe' are in a notation that has slipped through the cracks of the history of number-notations. These ciphers, of which the simplest form serving numbers from 1 to 99 was introduced in England from Greece in the 13th century, were developed by Cistercian monks in what are now the border regions of France and Belgium into a more viable system serving numbers from 1 to 9999, such as is used on this remarkable astrolabe. These ingenious Cistercian ciphers are attested in one form or another in manuscript and early printed sources from England to Italy and from Spain to Sweden. I have published elsewhere a detailed account of all attestations of this notation either for numbers or for alphabetic codes, following them from their origins as a stenographic code in ancient Greece to their fate and fortunes in Europe from the 13th to the 20th century.[5] This earlier article on the astrolabe appeared in a *Festschrift* for Prof. Gerard L'E. Turner of Oxford. See also VII below on the star-names on this astrolabe.

IV 'The medieval Catalan astrolabe of the Society of Antiquaries, London'. It was Kurt Maier who established the Catalan provenance of this charming little astrolabe. Our joint study of this piece attempts to consider all the astronomical, artistic and linguistic features of this piece. (Technological aspects are not considered in any Frankfurt studies.) Our joint article was published in a *Festschrift* for Prof. Juan Vernet of Barcelona.

The rete design on this piece is original, in the sense that it was clearly developed for it alone or for a batch of similar pieces. A derivative rete, obviously inspired by this design, but less competently fashioned, is found on a remarkable Andalusî astrolabe with inscriptions in Judaeo-Arabic, that is, Arabic written in Hebrew script. This piece is now published in detail.[6]

---

[4] See pp. 303–21 of the *Physis* publication and the accompanying 9 figures.

[5] See now *The Ciphers of the Monks* on this notation either for numbers or for alphabetic codes.

[6] On that instrument see now Mohamed Abu Zayed, David A. King & Petra G. Schmidl, 'From a heavenly Arabic poem to an enigmatic Judaeo-Arabic astrolabe', *Suhayl* 10 (2011).

V 'A remarkable Italian astrolabe from *ca.* 1300'. My analysis of the unusual miniature astrolabe in the Museum of the History of Science at Oxford raises as many problems as it solves. I argue that its inspiration is an Islamic type no longer extant. This paper appeared in a *Festschrift* for Dr Mara Miniati of Florence. When I published it, I did not know that an even smaller astrolabe from the same workshop is illustrated in the painting *St Jerome in his Study* associated with Jan van Eyck *ca.* 1425 (preserved in the Detroit Institute of Art).[7]

VI 'An astrolabe from Einbeck datable *ca.* 1330'. Few medieval astrolabes are signed or dated but this one bears an inscription mentioning an individual and his post: this information is sufficient to provide a rough date for its construction. It is possible that this piece derives from a workshop in Erfurt, about which nothing is otherwise known. Although it is the earliest surviving astrolabe from the German-speaking world, it exhibits features that are clearly developed from well-established traditions of instrument-making. Ideally, the vast majority of medieval European astrolabes should be subjected to this kind of analysis. This paper was published in a *Festschrift* for Prof. Menso Folkerts of Munich.

VII 'The stars-names on three 14th-century astrolabes from Spain, France and Italy'. The star-names on medieval European instruments are often full of surprises, even when one has at hand the writings of Paul Kunitzsch on astrolabe-star names in the manuscript tradition. The three instruments investigated here – see III and X on two of them – show what a mess we are dealing with. This paper was published in the *Festschrift* for Prof. Kunitzsch.

VIII 'A *vetustissimus* Arabic treatise on the *quadrans vetus*'. No medieval instrument has been so misunderstood in the modern literature as the universal horary quadrant, whose markings often appear on the backs of European astrolabes. The Islamic origin of the so-called *quadrans vetus*, both with and without a movable cursor for a solar-calendrical scale, is here established by the publication of an Arabic text from 9th- or 10th-century Baghdad. The markings, even though they are approximate, work extremely well in the latitudes where they were conceived and first used, but not in the latitudes of Northern Europe, where they were not understood but were used with abandon. In the early text on the quadrant, the markings are superposed on those of a shadow-square, also first attested in 9th-century Baghdad.[8] This paper is dedicated to the memory of Prof. Frans Bruin, formerly of the American University of Beirut.

---

[7] See now *In Synchrony with the Heavens*, XIIId–2.9.

[8] See now also *In Synchrony with the Heavens*, XIIa, in which the fate and fortune of this device is documented in Islamic and European sources.

IX: '14th-century England or 9th-century Baghdad? New insights on the origins of the elusive astronomical instrument called the *Navicula de Venetiis*'. Another medieval instrument that has had a chequered career in the modern literature is the *navicula*, a universal horary device for *timekeeping by the sun* in the form of a ship. The main evidence for my suspecting an Islamic origin to the underlying principle, a *proto-navicula*, is the fact that we have a related but more complicated universal instrument for *timekeeping by the stars* from 9th-century Baghdad.[9] Both universal instruments feature a movable, *i.e.*, mechanical, component for feeding in the latitude. (Tables and instruments that are universal – that is, serving all terrestrial latitudes – were favoured in Islamic astronomy, and I have dealt with them elsewhere.[10]) In that milieu, Baghdad *ca.* 900, there was an instrument called *lawh al-sâ'ât*, '(rectangular) board (for finding) the hours' (or 'Uhrtäfelchen'!) in circulation, an instrument on which we alas have no further information. Again, a remarkable and highly sophisticated plate for finding the hours (probably called *safîhat al-sâ'ât*) for the latitude of Baghdad, dating from *ca.* 900, has recently come to light.[11] Last, but not least, the universal horary quadrant that is on the back of the *navicula* is combined with markings for a shadow-square, a distinctive combination first attested in 9th-century Baghdad (see VIII).[12] All this having been said, the medieval European manuscripts relating to the construction and use of the *navicula*, now available for the first time, contain not a trace of any Islamic influence.[13] My 2003 paper appeared in a *Festschrift* for my teacher, Prof. Bernard R. Goldstein of Pittsburg.

X 'The astrolabe depicted in the intarsia of the *studiolo* of Archduke Federico in Urbino'. The astrolabe depicted in great detail in the *studiolo* of the Archducal Palace in Urbino *ca.* 1476 owes its inspiration to a class of Italian instruments with identical rete design, of which we have two examples. The first is medieval and the second was made in Urbino in 1462. See now VII above for more information on the star-names on these two astrolabes.[14] It is a curious coincidence that this astrolabe bears the same date as the one presented to Cardinal Bessarion by Regiomontanus (see XI).

---

9 See François Charette and Petra G. Schmidl, 'A universal plate for timekeeping by the stars by Habash al-Hâsib: Text, translation and preliminary commentary', *Suhayl* 2 (2001), pp. 107–59.

10 *In Synchrony with the Heavens*, VI.

11 See my study 'An instrument of mass calculation made by Nastûlus in Baghdad *ca.* 900', *Suhayl* 8 (2008), pp. 93–119.

12 See also *In Synchrony with the Heavens*, XIIb, in which the *navicula* is treated within the context of other universal horary dials.

13 See the valuable new study: Catherine Eagleton, *Monks, Manuscripts and Sundials – The Navicula in Medieval England*, Leiden & Boston MA: Brill, 2010, which presents and surveys all the known manuscript evidence relating to the *navicula*, mostly previously unpublished (my review is to appear in *Speculum – A journal of medieval studies*).

14 On the 1462 Urbino astrolabe see now also *Astrolabes and Angels, Epigrams and Enigmas*, pp. 275–84.

XI 'The astrolabe presented by Regiomontanus to Cardinal Bessarion in 1462'. This describes the most important and the most historically interesting instrument of the Renaissance, dedicated in 1462 by the young German astronomer Regiomontanus to his new patron, the Greek Cardinal Basileios Bessarion. The material context of this piece is provided by 10 other astrolabes with the same or related rete design from 15th-century Vienna (see also App. A to study I above).

In 1989, when this piece came up for auction in London, none of these other Viennese astrolabes was known in England. In this 1994 study, the 1462 astrolabe was reappraised in the light of all of these. Three questions remained: why did Regiomontanus depict an angel on the back of his instrument? What messages lie hidden in his enigmatic dedication? And why did he engrave his dedication, an elegaic couplet, in such a curious way? In 1994 I had no clue to the answers to any of these questions.

It emerges that the epigram is an acrostic with eight 'concealed' vertical axes. These conceal information that would have been of interest to Bessarion and Regiomontanus. In particular, the new astrolabe was made to 'celebrate' the 400th anniversary of an astrolabe made in 1062. This Byzantine astrolabe, which happily survives in Brescia, must therefore have been shown by Bessarion to his new protégé. It is immediately clear that the inscription on the Byzantine astrolabe inspired Regiomontanus' epigram. Furthermore, the young German angel depicted on the new astrolabe seems to be intended to represent the 5th-century Egyptian St Bessarion, who was venerated as an angel in the Byzantine liturgy and whose name was adopted by the young monk Basileios. The '*divus* Bessarion' referred to in the dedication would thus represent both the angelic saint and the Cardinal.

In addition to all this, groups of letters from left to right across the epigram, divided into eight sections, suggest in succession identities of five persons involved in the Passion and Flagellation of Christ and the downfall of the Byzantine Empire at the hands of the Turks, and of three others involved in the life of Bessarion. I maintain that these configurations somehow inspired the most enigmatic painting of the Renaissance, Piero della Francesca's 'Flagellation of Christ', with its five individuals in a Flagellation scene in the background on the left and three 'moderns' in the foreground on the right. Over 40 interpretations of this painting – mainly involving the identities of the three 'moderns' – have been ventured over the past 150 years, all without any textual foundation. My hypothesis yields double or multiple identities for each of the eight figures and classical god (statue atop the column behind Christ) in a reconstructed painting that corresponds very closely to that of Piero. It leads to the identification of the angelic blond-haired young man next to the Bessarion figure in the painting as, amongst others, the young astronomer Regiomontanus, who with his epigram and angel made the whole operation possible, and who here, as protégé of a cardinal, wears a tunic in cardinal red. His rather angelic image also represents three young men in Bessarion's closest circle who had recently died; even the Marian rosettes around his image – such as are also found on Regiomontanus' astrolabe – correspond to the letters ROSAE in the appropriate position in the epigram.[15] I believe that the whole business started when Bessarion showed Regiomontanus the Byzantine astrolabe of 1062 and helped him to read its Greek inscription.

XII Failing a catalogue of the entire corpus of medieval astronomical instruments, what needs to be done by future investigators is to prepare comparative studies of related groups of instruments. To facilitate such undertakings, I present here a list of European astrolabes from before *ca.* 1500, organized chronologically by provenance. This was prepared over 15 years ago and in the main has not been updated.[16]

DAVID A. KING

Frankfurt am Main
January 2011

---

15 The whole story is in *Astrolabes and Angels, Epigrams and Enigmas*, with a shorter version at: web.uni-frankfurt.de/fb13/ign/Code.htm (still available in 2010).

16 See also *In Synchrony with the Heavens*, XVIII, for a list of Islamic instruments up to *ca.* 1500 (labelled Parts 1–5).

ACKNOWLEDGEMENTS

I am grateful to my co-authors Kurt Maier (IV) and Gerard L'E. Turner (XI) for permission to publish again our joint articles.

For permission to reprint the various articles in this volume I thank the Österreichische Akademie der Wissenschaften, Vienna (I); Leo S. Olschki, Florence (II, V and XI); Prof. Julio Samsó, editor of *Suhayl – Journal for the History of the Exact and Natural Sciences in Islamic Civilisation*, Barcelona (IV); Deutsche Akademie der Naturforscher Leopoldina, Halle (VI); Otto Harrassowitz, Wiesbaden (VII); Science History Publications, Cambridge (VIII); Wiley-Blackwell, Oxford (IX); and the Accademia Raffaello, Urbino (X).

Above all, I am grateful to John Smedley of Ashgate Publishing Ltd for his tolerance and patience, and also to Lindsay Farthing for her editorial help.

I

ASTRONOMICAL INSTRUMENTS BETWEEN EAST AND WEST*

The astronomical instruments of the Middle Ages offer exciting new possibilities for the history of material culture as well as for the history of many other areas of human activity. They constitute a prime example of the transmission of material culture and the underlying ideas: in the main the various kinds of instruments were Greek in origin, developed subsequently by the Muslims, and then slowly but surely transmitted to or reinvented in Europe. Apparently, only the simplest forms of these instruments were transmitted from the early Byzantine milieu to the Muslims, and only the simplest and earliest forms of Islamic instruments were transmitted to the Latin West. But each category of instruments has a different history, which cannot be written solely from the available manuscript sources. We do well to let the surviving instruments speak for themselves. Many of them have individual histories which contribute to the overall picture.

---

* Acknowledgements: To Prof. Dr. Harry Kühnel and his colleagues in Krems my heartiest thanks for their kind invitation and generous hospitality during the conference. The invitation served a double purpose; it also enabled me to visit various collections in Austria and the South Tyrol during the week after the conference. My thanks go also to Dr. Meinrad Pizzinini of the Zeughausmuseum in Innsbruck; Dipl.-Ing. Karl Schwarzinger of Innsbruck, author of a survey of sundials in Austria; Dr. Alfred Auer of the Sammlungen Schloß Ambras near Innsbruck; Kustos Michl Rabensteuner of Schloß Prösels near Völs; Pater Paulus Kaimbacher O.S.B. and Mag. Eduard Pototschnig of the Stiftsgymnasium in St. Paul im Lavanttal; Dr. Eva Marko of the Landesmuseum Joanneum, Graz; Dr. Joseph Braunbeck of the Technisches Museum in Vienna; Dr. Franz Scharinger of the Uhrenmuseum in Vienna; and Dr. Rudolf Distelberger of the Kunsthistorisches Museum in Vienna. For kindnesses during previous visits to Austria I am grateful to Dr. Christa Svoboda of the Museum Carolino-Augusteum in Salzburg and Dr. Hermann Mucke of the Astronomisches Büro in Vienna.

I. The State of the Field

Over one thousand astronomical instruments of one sort or another, including globes, astrolabes, quadrants and sundials, survive from the Middle Ages, here generously defined as lasting until 1900 for the Islamic world and to ca. 1550, and occasionally even thereafter, for Europe.[1] An eighteenth-century Maghribi or Persian astrolabe is as "medieval" as one made in tenth-century Baghdad, and a German astrolabe with Arabic star-names or a Russian horary quadrant, both from the eighteenth century, also deserve our consideration as "medieval" instruments. In Europe generally after 1550 a new tradition of instrument-making developed, as attested by the instruments of, say, Humphrey Cole, Erasmus Habermel or the Arsenius brothers, and which included the telescope, but this later tradition lies outside the scope of my study.

The available corpus of medieval instruments represents only a small fraction of those actually made; nevertheless it bears witness to the scientific knowledge and interests of the divers societies in which the instruments were produced. They are often objects of beauty and elegance and reflect contemporaneous local artistic developments. Isolated instruments sometimes tell us just how little we know about the astronomical and artistic traditions from which they evolved. Thus the only surviving Byzantine astrolabe (#2), dated 1062 and made in Asia Minor, is more sophisticated than any Byzantine text on the astrolabe would lead us to believe, and it shows no trace whatsoever of Islamic influence. A spectacular astrolabe made by a leading Muslim astronomer of the tenth century (#111) bears a quatrefoil as decoration, a feature generally thought to be of European origin. The design influenced several later Islamic astrolabes (#3, #5, #2505, #2557 and #2708) and reappears on European astrolabes in the thirteenth century,

[1] The instruments mentioned in this paper are referred to in the text by numbers. Details of the instruments are given at the end of the paper. Each medieval instrument has been assigned a number, which I use with the American symbol #. Instruments with numbers up to about 300 are featured in Gunther, *Astrolabes*; those with numbers up to 4000 (not consecutively numbered!) are listed in *International Checklist* (currently being updated by Leonard Linton). Numbers between 4000 and 9999 have been assigned by the present writer (additional astrolabes 4001-4999, quadrants 5001-5999, globes 6001-6999, sundials 7001-7999, miscellaneous 8001-8999, and fakes 9001-9999). All of the instruments discussed in this paper have been described in the forthcoming catalogue - see note 16.

if not before, as well as thereafter (for example, #162). The earliest surviving European astrolabe (#3042), probably from the tenth century, contains enough information to force us to rewrite the history of the transmission of the instrument to Europe. An astrolabe with geared mechanism from fourteenth-century France (#198) has components which show that the earliest European astronomical clocks were inspired by the astrolabe. A miniature astrolabe (#169), probably from fourteenth-century Italy, attests to a forgotten tradition of making universal astrolabes of a kind not known from any texts.

The purpose of my paper is to present yet more examples of instruments which force a re-evaluation of many issues. It is inevitably rather provocative, and if it arouses interest or controversy I shall be delighted. I know that I am working in a gold-mine, and I shall be happy to share the spoils. There is unexploited material here for scholars from a multitude of disciplines. For the epigrapher there are inscriptions on the instruments in Greek, Arabic, Coptic, Judaeo-Arabic, Hebrew, Persian, Turkish, Armenian, Georgian, Sanskrit, Arabic transliterated in Gothic script with all manner of perversions, Latin of various degrees of sophistication, and old Catalan, French, German and Flemish. There is material for the historian of astronomy, mathematics, geography, technology and art. People possessed astrolabes in the Middle Ages, even if they did not or could not use them. Who made them? How and where were they made? Who owned them? How and when were they used? Were they ever used? We are far from being able to answer these questions, and I shall not deal with them here. It is better to begin with easy questions.

II. WHAT IS AN ASTROLABE?

By now this question will have occurred to some readers. To answer it one has to first consider the celestial globe.[2] The heavens are represented on a sphere which rests in a horizontal frame representing the horizon and is supported by a vertical frame representing the meridian. The axis of the sphere can be aligned so that the configuration corresponds to a specific terrestrial latitude. Rotating the sphere about

---

[2] In this paper I shall not deal with globes. Some 125 Islamic globes have been catalogued in SAVAGE-SMITH, *Islamicate Globes*, and a survey of early European globes is currently being conducted by Dr. ELLY DEKKER (see, for example, her contribution to the *Nuremberg 1992–93 Catalogue*).

its axis shows how the sun and stars rise and set over the horizon and how they culminate on the meridian. But with a standard globe one cannot determine the altitude of the sun and stars above the horizon or measure the passage of time; besides, the necessary additional components would merely clutter the globe. In fact, one needs a handier instrument altogether. There is a story in the Arabic sources that Ptolemy was riding on a donkey with a globe in his hand; he dropped it, the beast trod on it, and the result was an astrolabe.

The astrolabe is a two-dimensional model of the universe that one can hold in one's hands; its heavenly features include a star-map and the ecliptic (called the rete), and its terrestrial features (engraved on various plates) serve specific latitudes or localities.[3] The principal function of the astrolabe is to provide an image of the sky as it would be seen in a given location at any specific time. One determines the altitude of the sun or a given star with the sighting device on the back of the instrument and then sets the mark for the sun on the ecliptic or the selected star-pointer on the rete over the appropriate altitude mark on the plate. The instrument then displays the instantaneous configuration of the heavens with respect to the horizon and meridian of the observer. The astrolabe has obvious applications in timekeeping because rotating the rete over the plate simulates the apparent motion of the heavens about the observer, by which the passage of time can be measured. It also has applications in astrology because one can see at a glance the instantaneous configuration of the ecliptic relative to the local horizon and meridian, by which the astrological houses are defined. Its simplest use is in surveying, for calculating the height of inaccessible objects such as castles or the depth of a well.

Quadrants are mainly of three kinds: the horary quadrant, purely for timekeeping; the astrolabic quadrant, derived from an astrolabic plate and hence serving a single latitude; and the sine quadrant, for trigonometric calculations relating mostly to astronomy. Since an instrument in the form of a quarter-circle has two sides, two different kinds of quadrant-type markings could be incorporated on a single

---

[3] The best introductions to the astrolabe are MICHEL, *Astrolabe*; HARTNER, "*Astrolabe*"; NORTH, "*Astrolabe*"; and *Greenwich Astrolabe Booklet*. For an overview of Islamic instruments see KING, "*Strumentazione*". A useful introduction to medieval European instruments is in POULLE, *Instruments astronomiques*, and the most detailed study is ZINNER, *Instrumente*. See also A. J. TURNER, *Instruments*, and G. L'E. TURNER, *Strumenti*.

piece, or calendrical scales could be added on the back side. The astrolabic quadrant no longer functions as an analogue computer such as the astrolabe; in some regions, notably in Ottoman lands, the quadrant with astrolabic cum trigonometric markings replaced the astrolabe as the most popular instrument.

Sundials enable the user to reckon time from the position of the shadow cast by a gnomon. In the Islamic world a horizontal or vertical sundial became a feature of most mosques of consequence, although, alas, very few medieval examples have survived. Likewise few early non-trivial medieval European sundials have survived.[4] Also portable conical and cylindrical sundials were developed in the Islamic world and transmitted to Europe. In Vienna in the fifteenth century folding sundials evolved, and these became particularly popular in Nuremberg in the sixteenth century.

III. PREVIOUS STUDIES

The state of documentation of these important historical artefacts is catastrophic. The monumental survey of astrolabes by ROBERT T. GUNTHER (Oxford, 1932), still the standard work, has long been out of date, and its numerous errors remain uncorrected. It features but one-quarter of the astrolabes now known, and only about one-tenth of these in reasonable detail. About half of the instruments noted by GUNTHER were Islamic, but he was not an Arabist and was poorly advised by the Arabists at Oxford. Whilst his book is arranged chronologically by region, he began with "Persian astrolabes", most of which are late; the earliest "Arabian astrolabes" are hidden in the text of the book. The mistakes GUNTHER made in assigning general locations to early European astrolabes are actually more serious – they have held back scholarship on this subject for decades. Thus a German astrolabe is described as Italian because it looks nice and because contemporaneous German astrolabes – at least the ones featured by GUNTHER – were "plump und barbarisch". Or, an astrolabe is described as Sicilian because GUNTHER miscalculated the underlying latitude. But his

[4] The popular medieval scratch-dials, found all over Europe and consisting of a semi-circle divided into four, eight or twelve parts, barely qualify as scientific instruments. Yet no pan-European survey of these has been undertaken; such a survey might yield information on their transmission and their purpose, which are still not at all clear.

achievement was remarkable for the time, and we should not forget this. It is not his fault that his work was not continued, corrected or updated.

L. A. MAYER published in 1956 a valuable survey of Muslim instrument-makers and their works, adding much new material unknown to GUNTHER (but giving no descriptions of instruments).[5] In 1956 ERNST ZINNER produced a masterly survey of medieval and Renaissance instruments from the region which is now Austria, Germany, Poland, Belgium and the Netherlands, but his descriptions of individual instruments were inevitably rather short and unsigned instruments were excluded (or often incorrectly attributed).[6] In 1973 DEREK DE SOLLA PRICE took an important step forward by producing a checklist of astrolabes, signed and unsigned. More recently ALAIN BRIEUX and FRANCIS R. MADDISON have produced a repertory of Muslim astrolabe-makers and their works, the publication of which is awaited with anticipation.[7] The work is organized alphabetically by maker (as was MAYER's survey) and individual instruments are briefly described, with emphasis on their inscriptions; unsigned instruments will not be included.

Compared with Islamic instruments, early European instruments have not fared so well. Firstly, most of them are unsigned, so there can be no repertory of their makers. Secondly, they are more varied in style than Islamic pieces and it is often difficult to assign date and provenance. The result is that they have been ignored. Not even national interests have inspired any more recent studies of, say, all medieval or early Renaissance French or German or Italian or English astrolabes. In fact, most historians of astronomy seem to think that when one has seen one astrolabe, one has seen them all. Even popular surveys of the instrument by those most qualified to write about it do not bemoan the state of documentation of the numerous different varieties. Indeed, the history of the astrolabe has been written mainly by scholars who have worked only on the manuscript tradition and, in some cases, without ever properly looking at one.

---

[5] MAYER, *Islamic Astrolabists*.

[6] ZINNER, *Instrumente*.

[7] See PRICE et al., *Checklist*, and BRIEUX - MADDISON, *Répertoire*.

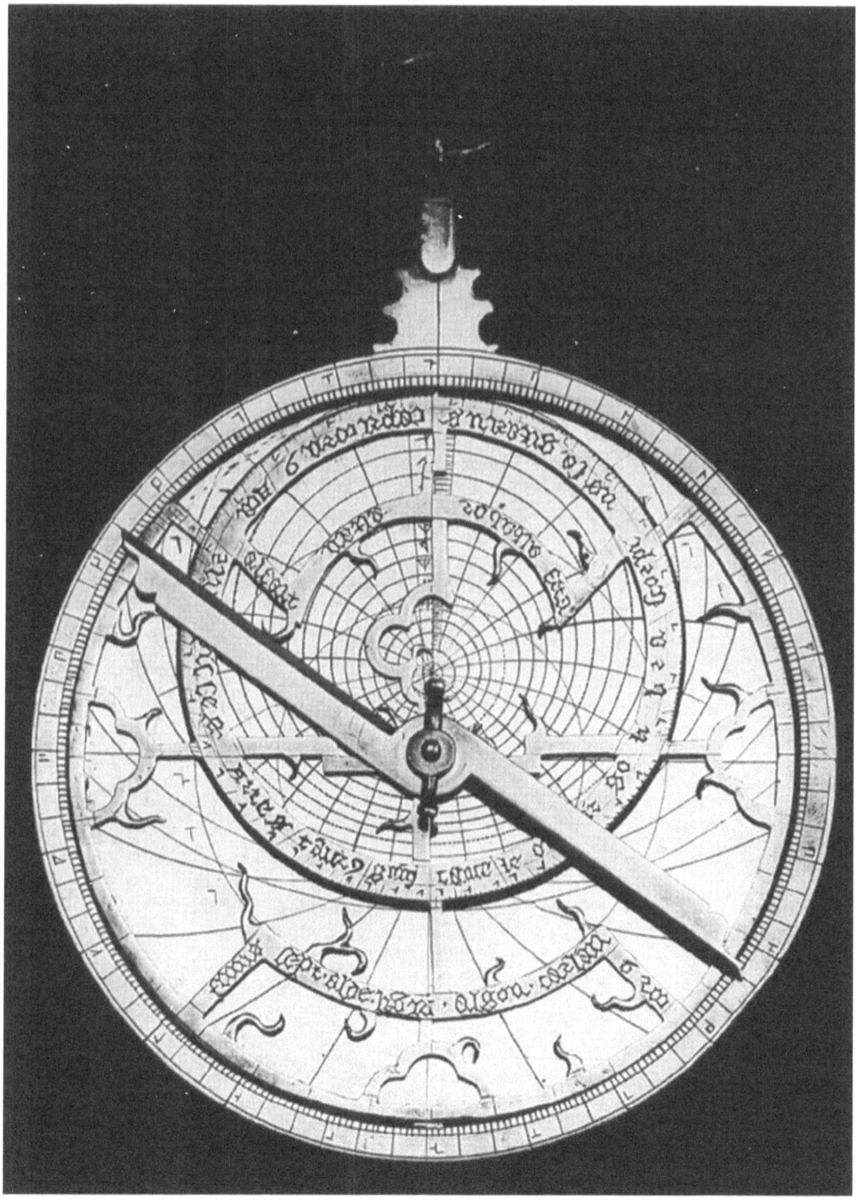

Pl. I (#202): The front of the "Berselius astrolabe", a fourteenth-century Picard astrolabe owned by Paschasius Berselius of Liège in the early sixteenth century. The rete shows various half-quatrefoil frames typical of medieval astrolabes. Courtesy of the owner (photograph courtesy of CHRISTIE'S of London).

Pl. II (#640): The back of the astrolabe presented by Regiomontanus to Bessarion in 1462. The angel looks over a device for time-keeping in any latitude known as the *organum Ptolemei*. This was invented by the Vienna school of the mid-15th century although the principle on which it is based was known to Ptolemy (ca. 125 A.D.). Photo courtesy of CHRISTIE'S of London and the owner.

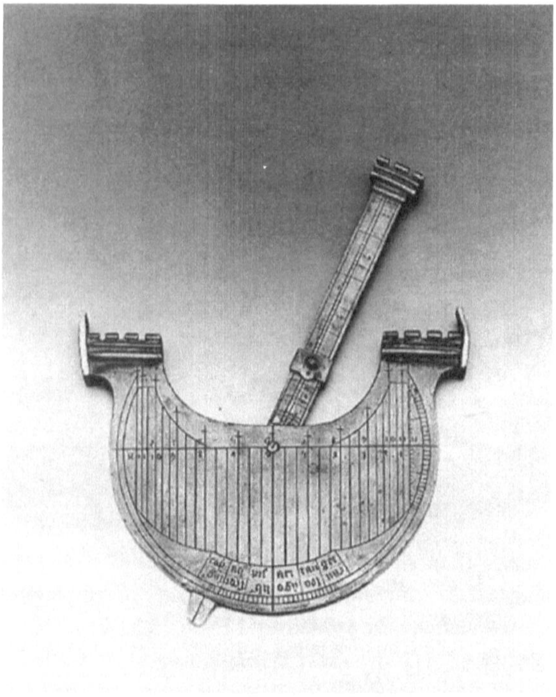

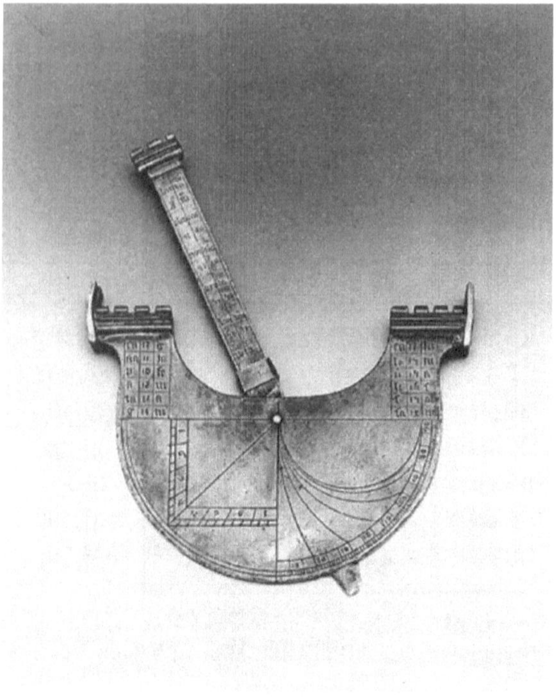

Pl. III a, b (#8537): The front and back of a *navicula de Venetiis*, a highly sophisticated medieval computing device for timekeeping, of which only a few other examples survive (see Brusa, "*Navicelle di Venezia*"). The medieval treatises on the use of the markings on the front of this instrument in Latin and Old English (see Price, "*Little Ship of Venice*") show that the Europeans did not unterstand it, a fact that points to an earlier origin for the underlying principle (later used by Regiomontanus for his universal rectilinear dial). The shadow-squares and universal horary quadrant on the back are known to have been invented in ninth-century Baghdad. Photos courtesy of Sotheby's of London.

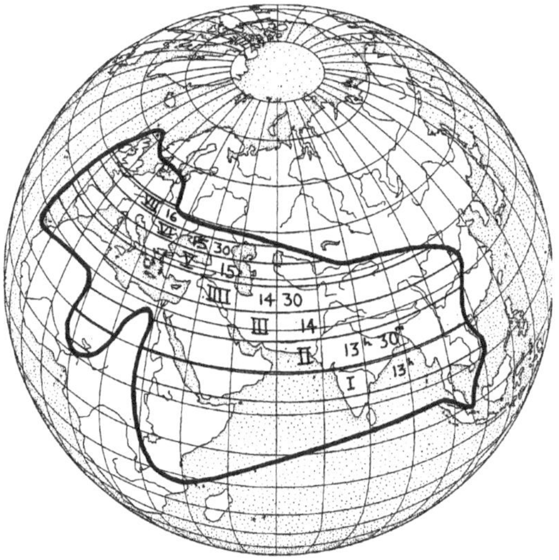

Pl. IV: The climates of antiquity, within the world known to Ptolemy. Their influence in astronomical instrumentation throughout the Islamic and European Middle Ages and the Renaissance is only now being investigated.
Courtesy of a student at Frankfurt University.

Table 1:
The latitudes of the middles of each of the seven climates (defined by the length of longest daylight) tabulated in terms of the changing value of the obliquity of the ecliptic

Climate	Daylight	Obliquity*			
		24°	23°51'	23°35'	23°30'
I	13^h	16 20	16 27	16 39	16 43
II	13^h30^m	23 40	23 49	24 5	24 10
III	14^h	30 10	30 21	30 40	30 46
IIII	14^h30^m	35 50	36 1	36 22	36 28
V	15^h	40 41	40 53	41 14	41 21
VI	15^h30^m	44 49	45 1	45 22	45 29
VII	16^h	48 19	48 31	48 53	48 59

* The Indian value of the ecliptic was 24°; Ptolemy's value was 23°51'; the value of some of the earliest Muslim astronomers was 23°35'; and that of Ulugh Beg (ca. 1425) was 23°30'.

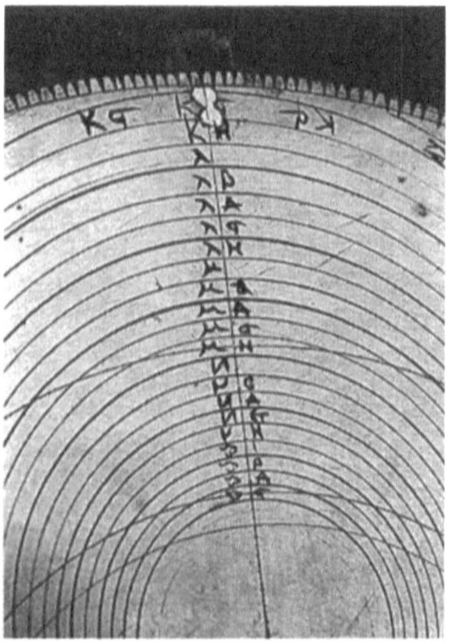

Pl. V a (#4509): Greek alphanumerical notation on a single Byzantine plate found in a composite medieval European astrolabe.
Courtesy of the owner (photograph by the author).

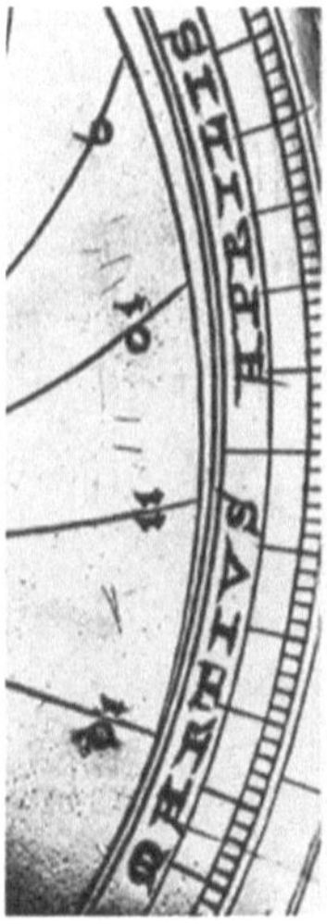

b (#548): On the back of this fourteenth-century Italian astrolabe there are additional markings scratched on by Henricus de Hollandia in Paris in the early fifteenth century and numerals punched on by a German, probably in Nuremberg, in the fifteenth or sixteenth century.
Courtesy of the Germanisches Nationalmuseum (photograph by the author).

c (#4509): Mistakes are of particular historical interest. On this plate from fourteenth-century Italy the maker first engraved the wrong information and then corrected it.
Courtesy of the owner (photograph by the author).

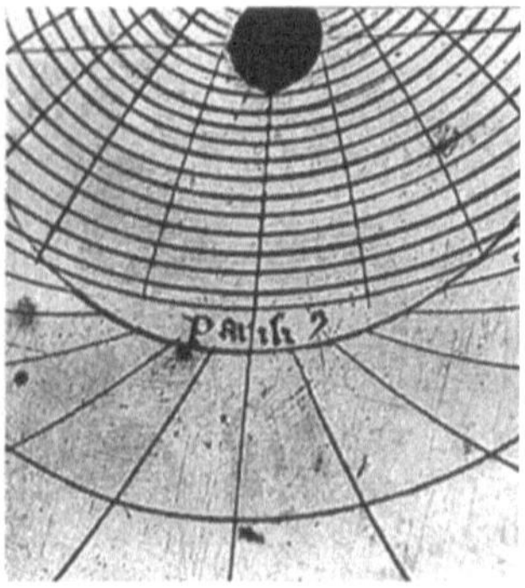

d (#4518): On this plate for a fourteenth-century English astrolabe the maker inscribed Paris on a plate intended for a latitude of ca. 60° serving the Island of Thule.
Courtesy of the owner (photograph by the author).

Pl. VI a (#111): A quatrefoil decoration on the rete of an astrolabe by Ḥāmid ibn Khiḍr al-Khujandī (Baghdad, 984/85).
Courtesy of the owner.

b (#162): A quatrefoil on the rete of one of the earliest surviving European astrolabes.
Courtesy of the Society of Antiquaries, London.

c (#158): A quatrefoil on the throne of a European astrolabe with Hebrew inscriptions.
Courtesy of the British Museum, London.

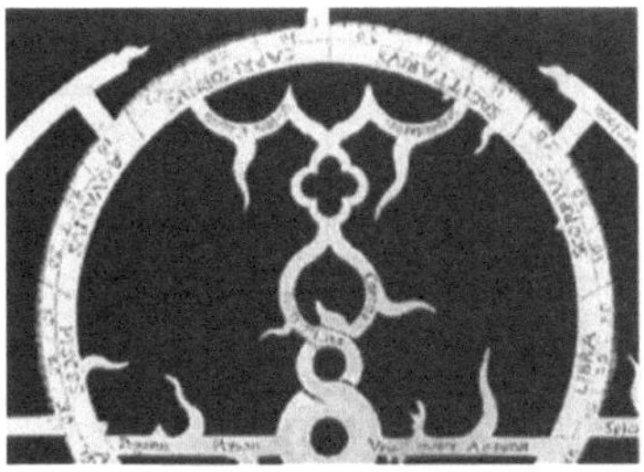

d (#544): A quatrefoil on the rete of an astrolabe by Bernadinus Sabeus, dated 1558.
Courtesy of Columbia University Special Collections.

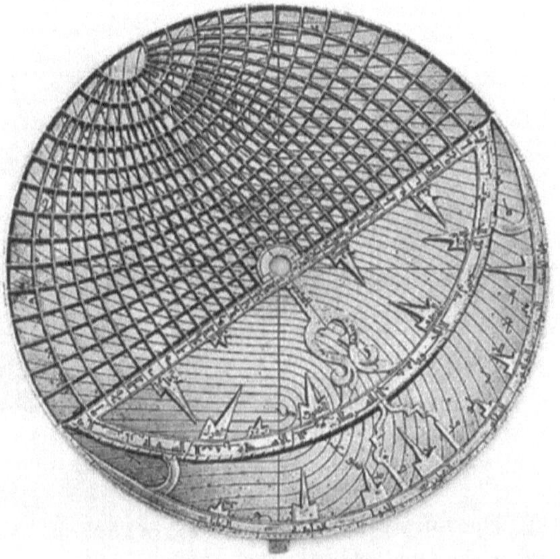

Pl. VII a (#140): The rete and one plate of the most sophisticated astrolabe ever made, which can be used universally, that is, for any terrestrial latitude, in five different ways. The maker was Ibn al-Sarrāj, who worked in Aleppo in the first half of the fourteenth century. Although the universal astrolabe was developed in Spain in the eleventh century and eventually became known in Europe, this instrument represents an independent development from the Andalusian prototype. Courtesy of the Benaki Museum, Athens.

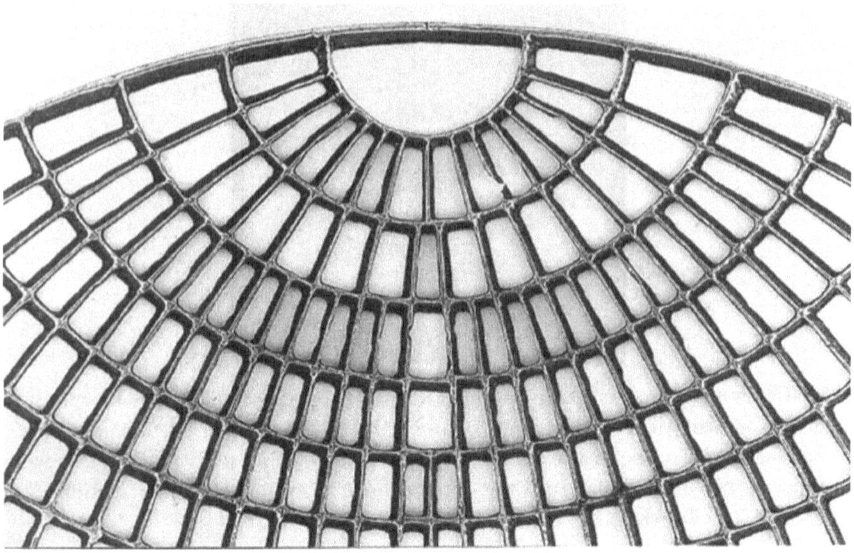

b (#140): A detail of the rete on the universal astrolabe of Ibn al-Sarrāj. Courtesy of the Benaki Museum, Athens.

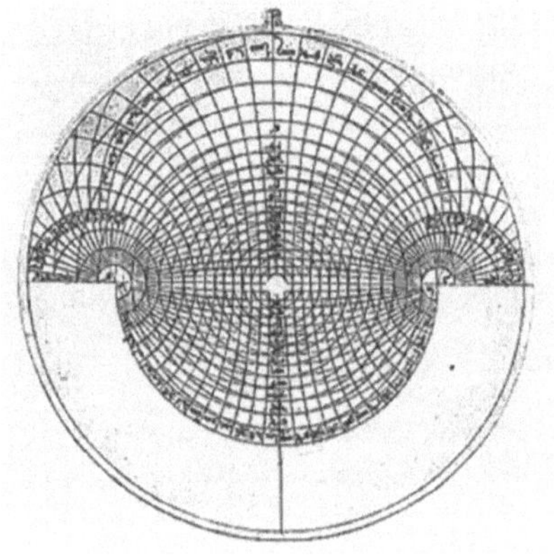

Pl. VIII a (#1203): The universal plate of Ibn Bāṣo of Granada ca. 1300, a highly ingenious device of considerable sophistication which has only recently been studied (see CALVO, *"Ibn Bāṣo's Plate"*).
Courtesy of the owner.

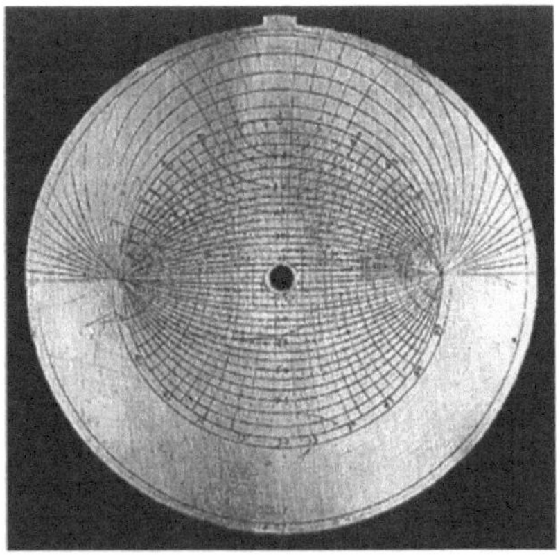

b (#545): A most unusual plate from a late-fifteenth-century astrolabe from either Vienna or Nuremberg. Only recently (see CALVO, *"Ibn Bāṣo's Plate"*) has it been realized that European astronomers had also developed a plate along the same lines as that of Ibn Bāṣo ca. 1300. This is the only known European example.
Courtesy of Columbia University Special Collections.

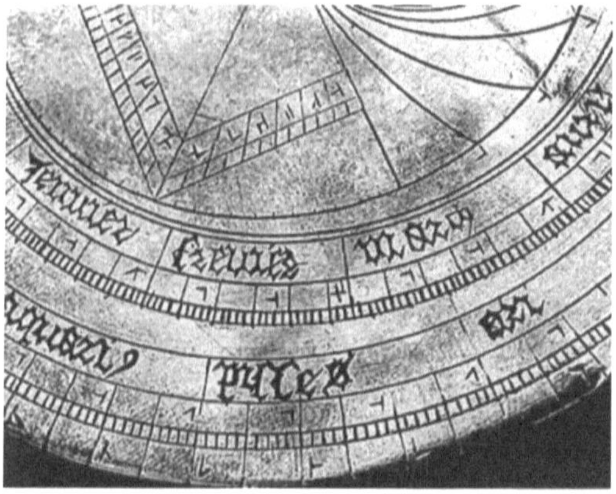

Pl. IX a (#202): Detail of the back of the fourteenth-century Picard astrolabe with month-names in Picard dialect and all numerals in ciphers.
Courtesy of the owner (photographs by the author).

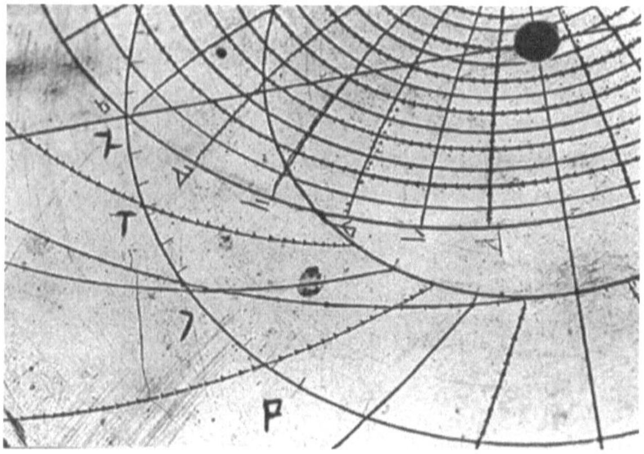

b (#202): Detail of a plate of the same astrolabe showing numerals in ciphers.

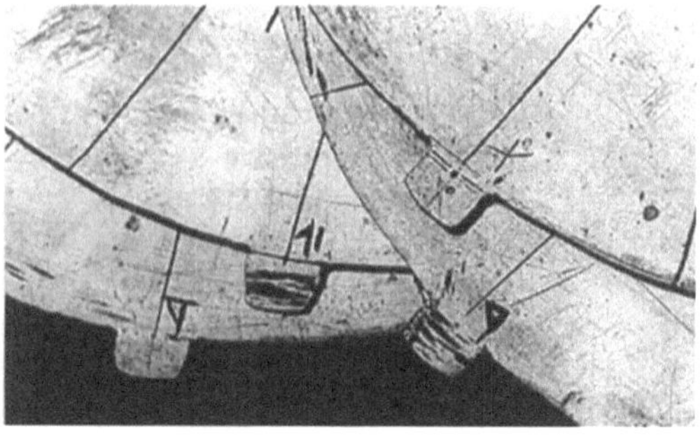

c (#202): Detail of the latitudes engraved on the plates of the Picard astrolabe.

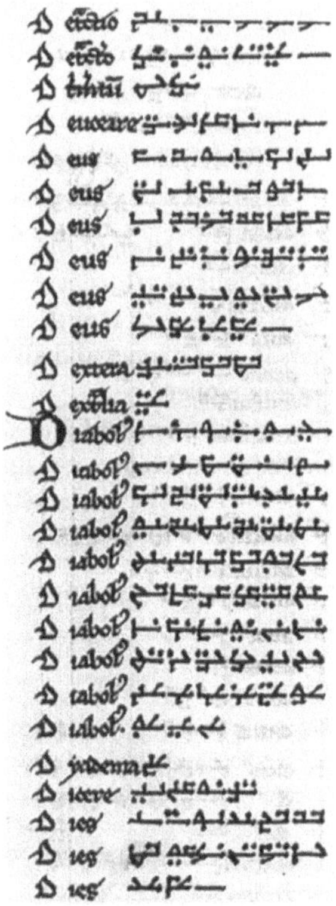

Pl. X a: Ciphers used in the index of a late-thirteenth-century Cistercian manuscript. These are of a different, earlier variety than those on the Picard astrolabe.
From MS Brussels BR II.1051, courtesy of the Bibliothèque Royale.

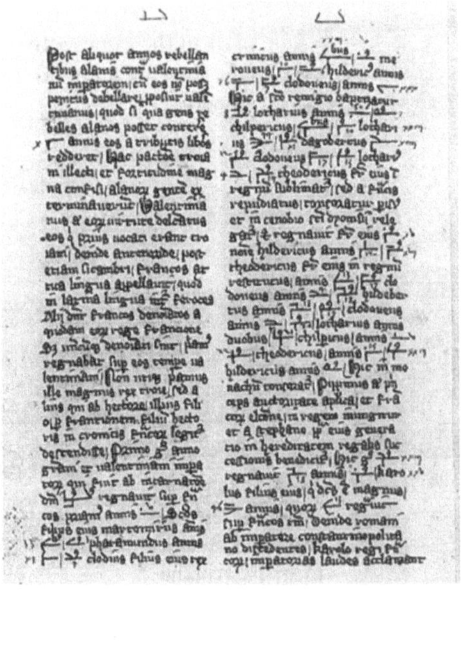

b: Horizontal ciphers used in a fifteenth-century French Cistercian manuscript of the "Chronica" of Martin of Troppau called Martinus Polonus.
Taken from MS Turin B.N. 1056 (G-IV-26), courtesy of the Biblioteca Nazionale.

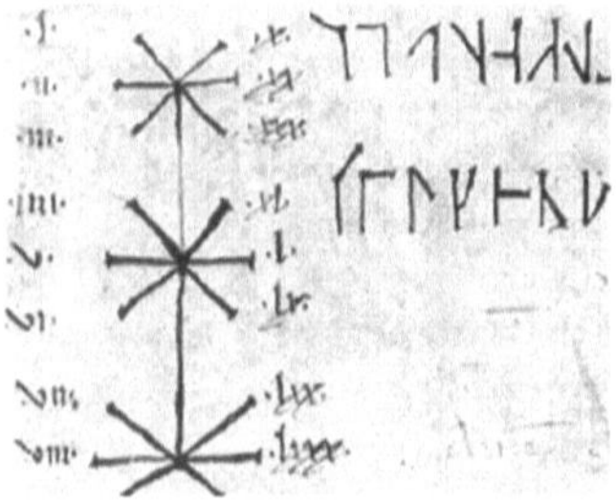

Pl. XI a: The ciphers of John of Basingstoke, reportedly brought from Athens in the first half of the thirteenth century. The first row represents the ciphers for 1–9, the second those for 10–90. When combined on a single vertical stem these ciphers can represent any number from 1 to 99. Taken from MS Cambridge, CCC, MS 468, fol. i^{v}, courtesy of Corpus Christi College.

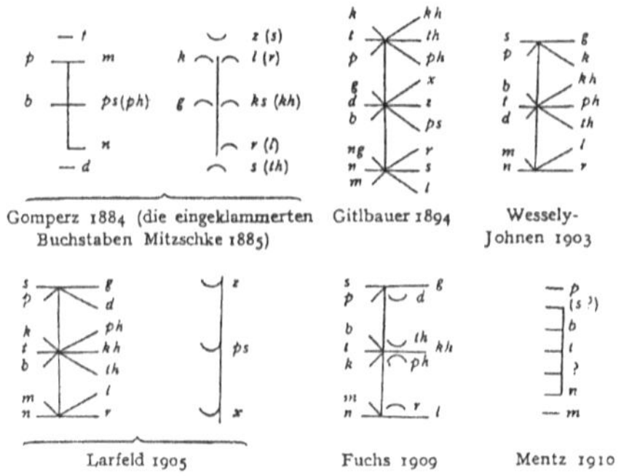

b: Various interpretations of the shorthand code recorded on the Acropolis stone (fourth century B.C.).
Taken from JOHNEN, *Geschichte der Stenographie*, p. 116.

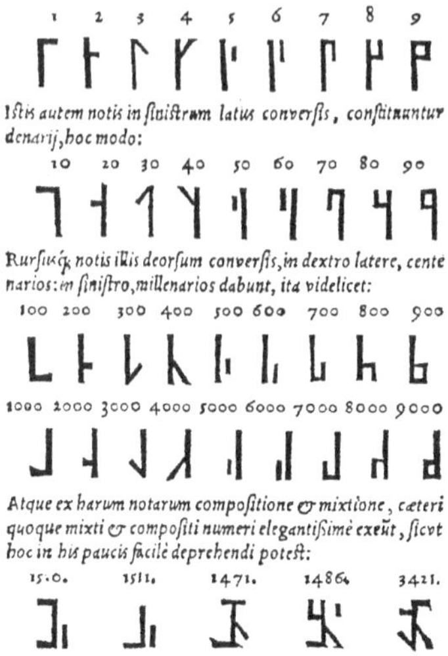

c: The vertical French ciphers, as found on the Picard astrolabe, recorded by Agrippa of Nettesheim in his "De occulta philosophia" (Lib. II, Cap. XIX). The ciphers could not survive in the new world of the printed book; already Agrippa's printer could not print them properly.

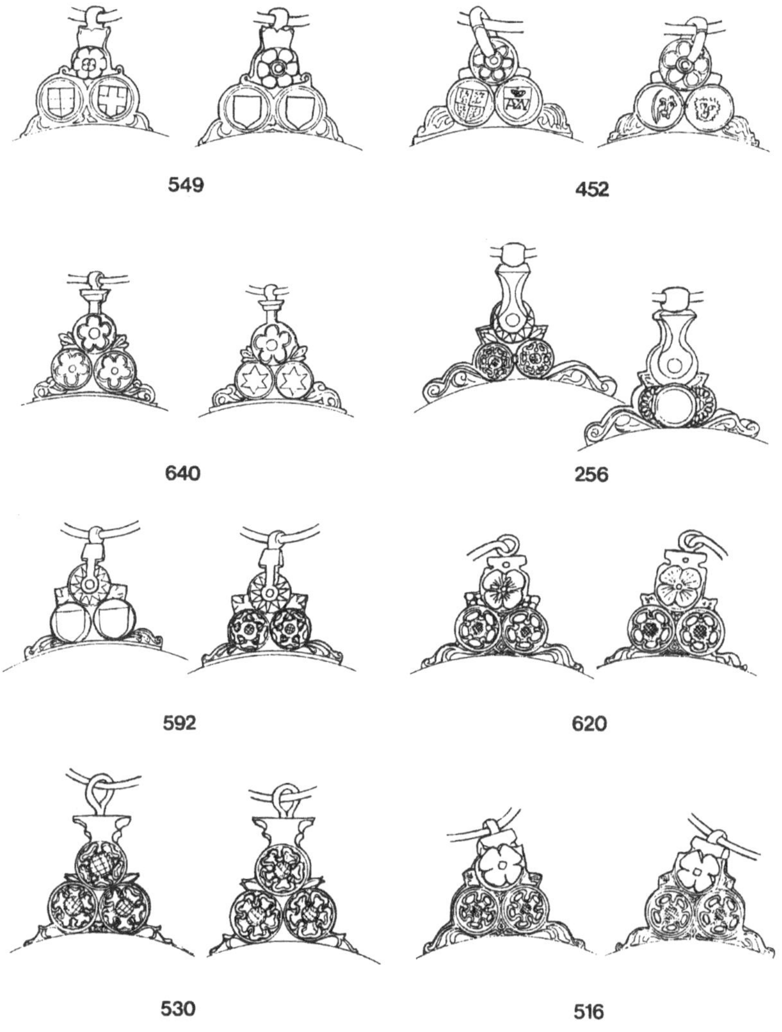

Pl. XII: Details of the thrones of various astrolabes from the fifteenth-century Vienna school.
Drawn by HELEN WALDMAN.

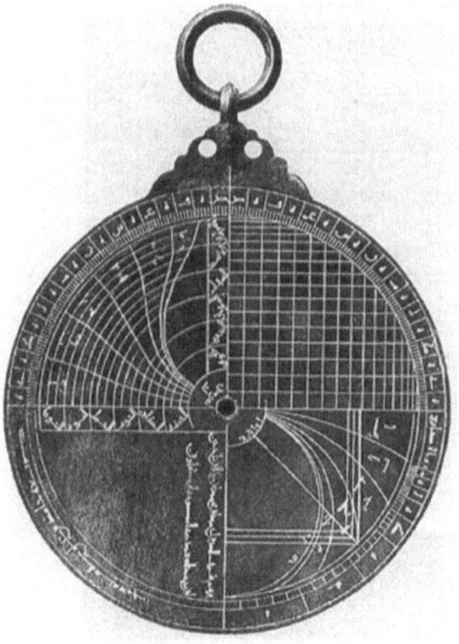

Pl. XIII a (#100): An horary quadrant for the latitude of Baghdad (upper left) on an astrolabe by Ḥāmid ibn 'Alī al-Wāsiṭī, dated 959. There is also a universal horary quadrant for the seasonal hours (lower right) and a trigonometric grid (upper right). The instrument has been stolen from the Museo Nazionale di Palermo, and no better photographs are available. Taken from Caldo, *"Astrolabi"*.

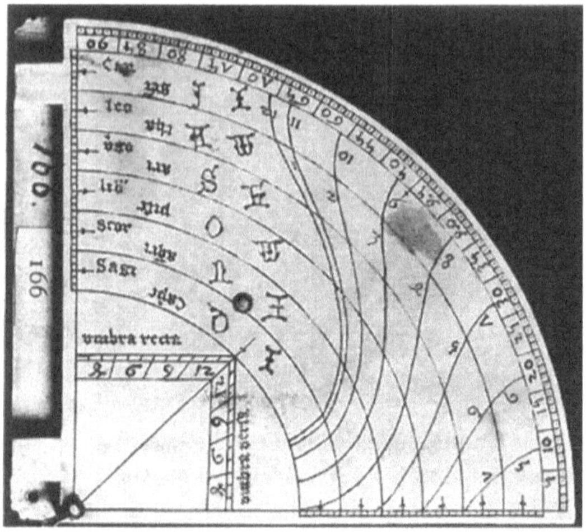

b (#5510): A Viennese horary quadrant dated 1438. Courtesy of the Kunsthistorisches Museum, Vienna.

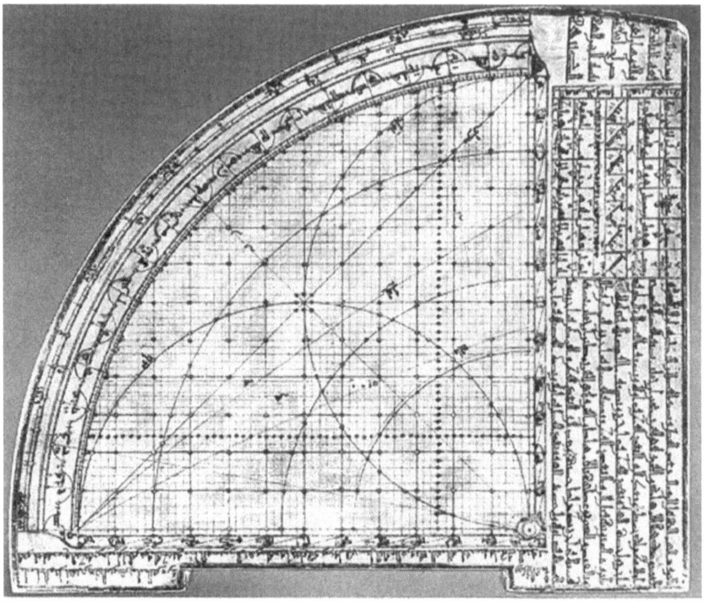

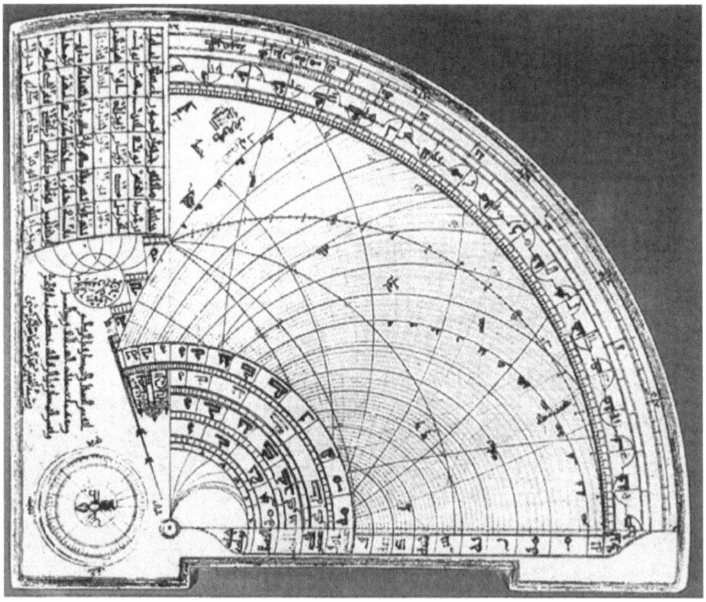

Pl. XIV (#5012): A quadrant made in Damascus about 1800. The makings on the front are essentially half of an astrolabic plate for the latitude of that city. Those on the back represent the culmination of a millennium of activity in the graphical solutions to problems of spherical astronomy which started in Baghdad in the early 9th century.

Photos courtesy of Mr. LUIS MARDEN, Washington/D.C., and the owner.

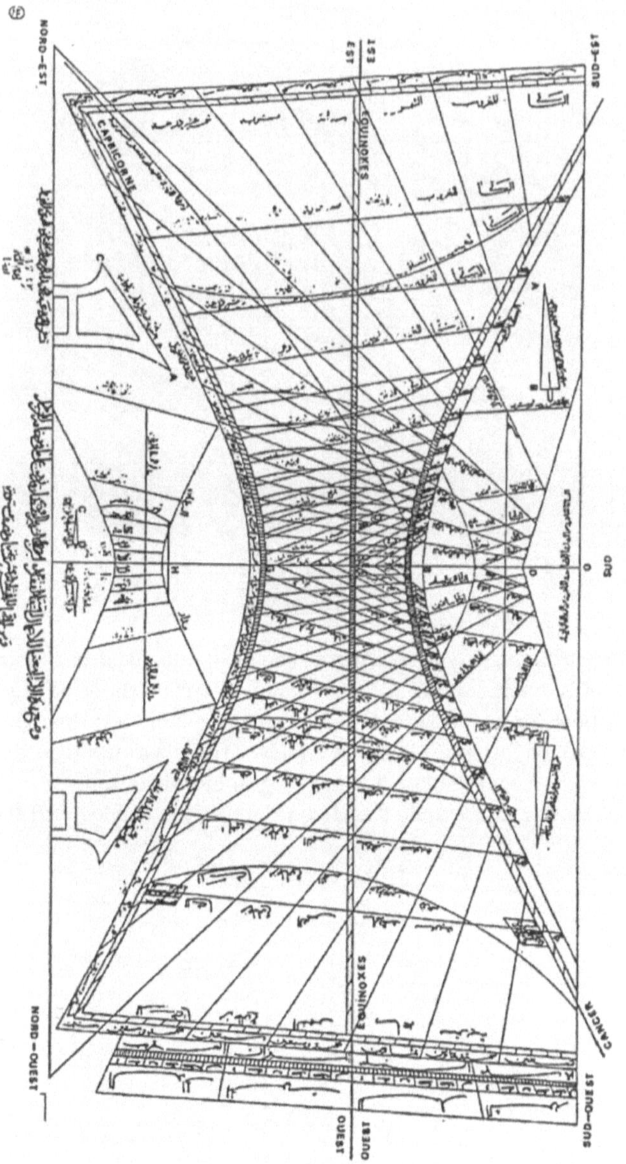

Pl. XV (#7319, after #7318): A nineteenth-century copy of the sundial constructed in 1371 by the astronomer Ibn al-Shāṭir for the main minaret of the Umayyad Mosque in Damascus. The markings enable the user to read the time of day in seasonal or equinoctial hours relative to each of the five astronomically-defined times of Muslim prayer.

Photo courtesy of the late Alain Brieux, Paris.

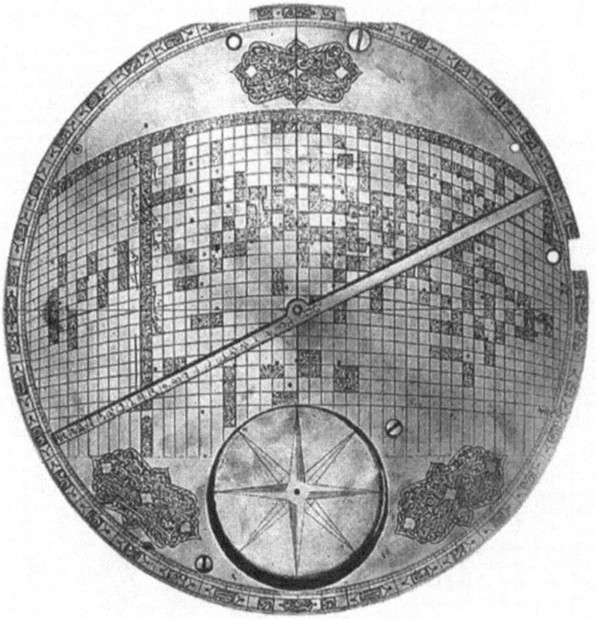

Pl. XVI a (#8024): A highly sophisticated cartographical grid from Isfahan, ca. 1710. The *qibla* of any locality can be read directly from the scale around the grid, and the distance from Mecca from the scale on the diametral rule. The instrument shows no trace of European influence and is entirely within the medieval Islamic mathematical and cartographical tradition.
Courtesy of the Museum of the History of Science, Oxford, and the owner.

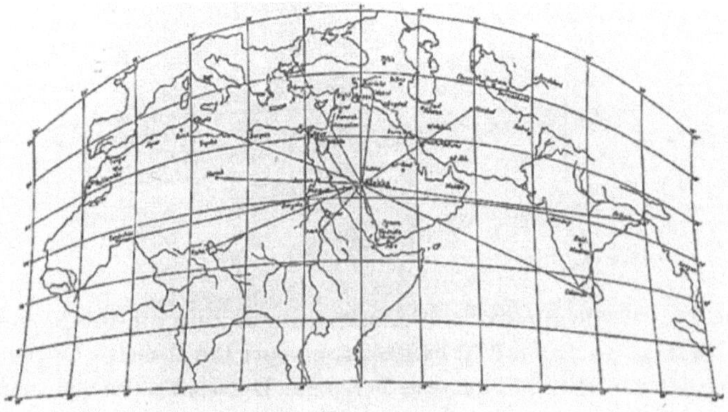

b: A cartographic grid developed in the 1920's by the German historian of Islamic science CARL SCHOY. The grid preserves directions towards Mecca and distances from Mecca.

IV. QUADRANTS

There are no surveys at all of medieval quadrants or sundials. The quadrant in particular has suffered from distortions of its history which have percolated through the literature. Two examples: Firstly, the horary quadrant with and without cursor, popular in medieval Europe, is an Islamic invention of the ninth century, not a European development. No Islamic examples with cursors have survived, but the texts do. Secondly, the *quadrans novus* of Prophatius, popular in medieval Europe, is a bastardized version of this instrument; it bears no relation to the astrolabic quadrant as is often stated in the literature. The astrolabic quadrant was developed not by Prophatius, but in the Islamic world during the eleventh or twelfth century. The quadrant for trigonometric calculations, one variety of which is the precursor of modern graph-paper, is likewise an Islamic invention, the first examples of which are described in texts from ninth-century Baghdad. Very few adequate descriptions of Islamic or European quadrants have been published.

V. SUNDIALS

Islamic sundials still in situ have suffered a neglect far worse than that which has afflicted the buildings they once graced.[8] European sundials have been blessed by a painstaking survey by ERNST ZINNER (1964), arranged, however, alphabetically by location.[9] The medieval ones have to be carefully extracted from his writings and documented afresh. Many have suffered neglect and incompetent restoration. A sundial dated 1487 was rediscovered by the present writer and his wife in Schloß Prösels near Völs in the South Tyrol the week after the Krems conference. ZINNER and his wife did not see it decades ago because it was under the roof of a nineteenth-century extension to the Castle.[10] But many of the medieval sundials which ZINNER did see in the thirties, forties and fifties alas no longer look as they did then. Portable sundials are fortunately now mainly in museums, and but for a splendid

[8] See the summary in my article "Mizwala (sundial)," in: *Enc. Islam*, VI, pp. 210–211.

[9] ZINNER, *Sonnenuhren*. A recent survey of sundials in Austria is in SCHWARZINGER, *Sonnenuhren*.

[10] See ZINNER, *Sonnenuhren*, p. 161.

overview of the sixteenth-century Nuremberg sundials,[11] the state of their documentation is so poor that, for example, the mid-fifteenth-century Vienna sundials are generally regarded as fakes. At least eight out of the ten known pieces are genuine. (I have not yet been able to inspect the other two.)

The major collections of medieval instruments are in Oxford, London, Greenwich, Florence, Washington, Nuremberg, Munich, Chicago, Paris, Istanbul and Cairo. Only catalogues of the Washington collection[12] and the Nuremberg collection[13] have been published. Another for the Chicago astrolabes is being prepared for publication. Otherwise, none of these major collections has been catalogued, a situation particularly serious for Oxford, which houses the largest collection in the world. Catalogues on a national level, featuring all of the instruments in one country, exist only for Spain (by S. GARCÍA FRANCO).[14] Smaller, but important collections are in Rockford, Illinois (catalogued by ANTHONY J. TURNER, but the collection is now being dispersed);[15] Cracow (catalogued by the present writer but still unpublished); and Milan (catalogued by TULLIO TOMBA). Numerous astrolabes are in private collections; some owners will provide access and photographs, others, less enlightened, will not. Not all museums will provide photographs. In some museums the situation borders on the hopeless: the present writer was actually refused permission to catalogue the Cairo collection in January, 1993.

In addition to this general situation it should be mentioned that some of the most visually and technologically spectacular and historically important instruments have never been published. Several of these deserve to be published in the minutest detail; all instruments merit more than the one line the astrolabes get in the *International Checklist*, and at least a basic description or comparison with others from the same workshop.

---

[11] See GOUK, *Nuremberg Sundials*. This is a model study of all aspects of the production of these pieces, the first of its kind, dealing with questions: who made them, for whom were they made, how were they made, how were they used, etc.

[12] See *Washington NMAH Catalogue*.

[13] See *Nuremberg 1992–93 Catalogue*.

[14] See GARCÍA FRANCO, *Astrolabios*.

[15] See *Rockford TM Catalogue*, which has valuable introductory sections providing historical overviews.

VI. NEW APPROACHES

Historical scientific instruments indeed demand the same kind of respect as manuscripts and incunabula. They call out for attention to their basic design (equivalent to the text, literary style and geographic milieu) as well as to details like inscriptions (author, location, date and title), special markings (colophon, copyist and script), scales and numeral forms (foliation, chapter titles, indexing), and not least additional markings, replacement parts and repairs (marginalia, flyleaves, binding, owner's marks). Too often amateurs have been given free rein in the delicate field of historical instruments, often with disastrous consequences. Also, one cannot work on these instruments from photographs alone; one has to hold the instrument in one's hands, take it apart, look carefully at the engraving and the markings. This is no armchair science. The information provided by the instruments themselves should always be viewed in the light of that which can be extracted from contemporaneous texts, the documentation of which also leaves much to be desired.

In an attempt to remedy the unhappy state of basic documentation of medieval instruments I have begun to prepare a critical catalogue of all surviving examples.[16] A large proportion of the earliest instruments (Islamic and European to ca. 1500) have now been inspected at first hand. Descriptions are being prepared that are as detailed as is feasible. Also, where possible, related texts are being consulted. The preparation of this catalogue has already revealed a wealth of new information on medieval instrumentation. The first phase of the project (1989–1992) produced a preliminary draft of 3000 pages; the second phase (1992–1994), funded by the German Research Organization, will attempt to move closer to the first goal, namely, to produce a publishable version for the earliest instruments (Islamic and European to ca. 1500). Progress is hampered by the fact that many museums are incapable of producing even the most commonplace of photographs necessary for the eventual publication, and such pictures often fail to reveal some of the most important features anyway. The preparation of photographs of significant details is now a standard part of the cataloguing procedure, at least in those collections where photography is allowed. With one or two notable exceptions all museums have been extremely cooperative with the project to date.

---

[16] See KING, "*Report*," for details.

Inscriptions, star-names, geographical latitudes and localities, variants for names of zodiacal signs and months, must all be recorded and eventually indexed. The star-names featured on instruments, for example, provide a vast corpus of data extending far beyond that in the manuscript tradition so laboriously documented by Prof. PAUL KUNITZSCH of Munich.[17] Prof. KUNITZSCH is graciously assisting the project. Unusual Latin or vernacular month-names have already served to identify the provenance of several items. We have astrolabes with Catalan month-names, nay, month-names in the Catalan dialect of Valencia, and others with Old French month-names, or rather, month-names in the dialect of Picardy and another more difficult to pin-point. KURT MAIER of Frankfurt, assistant to the project and versed in Romance linguistics as well as Arabic and Islamic Studies, was able to pin-point the provenance of these examples. An English astrolabe preserved in Innsbruck bears the following place-names and latitudes on the plates:

Londonie	(London)	51°34′
Northāptō	(Northampton)	52°50′
Eborac⁹	(York)	53°40′
Nouūcastrū	(Newcastle)	55° (?)

What does this information tell us? Alas there is no index of medieval European geographical coordinates analogous to the one available for Islamic coordinates[18] to which one can turn in order to identify a textual source for this information; it is to contemporaneous English astronomical texts that we should turn, but here again the state of documentation is deplorable. However, other medieval English instruments feature some of these distinctive latitude values (#4521 and #4545). It is clear that some medieval astrolabists had only a vague idea of what they were doing: two English astrolabes (#299 and #4518), otherwise quite respectable pieces, both name Paris on a plate intended for Thule, an island thought to be north of the British Isles.

An Italian astrolabe (#169, misidentified by GUNTHER as Sicilian) serves a single latitude 24°, so that the instrument could be used, for example, in Aswan. It is therefore useless in Europe from a practical point of view, but it nevertheless has a message for us: 24° is also the

---

[17] See KUNITZSCH, *Sternverzeichnisse*, and numerous other publications by the same author.

[18] See KENNEDY - KENNEDY, *Geographical Coordinates.*

latitude of the second of the seven climates of antiquity. Greek astrolabes had seven sets of geographical markings for the climates (progressing non-uniformly from 16° to 48°), thus making the instruments "universal". Whilst the earliest Islamic astrolabes had the same seven sets, a trend developed to produce plates for integral latitudes corresponding to a series of localities from Spain to Central Asia or for latitudes serving specific locations. The earliest European astrolabes, however, were by no means representative of the developments which had taken place in the Islamic East. So they recommenced with the climates, often in disguise. This Italian astrolabe (#169) was copied from another probably fitted with plates for all of the climates. When any European astrolabe has a series of plates for latitudes between the Yemen and Northern France or Germany this generally means that the climates underlie the plates. But the latitudes of the climates changed slowly as a result of the decreasing value of the obliquity of the ecliptic; astrolabe plates often reflect these small changes. Invariably tradition reigned supreme. When on the plate of a late Persian astrolabe the length of longest daylight is stated alongside the latitude, this has usually been calculated using the Ptolemaic value of the obliquity (23°51′) rather than the parameters of the early Muslim astronomers (for example, 23°35′) or even contemporaneous ones (23°30′). Alternatively, the latitude of the locality may have been calculated from the assumed (often erroneous) length of maximum daylight. For example, some Byzantine astronomers used 41° for the latitude of Constantinople (which happens to be reasonably accurate), but the latitude was not measured; rather it was rounded from the latitude of the middle of the fifth climate. Others used 45° because they thought their metropolis was in the middle of the sixth climate!

In the new catalogue it is proposed to provide graphic indexes of the shape of throne (the protrusion at the top of the disc), star-pointers and significant decorative features, and also of numeral and epigraphic forms. With these to hand it will become clear, for example, that (but not necessarily how) the quatrefoil design was transmitted from Islamic to European astrolabes. Another example: The thrones of the astrolabes of the three generations of the Kirmānī family, active in Iran in the fourteenth century, all share the same feature; one that does not is a fake. Yet another: The numeral forms on the dates of some of the dated astrolabes of the fifteenth-century Vienna school are very similar to those on the scales and dates of the portable sundials from the same school, so similar, in fact, that they must stem from the same person.

VII. Some surprises

Some preliminary results are surprising. The star-pointer for Vega shaped like a bird and common on late Eastern Islamic astrolabes has invariably been thought to be an Islamic contribution, the star being called *al-nasr al-waqi'*, "the falling eagle", in Arabic. But the star-pointer for the same star, Lyra in Greek, is in the form of a bird on the Byzantine astrolabe of 1062. It is also in the form of a bird on a copy of the oldest known Islamic astrolabe (#3702), the original of which is dateable to about 800 and modelled after the simplest early Byzantine variety of astrolabe. (The lyre of the Greek constellation Lyra was held in the beak of a bird.) But no surviving Islamic astrolabes from the next few centuries, not even those with other zoomorphic features on the retes (for example, #111), feature the bird. Muslim astrolabists in the ninth century clearly abandoned the bird; a few centuries later the Byzantine bird was reintroduced.

The frames on astrolabe retes were originally designed to accommodate star-pointers of a reasonable length. However, as a result of the precessional motion of the stars, the shapes and arrangements of the various components were modified from time to time because a rete produced in an old style would have some pointers disproportionately long or short. A new design was then introduced to solve this problem. If a quatrefoil or trefoil appears on a rete without a star-pointer attached this means that it is late in the tradition of that particular rete model.

The star-name Vega has lost the definite article of the Arabic name *al-waqi'* as well as the component *al-nasr*. But Altair, from *al-nasr al-ṭa'ir*, "the flying eagle" has retained the article. The reason is found in the positions of the stars on the astrolabe rete relative to the centre and to the outer rim, and the resulting possibilities for the positions and lengths of the star-pointers. On European astrolabes there was invariably space on the longish star-pointer for Altair but less space on the cramped central disc and short pointer for Vega; on neither was there space for a Europeanized version of *al-nasr*. The problem does not arise on Islamic astrolabes because of the flexible nature of Arabic script. Sometimes the nature of Gothic script interfered: Altair sometimes came to be written Alcair (or Alcayr). This was Germanized to Alkair (or Alkayr), and German astrolabes of the fifteenth and sixteenth centuries all have this form.

Particular attention is being paid to astrolabes with different layers of inscriptions, markings or components of varied provenance. One such is the 1062 Byzantine astrolabe (#2): The shadow scales on the back are a later addition, interfering with the elegant inscription around the rim. Another is a late-tenth-century Persian astrolabe (#3): On this the rete is a later replacement from ca. 1200. Yet another (#643 or #644), now in Cracow, is clearly fourteenth-century Italian, but the rete and plates were made by the same Viennese who made the 1457 astrolabe now in Nuremberg (#549). The Gothic inscriptions were made by the same person (Peuerbach?); for a miniscule "a" the maker used the same form, resembling the Russian letter Я. That Nuremberg astro labe is reputed to have been in the possession of Regiomontanus and hence to have been in Nuremberg ever since; but there are markings in Renaissance French on the plate for 48°, which was originally intended to serve Vienna: These imply that the instrument was modified in Paris.

Another fourteenth-century Italian astrolabe (#548) bears additional markings and a signature added in France in the fifteenth century by Henricus de Hollandia. A century later it was in Nuremberg; a third layer of inscriptions was added by a German.

Cross-cultural influences are particularly interesting. An early-fifteenth century astrolabe (# 460), made in Paris and now preserved in Antwerp, has had its original inscriptions removed and substituted with Arabic ones in nineteenth-century Egypt. A fourteenth-century Andalusian astrolabe (#1100) now in Vienna bears markings by an Italian in about the sixteenth century: Not only the month names but also the numeral forms prove this. That Italian was probably active in Naples or Rome because he added scales to the plate for latitude 42°. It is not difficult to set the general scene for the transmission of the instrument from Spain to Italy and thence to Vienna. A myth surrounding the instrument asserts that it was (or rather, must have been) found in an abandoned Turkish military camp outside Vienna.

Most of the nice folklore associated with instruments which have never been properly studied can be disproved. The "Astrolabe of Alfonso X", otherwise known as the "Astrolabe of Pope Sylvester" (#101), a piece that has enjoyed the limelight of several exhibitions, is but one example. It is nothing more than a tenth-century Iraqi astrolabe with no original markings on the back. The incompetently-executed Europeanized Andalusian-type scales on the back were added by a faker in the nineteenth century.

Various early Islamic astrolabes reflect the innovations to the standard astrolabe effected by astronomers in the ninth and tenth centuries. Most of these were not transmitted to Europe in the Middle Ages; some of the more interesting innovations were not introduced in Europe until the early seventeenth century.[19] A spectacular (unpublished) piece (#111) from the late tenth century constructed by a leading astronomer shows the sophistication in instrumentation in the Islamic East at that time. Various later Islamic instruments representing the major schools show that the importance of the instruments far exceeds the boundaries of the history of science. A twelfth-century Syrian astrolabe (#137) has a rete decorated with human figures resembling clowns from a circus troupe; on this spectacular piece the star-positions correspond to about 600 A. D. The figures are typical of decoration on contemporaneous Syrian metalwork. Preliminary investigations by BURKHARD STAUTZ, assistant to the Frankfurt project, reveal that the star positions were actually computed for about the year 800 from Ptolemy's positions using his own incorrect value of precession. (If one uses modern positions projected backwards in time or positions calculated from those of Ptolemy with a better value for precession, such as that found by Muslim astronomers in the ninth century, the star-positions appear to correspond to about 600.) This means that the positions of the pointers for the astrolabe are based from one of the earliest Arabic adaptations of Ptolemy's star positions. Why the maker of the astrolabe did not bother to use updated star-positions we shall never know.

Another example: A fourteenth-century Tunisian sundial (#7341) bears markings for the times of day with religious significance. In Islam the times of the five canonical prayers were astronomically defined, using shadows for the two daylight prayers and horizon and twilight phenomena for the others. Now this sundial shows markings for a prayer at mid-morning as well as at the standard times shortly after midday and in the afternoon. A study of this sundial has led to the investigation of Islamic manuscript sources never studied before in

---

[19] See MICHEL, *Astrolabe*, pp. 69–72; and CALVO, *"Ibn Bāṣo's Plate,"* especially p. 69. The mixed northern and southern astrolabe retes devised by Muslim astronomers in the ninth and tenth centuries were far more than a "coquetterie géométrique" as supposed by MICHEL. They are currently being studied by Dr. RICHARD LORCH, Munich and Frankfurt/M. The reinvention in Europe in the early seventeenth century of various mixed northern and southern astrolabe projections on retes as well as of the astrolabic quadrant is not yet properly documented in the modern literature.

modern times and thence to a new interpretation of why there are five prayers in Islam and why their prescribed times are astronomically determined.[20]

Yet another example. A cartographic dial (#8024) from Isfahan ca. 1710 bears a grid from which one can read the direction and distance of Mecca from any locality in the Islamic world between the Maghrib and China. This instrument came to light only in 1989 when it was auctioned at SOTHEBY'S in London. In the catalogue it was misdated to ca. 1800 and the projection was said to be inspired by a European. In fact, the instrument belongs to the well-known Isfahan school of ca. 1710 and the projection is fully within the tradition of Islamic cartography, inspired by achievements in mathematical astronomy. The geographical coordinates used for the major cities displayed on the grid are taken from the astronomical handbook of the early fifteenth-century astronomer-prince Ulugh Beg of Samarqand. The *qibla* for each point of the grid was probably taken from a table displaying the *qibla* for each degree of latitude and each degree of longitude. This ingenious instrument is a cartographical solution to one of the most important problems of Islamic science.[21] Twenty years ago we had little inkling of the sophistication of Muslim mathematical solutions to this problem; these have since been recovered and studied from the vast corpus of available manuscript sources.[22] Now we have to get used to the idea that Muslim cartographical solutions reached the same level of sophistication.

VIII. FORGOTTEN MEDIEVAL NUMERAL NOTATIONS

Greek, Islamic and Jewish astrolabes use an alphanumerical notation for marking numbers; the same notation was used for astronomical tables in general in those three traditions. The so-called Arabic numerals, of Indian origin, were used only in arithmetic and in common practice. The numeral notations used in medieval Europe were

---

[20] See my forthcoming paper "On the Times of the Prayers in Islam".

[21] A detailed description of this instrument is in *Nuremberg 1992–93 Catalogue*, I, pp. 167–171, and II, pp. 686–691 (no. 1.43). The second function of the instrument, namely, to provide the distance of Mecca, is here incorrectly interpreted as approximate.

[22] See KING, *Studies*, A-XIII and C-XIV, and the overview in the article "Ḳibla," in: *Enc. Islam*, V, pp. 83–88, reprinted in KING, *Studies*, C-IX.

the cumbersome Roman system and the new Hindu-Arabic numerals in their Gothic forms.[23]

Now the earliest known European astrolabe (#3042), important for numerous reasons, has, in addition to its Latin inscriptions, all numbers in an alphanumerical notation. Such notations were used by Greek and Muslim astronomers in general and instrument-makers in particular. Some prominent historians have, however, branded the instrument - described by MARCEL DESTOMBES as tenth-century Carolingian[24] - a fake or at least as highly dubious. The fact is that it simply tells us things that are not recorded in the modern literature. It is trying to speak to us, and fortunately there are some of us willing to listen: This single instrument was the focus of a series of papers at the International Congress on the History of Science held at Saragossa in 1993. I maintain that it is trying to tell us that the history of the transmission of the astrolabe to Europe needs to be, and can be, rewritten.

Another medieval instrument actually addresses us in the first person. This is an astrolabe made in Picardy in the fourteenth century (#202). The month-names are in the Picard dialect and the plates for latitudes 50° and 51° are more fully worked than the others. A later dedication says: "Berselius gave me as a gift (*me dono dedit*) to Hadrianus Amerotius in 1522". When this astrolabe was auctioned at CHRISTIE'S of London in 1991 the identity of these two individuals was not known. Berselius was a humanist Benedictine monk from Liège, correspondent of Erasmus and familiar of Vives and Budé. Amerotius, a Picard, was the second professor of Greek at the newly-founded Collegium Trilingue in Louvain. Berselius left Louvain about 1522 to return to his monastery in Liège; this was in all probability a parting gift.

Now it is significant that Berselius was a monk. The numerals on this instrument are in a highly ingenious cipher notation not found on any other known instrument but attested in about 20 medieval manuscripts, mainly of monastic provenance. The earliest form of ciphers (serving numbers from 1 to 99) was introduced in England in the early thirteenth century by the monk John of Basingstoke, whose biographer Matthew

---

23 On the latter see for example HILL, *Arabic Numerals*, based mainly on manuscript sources.

24 See DESTOMBES, "*Astrolabe carolingien*". DESTOMBES was a scholar who knew how to look at instruments. His views on this instrument have been criticized mainly by experts in the manuscript tradition, this notwithstanding the fact that manuscripts relating to this kind of astrolabe - if there ever were any - are not known to have survived.

Paris wrote that he learned of them in Athens. But the shapes of these ciphers are not related to known Greek or Byzantine numeral forms. However, the same forms were known in England already in the late twelfth century, introduced in a manual of shorthand developed from the Tironean notes by the monk John of Tilbury. Indeed the basic forms also feature in a stone inscription on shorthand from the fourth century B. C. found on the Acropolis towards the end of the last century. Maybe it was just the idea of using such symbols for numbers which Basingstoke brought back from Athens. He was certainly not the first North European to find out that the Greeks used letters for numbers (numerous earlier European manuscripts list the Greek letters and their numerical equivalents). In any case the Basingstoke ciphers were not widely used outside the circle of Grosseteste for numbering items in lists and writing arguments in Easter tables. In the late thirteenth century Cistercian monks in what is now Belgium developed a second system of horizontal ciphers (serving numbers from 1 to 9999). These were used in limited circles for foliating pages in manuscripts, numbering sermons, compiling indexes and writing year-numbers. In the fourteenth century the same ciphers in a vertical position were favoured in France. Of particular interest is an anonymous Norman treatise on arithmetic which introduces the ciphers as an alternative to the Gothic numerals. In the same manuscript is a copy of the treatise on astrolabe construction by Jean Fusoris.

The Picard astrolabe proves that these ciphers were used outside the *scriptorium*. The later history of the ciphers in the printed literature is almost as colourful as their history between the thirteenth and fifteenth centuries. So the Berselius astrolabe really is speaking to us, telling us about a forgotten numeral system of the Middle Ages.[25]

IX. MORE SURPRISES

Another astrolabe with a dedication in the first person singular (#640) was presented by the German astronomer Regiomontanus to his patron, the Cardinal Bessarion, in Rome in 1462. Regiomontanus dedicated

[25] See my forthcoming study listed as KING, *Ciphers*, for a detailed account of the ciphers and the Picard astrolabe, as well as for new insights on other aspects of medieval instrumentation. Brief notices are contained in the papers listed as KING, "*Ciphers / Berselius Astrolabe*".

several of his works to Bessarion, signing some of them simply *Ioannes*. The inscription in elegant Latin reads: *SVB DIVI BESSARIONIS DE CARDINE DICTI PRAESIDIO ROMAE SVRGO IOANNIS OPVS: - 1462* which may be rendered: "Under the protection of the divine Bessarion, called ‚de cardine' (an allusion to his rank as Cardinal), I see the light of day as the work of Johannes in Rome (in the year) 1462."

This astrolabe was brought from Rome to England in the nineteenth century and put on display at the National Maritime Museum in Greenwich in 1959. In 1989 it was auctioned in London. At that time questions were raised about its authenticity; compared with the fifteenth-century German astrolabes featured by GUNTHER, it does indeed "stick out like a sore thumb", to use a colloquial English phrase. But during the course of the cataloguing project I came across ten other astrolabes from the same workshop. Only the rete of one other piece was made by Regiomontanus, but the others clearly stem from the same workshop(s) which survived from the time of Peuerbach and inspired the work of Hans Dorn in the 1480's. The same astrolabes also inspired those of Georg Hartmann of Nuremberg in the first half of the sixteenth century. The reason the Regiomontanus instrument was "suspicious" was that in 1989 we knew virtually nothing about the astrolabes of the Vienna school. "Nothing" is a bit strong, for one piece dated 1457 preserved in Nuremberg (#549) was "known", and one or two others are illustrated by ZINNER. But those pieces had at the time never been properly published, and the world outside Germany and Austria, alas, does not read ZINNER.

Fortunately this charming astrolabe by Regiomontanus has been reinstated to its rightful place in the history of instrumentation. It was exhibited at the Germanisches Nationalmuseum in Nuremberg in 1992–93 and at the Biblioteca Marciana in Venice in 1994. What more appropriate present for Regiomontanus to give his friend and patron: a model of the universe which one can hold in one's hands, with special engraving on the plate for Rome and an original feature on the back devised by the Vienna school?

X. Fakes – a Pain in the Neck

A word about fakes, of which there are plenty around. Some are more appropriately labelled degenerate. Most Islamic degenerate instruments can be easily identified, having been made for tourists in Iran or India in past decades and up to the present. Such pieces, once identified for what they are, can – if their owners so wish – be offered at appropriate prices as decorative art. But when an instrument purports to look medieval and is sold for an astronomical price befitting a genuine medieval instrument, it is a fake.

The most remarkable fakes are perhaps those of what I used to call "the Bombay school". The makers have a singularly eclectic taste. One piece I have inspected bears a rete in standard Maghribi style, a scale on the rim copied from a thirteenth-century Yemeni astrolabe (#109 = #3549, published by myself), an unusual plate copied from an illustration in an eleventh-century Central Asian treatise (published by S. H. NASR), a back with scales found only on Indian Sanskrit astrolabes, and a distinctive alidade copied from one on a Persian astrolabe (#2505, preserved in Oxford). For $ 5000 – a bargain price for a genuine medieval instrument – you can get pieces from five different traditions; not, though, five for the price of one, but rather one for the price of fifty. Clearly the "Bombay school" has been advised by someone, presumably, but not necessarily, a European or an American, who knows the scholarly literature. But they are also producing fake European astrolabes. A distinguishing feature of all their productions that I have seen is the thickness of the back, which makes the instruments unduly heavy. Some of their pieces have reached the leading auction houses in Europe.

The other known fake European instruments were in the main put into circulation by European private collectors of genuine instruments in the late nineteenth century. Many fakes will be included in a special section of the new catalogue, not, however, to honour their makers.

XI. CONCLUDING REMARKS

So much for a brief introduction to the field of medieval astronomical instrumentation and the work currently in progress. A sceptic may still ask "So what?" (This question was actually posed after my paper in Krems, but I had heard it before from my teenage son.) I have not only outlined the scope of the field for numerous disciplines, but have hinted at forgotten national heritages and mentioned single instruments whose market value can be anything up to ATS 10,000,000. What else must I do to attract to these instruments the attention they deserve?

Our culture is benighted by the unnatural division between the sciences and the humanities. Scholars of the sciences will produce astrolabic components using computer graphics without giving thought to the 2000-year-long history of the underlying notions. Scholars of the humanities will always say they cannot understand what an astrolabe is (even though it be, as Chaucer said "bread and milk for children"). People write books about medieval metalwork and leave out instruments. They write books about medieval astronomy and dismiss the astrolabe in a few sentences. They write on medieval civilization and ignore astronomy and the astrolabe altogether. (A recently-published respectable encyclopaedia on the European Middle Ages includes an illustration of an Islamic astrolabe wrongly assembled and with the charming caption "A 9th-century astrolabe from Iraq used to decipher the celestial mechanism that was said to govern life on earth.") People write books about Schloß Prösels and do not mention its sundials. Others restore medieval buildings and plaster over the sundials.

Our museums are dominated by works of art (to which category, according to the art historians, these instruments do not belong). A potsherd stands more chance of being properly exhibited in a museum than an astrolabe. Even museums of science and technology sometimes do not know what to do with anything predating the Industrial Revolution. The Science Museum in London has found a solution: Their important astrolabe collection, together with all their other historical astronomical instruments, has been put in storage to make room for a long-term exhibition on nutrition sponsored by a leading British food chain. We have a problem here, and it is not going to be solved by new generations totally ignorant of the most basic knowledge of astronomy, which is not a subject to teach in schools simply because it is usually not on the curriculum. I once had students at New York University ask people on Washington Square, for example, "On which days of the year does the sun reach the point directly overhead?" ... Only a few colleagues involved in secondary and university education realize that the study of historical instruments in a class-room can open both eyes and minds. This is a field which covers history, science and art; indeed it is a field to itself, perhaps the most interdisciplinary field within the entire disciplines of the history of material culture and the history of ideas.

We are now at a turning point, and there is hope for the future. There is also an obvious role for Austria, for which there is no national inventory of historical scientific instruments. One could start by preparing an exhibition on the newly-discovered Viennese astrolabes of the fifteenth century and the associated sundials and manuscripts. Half of the astrolabes in Austrian collections before the Second World War were lost, and not destroyed – in other words, they were sold or stolen. Some of these can be identified with instruments now in museums in other countries, but only if the cataloguing project receives sufficient support and is allowed to continue. To whet the appetite I append two lists of instruments, one of those made in Vienna in the fifteenth century and another of medieval instruments of varied provenance but now or formerly in Austrian collections.

In short, all instruments can speak to us if we are prepared to listen. They constitute an untapped source for the history of human preoccupation with the celestial environment and man's attempts to understand it. There are enough instruments available and in sufficient variety that even the most hardened sceptic could find at least one of interest for one reason or another. And one instrument will suffice for a start because enthusiasm for scientific instruments is progressive, incurable, and also – I hope – contagious.

ABBREVIATIONS USED FOR COLLECTIONS

AP = Adler Planetarium (Chicago)
BM = British Museum (London)
BNM = Bayerisches Nationalmuseum (Munich)
CU = Columbia University Special Collections (New York)
DM = Deutsches Museum (Munich)
GNM = Germanisches Nationalmuseum (Nuremberg)
IMA = Institut du Monde Arabe (Paris)
JUM = Jagellonian University Museum (Cracow)
KHM = Kunsthistorisches Museum (Vienna)
LMJ = Landesmuseum Joanneum (Graz)
MCA = Museum Carolino-Augusteum (Salzburg)
MHS = Museum of the History of Science (Oxford)
MKG = Museum für Kunst und Gewerbe (Berlin)
MMA = Metropolitan Museum of Art (New York)
MSS = Museo di Storia della Scienza (Florence)
PC = Private collection
SM = Science Museum (London)
TLMF = Tiroler Landesmuseum Ferdinandeum – Zeughausmuseum (Innsbruck)
TMKG = Technisches Museum für Kunst und Gewerbe (Vienna)
UM = Uhrenmuseum (Vienna)
WLM = Württembergisches Landesmuseum (Stuttgart)

List of Instruments Mentioned in The text and Featured in the Illustrations*

1. Astrolabes

#2 - Brescia, Museo dell'Età Cristiana, inv. no. 36 - Byzantine astrolabe dated 1062. See GUNTHER, *Astrolabes*, no. 2, and the earlier work by O. DALTON on which GUNTHER'S description is based.
#3 - Oxford, MHS - astrolabe by the brothers Aḥmad and Muḥammad sons of Ibrāhīm al-Iṣfahānī, dated 984, with replacement rete. See GUNTHER, *Astrolabes*, no. 3 (it has not been previously noted that the rete is a later replacement).
#5 - Oxford, MHS - astrolabe with luni-solar gear mechanism by Muḥammad ibn Abī Bakr al-Iṣfahānī, dated 1223/24. See GUNTHER, *Astrolabes*, no. 5.
#100 - Museo Nazionale di Palermo, inv. no. 2131, stolen - mater of an astrolabe by Ḥāmid ibn ʿAlī al-Wāsiṭī of Baghdad, dated 959. See CALDO, "*Astrolabi*," pp. 6–9, and figs. 1 and 2.
#101 - Florence, MSS, inv. no. 1113 - unsigned tenth-century Iraqi astrolabe with additional markings on the back by a European, purporting to be medieval but actually nineteenth-century. The best description is GARCÍA FRANCO, *Astrolabios*, pp. 131–160.
#109 (identical with #3549) - New York, MMA, inv. no. 91.1.535 - astrolabe by the Yemeni Sultan al-Ashraf dated 1291. See GUNTHER, *Astrolabes*, no. 109, and KING, *Studies*, B–II.
#111 - Kuwait, PC - astrolabe by Ḥāmid ibn Khiḍr al-Khujandī dated 984/85. Unpublished! See GUNTHER, *Astrolabes*, no. 111 (GUNTHER missed a golden opportunity to acquire this piece for Oxford; his description is a mess) and KING, "*Strumentazione*," pp. 158, 159, and 162, for some illustrations.
#137 - Nuremberg, GNM, inv. no. WI 20 - eleventh- or twelfth-century Syrian astrolabe by al-Sahl al-Nīsābūrī with circus figures on the rete. See *Nuremberg 1983 Catalogue*, pp. 33–35 (no. 2), with superb illustrations; *Linz 1990 Catalogue*, 2, p. 22 (no. 18); KING, "*Strumentazione*," p. 160; *Nuremberg 1992-93 Catalogue*, item 1.71; and *Paris IMA 1993–94 Catalogue*, pp. 432–434. #140 - Athens, Benaki Museum, inv. no. 13178 - universal astrolabe by Ibn al-Sarrāj of Aleppo, dated 1328./29. See ibid., pp. 432–433; GUNTHER, *Astrolabes*, no. 140, and KING, *Studies*, B–IX.

---

* For reasons of space only brief details are given here. All of the following instruments are described in detail in the new catalogue.

#158 - London, BM, inv. no. MLA 1893 6-16 1 - unsigned European astrolabe with inscriptions in Hebrew, datable ca. 1400. See GUNTHER, *Astrolabes*, II, p. 304 (no. 158), and GOLDSTEIN, "*Hebrew Astrolabe*".
#162 - London, Society of Antiquaries - unsigned fourteenth-century European astrolabe of uncertain provenance. See GUNTHER, *Astrolabes*, no. 162.
#169 - Oxford, MHS - unsigned fourteenth-century Italian (?) astrolabic plate with a special rete and serving the latitude of the second climate. See GUNTHER, *Astrolabes*, no. 169 (confused).
#198 - London, SM, inv. no. 1880.32 - unsigned fourteenth-century Northern French astrolabe with luni-solar gear mechanism. See GUNTHER, *Astrolabes*, no. 198.
#202 - PC - fourteenth-century Picard astrolabe with numerals in cipher notation and a dedication dated 1522 by Berselius. See GUNTHER, *Astrolabes*, no. 202, and KING, "*Ciphers / Berselius Astrolabe*".
#299 - Oxford, MHS, inv. no. IC 299 - unsigned fourteenth-century English astrolabe. See GUNTHER, *Astrolabes*, no. 299.
#460 - Antwerp, Nationaal Scheepvaart Museum, inv. no. A. S. 43.9.127 - early-fifteenth-century Parisian astrolabe of the "Fusoris" type with nineteenth-century Egyptian replacement inscriptions. See *Brussels 1984 Catalogue*, pp. 37-38 (no. 8).
#544 - New York, CU, inv. no. 27-255 (14) - astrolabe by Bernardinus Sabeus dated Pavia, 1558. See *New York IM Catalogue*, p. 59 (no. 14) and pp. 3, 5, 7 and 9 (illustrations).
#545 - New York, CU, inv. no. 27-256 (13) - unsigned, undated fifteenth-century astrolabe from Vienna or possibly Nuremberg. Unpublished, though see *New York IM Catalogue*, p. 59 (no. 13).
#548 - Nuremberg, GNM, inv. no. WI 6 - fourteenth-century Italian astrolabe with later additions by Henricus de Hollandia. See *Nuremberg 1992-93 Catalogue*, item I.74.
#549 - Nuremberg, GNM, inv. no. WI 129 - astrolabe from the Vienna school dated 1457. See *Nuremberg 1992-93 Catalogue*, item I.75.
#640 - Private Collection - Astrolabe dedicated by Regiomontanus to Cardinal Bessarion in 1462. See PRICE, "First Renaissance Instrument"; G. L'E. TURNER, *The Regiomontanus Astrolabe*; *Nuremberg 1992-93 Catalogue*, item I.76; and KING - TURNER, "*Bessarion's Astrolabe*".
#643 or 644 - Cracow, JUM, inv. no. 4042-40/V - fourteenth-century Italian mater with replacement plates and rete by the maker of #549, later owned by Broscius. See ZINNER, *Instrumente*, p. 296; BURCZYK-MARONA, "*Astrolabium Brożka*".

#1100 - Vienna, TMKG, inv. no. 15144 - unsigned fourteenth-century astrolabe from Granada in the tradition of Ibn Bāṣo, with later additions by an Italian, probably in Naples or Rome. Some features are discussed in FIRNEIS, *"Moorish Astrolabe"*.
#1203 - PC - astrolabe made by Aḥmad ibn Ḥusayn ibn Bāṣo of Granada in 1309/10. See *Amsterdam 1990 Catalogue*, pp. 68–69 (no. 79) and the illustrations on pp. 66–67.
#2505 - Oxford, MHS, inv. no. Billmeir 158 - astrolabe by Muḥammad Ṣ affār dated 1505/06. See *Billmeir Supplement Catalogue*, pp. 19–20 (no. 158).
#2557 - Chicago, AP, inv. no. A84 - astrolabe in the tradition of al-Khujandī signed by Badr and dated 1130/31. Unpublished, though see the forthcoming catalogue of the Chicago collection.
#2708 - LEONARD LINTON Collection - anomalous astrolabe in the tradition of al-Khujandī (see #111) signed by Wafā'ī-yi Munajjim, dated 1608/09, with later inscriptions in Sanskrit. See *Linton Catalogue*, pp. 121–122 (no. 179) (description confused).
#3042 - Paris, IMA, inv. no. AI 86–31 - unsigned, undated astrolabe, probably tenth century, provenance Catalonia. See DESTOMBES, *"Astrolabe carolingien"*.
#3549 - see #109.
#3702 - Baghdad, Archaeological Museum, inv. no. 9723 - unsigned astrolabe signed by an Ottoman, Ahmad ibn Kamāl, but copied from (or reworked from) the earliest variety of Islamic astrolabe from the eighth or ninth century. See FRANSIS - NAQSHBANDI, *"Astrolabes"* (in Arabic), pp. 12–13 and plates 2–3.
#4509 - PC. See *Amsterdam 1990 Catalogue*, p. 101 (no. 186) and the illustration on p. 106.
#4518 - PC. See *Brussels 1984 Catalogue*, p. 37 (no. 7).
#4521 - Milan, PC. Unpublished.
#4545 - Oxford MHS, "GUNTHER loan" - fourteenth-century English astrolabe. Unpublished.

2. Quadrants

#5012 - PC - a quadrant by Muḥammad al-Ṣakāsī al-Jirkashī ca. 1800. See *Paris IMA 1993–94 Catalogue* for a description.
#5510 - Vienna, KHM, inv. no. 166 - ivory quadrant from Vienna, dated 1438. See *Linz 1990 Catalogue*, II, p. 36 (no. 34) (brief account), with a colour plate of the back only.

3. Sundials

#7318 and #7319 – fragments of the original sundial in the Archaeological Museum, Damascus, and nineteenth-century copy on the main minaret of the Umayyad Mosque. See JANIN, "*Cadran solaire damascène*" and *Paris IMA 1993–94 Catalogue*, p. 439.
#7341 – Tunis, Musée de Carthage – sundial for the latitude of Tunis by Abu l-Qāsim ibn Ḥasan al-Shaddād, dated 1345/46. See KING, *Studies*, B-XVIII.

4. Miscellaneous

#8024 – PC – unsigned cartographic grid for determining the *qibla*, Isfahan, ca. 1710. See *Nuremberg 1992–93 Catalogue*, item II.43.
#8537 – auctioned at SOTHEBY's of London on 25. 2. 1993 – a *navicula de Venetiis* (fourteenth – if not thirteenth-century English). Unpublished.

APPENDIX A
ASTRONOMICAL INSTRUMENTS MADE IN VIENNA IN THE FIFTEENTH CENTURY

1. General notes

The expression "made in Vienna" really means that I currently think that these instruments were made in Vienna, or that, in the case of the Regiomontanus astrolabe, it was devised and probably made in Vienna, or that, in the case of the Dorn instruments, the maker was essentially based in Vienna but also made instruments in Hungary. Whether the later "Regiomontanus-type" astrolabes were made in Nuremberg rather than Vienna is an open question. Also I cannot claim to understand the concepts behind the expressions "made by" and "associated with"; I use them for want of any better terminology.

Another problem is this: If Regiomontanus made astrolabes in Nuremberg, as he wrote that he did, why is not one single piece that one can attribute to Regiomontanus from his Nuremberg years preserved for us? The fact that some of his instruments were deliberately destroyed after his death is not a complete explanation because his Nuremberg workshop was surely as prolific as the earlier Vienna workshop. Are some of the "sixteenth-century" German astrolabes recently listed by the present writer and G. L'E. TURNER, and also mentioned below, actually the lost Nuremberg productions of Regiomontanus?

Yet another problem is this: how to explain the apparently continuous production of instruments in Vienna from the 1430's to the 1480's, and the subsequent appearance of a completely different class of instruments from the 1480's associated with Hans Dorn? Some, but not all of the problems could be solved by investigating all the instruments at the same time - a situation surely possible in modern times, if never before. Others could be solved by renewing research on the abundant relevant manuscript sources which have not been looked at properly since ZINNER took the trouble to list them. Poor ZINNER prepared catalogues of the astronomical instruments in both Nuremberg and Vienna and either the conditions of the time or sheer inertia on the part of the museums (or both) prevented their ever being published.

2. Bibliographical notes

On the early Vienna school see most recently the papers by PAUL UIBLEIN, MARIA G. FIRNEIS and HANS K. KAISER, in: HAMANN, *J. von Gmunden*. On Peuerbach and Regiomontanus and Dorn see, for example, ZINNER, *Instrumente*, pp. 463–465, 479–483, and 292–297. On the astrolabes see *Nuremberg 1992–93 Catalogue*, items I.75 and I.76; and KING - TURNER, "*Bessarion's Astrolabe*".

ASTROLABES

Group 1: Astrolabes probably to be associated with Georg Peuerbach

#549 - Nuremberg, GNM, inv. no. WI 129, dated 1457, Gothic script as on #643/644, with a lunar volvelle on the back. See ZINNER, *Instrumente*, pp. 150 and 465; *Linz 1990 Catalogue*, II, p. 32 (no. 29); *Nuremberg 1992–93 Exhibition Catalogue*, item I.75.
#643 or #644 – Cracow, JUM, inv. no. 4042-40/V, undated rete and plates, engraved in the same Gothic hand as #549, fitted to a fourteenth-century Italian mater. See ZINNER, *Instrumente*, p. 296, and BURCZYK-MARONA, "*Astrolabium Brożka*".

Group 2: An astrolabe probably to be associated with Peuerbach and Regiomontanus

#452 - Paris, PC, undated, Gothic script as on #549 on the plates and back and Antiqua as on #640 on the rete. See *Greppin Catalogue*, p. 32 (no. 3122).

Group 3: An astrolabe signed by Regiomontanus

#640 - PC, dated 1462, signed by Ioannes (Regiomontanus), dedicated to Cardinal Bessarion, Antiqua script as on the rete of #452; with an *organum Ptolemei* on the back. See ZINNER, *Regiomontanus*, p. 137; IDEM, *Instrumente*, p. 481; PRICE, *"First Renaissance Instrument"*; G. L'E. TURNER, *The Regiomontanus Astrolabe*; *Nuremberg 1992–93 Catalogue*, item I.76; and KING - TURNER, *"Bessarion's Astrolabe"*.

Group 4: Related astrolabes with mainly Gothic script

#256 - Paris, PC - undated (with some early-sixteenth-century additions on the back of the movable plate), Gothic script resembling that on #549, with an *organum Ptolemei* on the back. See GUNTHER, *Astrolabes*, II, p. 434 (no. 256), and also KING, *"Report,"* p. 7, fig. 5 (illustration of throne and inscription on mater).
#592 - Darmstadt, Hessisches Landesmuseum, inv. no. Kg. 25:15, dated 1466, Gothic script. Unpublished.
#620 - Munich, DM, inv. no. 70678, undated, Gothic script. Unpublished, though see ZINNER, *Instrumente*, Pl. 41, fig. 1 ("Gotisches Astrolab").
#530 - Strasbourg, Musée Historique, inv. no. IL 74, dated 1481, Gothic script as on #4501. Unpublished, though see POULLE, "Fabrication des astrolabes," p. 118 (illustrations); ZINNER, *Instrumente*, pp. 150 and 421 (sub Arnold Kunig [!]); *Strasbourg 1987 Exhibition Catalogue*, (illustrations).
#4501 - Cracow, Muzeum Narodove, Czartoryski Collection, inv. no. XIII/753, dated Schmalkalden, 1482, Gothic script as on #530. Unpublished.

Group 5: Astrolabes with unusual plates and three different contemporary styles of script (all probably by the same individual)

#516 - Paris, Conservatoire National des Arts et Métiers, inv. no. 3882/1, undated, same three hands as #545.
#545 - New York, CU, inv. no. 27-256, undated, same three hands as #516.
Some other early-sixteenth-century German astrolabes continued the tradition of the fifteenth-century Vienna astrolabes, namely:
Four undated astrolabes by the same maker with distinctive Y-shaped frames inside the ecliptic and below (one - the only one known to GUNTHER - was misidentified as Italian and dated ca. 1500; a later dating may be more appropriate).
#173 - Oxford, MHS, IC173, diameter 197 mm. See GUNTHER, *Astrolabes*, II, pp. 327–328 (no. 173).

#453 - LEONARD LINTON Collection, diameter 197 mm. See *Greppin Catalogue*, p. 33 (no. 3124: "français vers 1500"); *Linton Catalogue*, p. 133 (no. 186).
#4505 - Munich, DM, inv. no. 1991-229,1 (astrolabe) + 2 (case), diameter 197 mm, with a leather case, possibly original, dated 1542. Unpublished.
#4543 - Rome, Osservatorio Astronomico, diameter not known.
Two undated astrolabes by the same maker with a vaguely rectangular frame on the rete developed from #643 or 644:
#491 - Florence, MSS, inv. no. 1110, diameter 134 mm. See POULLE, "*Instruments astronomiques,*" pp. 20–21 (illustration of front).
#4544 - Stuttgart, WLM, inv. no. L1985-200, diameter 187 mm. Unpublished.
Two astrolabes attributable to Hans Dorn:
An unsigned astrolabe dated 1483, fitted with an *organum Ptolemei* and an eclipse calculator:
#492 - Florence, MSS, inv. no. 1096, diameter 460 mm. Unpublished, though see G. L'E. TURNER, *Strumenti*, p. 73 (illustration of front).
An unsigned astrolabe dated 1486:
#642 - Cracow, JUM, inv. no. 4040-38/V, diameter 451 mm. See ZINNER, *Instrumente*, p. 296; A. J. TURNER, *Instruments*, pls. 25 and 26 on pp. 36–37; *Washington 1992 Catalogue*, pp. 222–223.

See also the astrolabic plate on the celestial globe dated 1480 and attributable to Hans Dorn, listed below. Two astrolabic plates for a single latitude in Florence and Salzburg are closely related to Dorn's work, if not actually by him. A detailed investigation of all of his instruments would be a useful contribution. A search for more biographical information on him could also be worthwhile.

PORTABLE FOLDING SUNDIALS

Remark: The following ten instruments are probably all from the same workshop as the astrolabes listed above. It is traditional to divide them as productions of Peuerbach (d. 1461) and Regiomontanus; they were, however, probably all made by the same person, that is neither Peuerbach nor Regiomontanus. An overview follows:

Ten folding sundials, all of of similar design, with varying decoration, between 44 and 80 mm long:
#7511 - Innsbruck, TLMF, inv. no. U5, dimensions (length/width) 79/58 mm, dated 1451.

#7512 - London, BM, inv. no. 93 6-16 9, dimensions 79/59, dated 1453.
#7513 - Graz, LMJ, inv. no. 4525, dimensions 80/58, dated 1455.
#7514 - Ecouen, Musée de la Renaissance, inv. no. ? (formerly Cluny 18546), dimensions 79/59, dated 1455.
#7515 - Munich, BNM, inv. no. Phys. 22, length 70 mm, dated 1456.
#7516 - Vienna, KHM, inv. no. 122, dimensions 44/29, dated 1463.
#7517 - Pécs (H), PC, dimensions 55/42, dated 1465.
#7518 - Nuremberg, GNM, inv. no. WI 7, dimensions 62/44, datable ca. 1465.
#7519 - Vienna, UM, dimensions 55/43, undated.
#7520 - Berlin, MKG, inv. no. ?, dimensions ? undated?
In addition we should mention the folding sundial on the horizontal frame of the celestial globe attributable to Hans Dorn - see below.
#7521 - Cracow, JUM, dated 1480.

On some of these eleven pieces see ZINNER, *Instrumente*, pp. 464–465 and 481; *London BM Catalogue*, pp. 21–22; and WARD, "*Sundials*" (two examples); *Linz 1990 Catalogue*, pp. 35, 37, and 39 (three examples). I have not seen the Ecouen and Berlin pieces yet; all of the others are genuine.

MISCELLANEOUS SMALL INSTRUMENTS (QUADRANTS, RING-DIALS, CYLINDRICAL DIALS, COMPENDIA)

An unsigned ivory horary quadrant dated 1438:
#5510 - Vienna, KHM, inv. no. 166, radii 90 / 80 mm, thickness 5 mm. See *Linz 1990 Catalogue*, II, p. 36 (no. 34) (brief account), with a colour plate of the back only.
An unsigned, undated ivory cylindrical dial dated 1455:
#7524 - Munich, BNM, inv. no. Phys. 64. Height ca. 150 mm. See GOUK, *Nuremberg Sundials*, p. 28, for an illustration of one side.
An unsigned, undated ring-dial:
#7525 - Chicago, AP, inv. no. M307, diameter 143 mm. See ZINNER, *Instrumente*, pp. 480–481, and the forthcoming catalogue of the Chicago collection.
A compendium (multi-functional instrument) with an astrological astrolabic plate signed by Hans Dorn:
#250 = #8611 - London, BM, inv. no. 94 6-15 1, overall length 226 mm, dated 1491. See ZINNER, *Instrumente*, p. 293; *London BM Catalogue*, p. 74 (no. 207), and pl. XXVIII (a); G. L'E. TURNER, *Strumenti*, p. 120 (illustrations).

Four other unsigned compendia associated with Dorn need to be compared with the London piece:
#8612 - Stuttgart, WLM, inv. no. 12364, dated 1476.
#8613 - Chicago, AP, inv. no. M288, dated 1479;
#8614 - Hamburg, MKG, inv. no. 1893.2, undated;
#8615 - Oxford, MHS, inv. no. G425, undated.
On all five compendia see already ZINNER, *Instrumente*, pp. 293–295.

MISCELLANEOUS LARGER INSTRUMENTS ASSOCIATED WITH HANS DORN

A celestial globe dated 1480 attributable to Hans Dorn, with an astrolabic plate attached to the top and a folding sundial on the horizontal frame:
#4510 - Cracow, JUM, inv. no. 4039-37/V, total height 1240 mm. See ZINNER, *Regiomontanus*, pl. 45, figs. 89–90; IDEM, *Instrumente*, p. 295; AMEISENOWA, *Bylika Globe*; WARD, "*Sundials*"; and *Washington 1992 Catalogue*, p. 222.
A torquetum, undated (datable to ca. 1485), attributable to Hans Dorn:
#8503 - Cracow, JUM, inv. no. 4038-38/V, dimensions of rectangular base (reconstruction): 560 × 420 mm, overall height 710 mm. See ZINNER, *Instrumente*, pp. 296–297, and *Washington 1992 Catalogue*, pp. 223–224.

FIXED SUNDIALS

Various sundials in Austria predate 1550. As this paper was going to press, Dipl.-Ing. KARL SCHWARZINGER (Am Tigls 76A, A-6073 Sistrans) kindly sent me a list, which it is not possible to include here. For the present the interested reader may consult ZINNER, *Sonnenuhren* (Europe) and SCHWARZINGER, *Sonnenuhren* (Austria).

APPENDIX B
MEDIEVAL ASTRONOMICAL INSTRUMENTS IN OR LOST FROM AUSTRIAN COLLECTIONS

Notes: Fortunately ERNST ZINNER'S notes on various instruments that he inspected in Austria in the 1930's, 40's and 50's still exist. His information on astrolabes in Austrian collections before World War II was used by DEREK DE SOLLA PRICE in his *International Checklist* of astrolabes. Presumably PRICE did not check that the instruments were still in situ in the mid 50's; hence they appear in his 1973 *Checklist* even though they had already disappeared in the mid 40's. Many stolen Austrian art treasures were recovered after the War, but, for some reason which is as yet not at all clear to me, apparently no one pursued the missing astrolabes. It seems probable that some of these astrolabes were actually sold (see the remarks in ZINNER, *"Forschungen"*, kindly brought to my attention by KARL-HEINZ SCHALDACH of Schlüchtern): yet only one has been rediscovered in a museum elsewhere, and, as far as I am aware, none has passed through any of the leading auction houses in the past 50 years.

The following list is not necessarily complete, and I shall be grateful for information on any pieces omitted. The reader is reminded that "medieval" means here Islamic to 1900 and European to ca. 1550. Also it should be pointed out that ZINNER often misdated unsigned astrolabes, so pieces dated by him to, say, 1500, may well be from the mid-fifteenth-century Vienna school. He also, alas, sometimes rounded values of dimensions of instruments. An asterisk denotes that the instrument is mentioned in Appendix A.

Graz, Landesmuseum Joanneum:

#7513 – Vienna sundial dated 1455*, inv. no. 4525, dimensions 80/58 mm.

Another instrument is missing:

#8025 – Islamic plate with a list of localities and their qiblas around the rim (Isfahan, ca. 1710), unique of its genre, inv. no. 0330.

Innsbruck, Tiroler Landesmuseum Ferdinandeum:

#4546 – fourteenth-century English astrolabe, inv. no. 2957, diameter 253 mm.

#7511 – Vienna sundial dated 1451*, inv. no. U5, dimensions 79/58 mm.

Innsbruck, Schloß Ambras:

#4547 - single astrolabic plate from ca. 1600.

Several other late-sixteenth-century pieces are preserved in this collection.

Kreuzenstein, Burg:

Three astrolabes, all missing:

#604 - diameter 119 mm, dated 1487 and signed by Hans Bamann of Schmalkalden. See ZINNER, *Instrumente*, p. 244.

#605 - diameter 190 mm. No information available in the published sources.

#606 - diameter 165 mm. Thought by ZINNER to be fifteenth century.

Salzburg, St. Peter-Stift:

Two astrolabes, both missing from the Stift:

#625 - diameter 155 mm, dated by ZINNER to the fifteenth century, in fact probably fourteenth century.

#626 - diameter 305 mm, dated by ZINNER to the fifteenth century.

Salzburg, Museum Carolino-Augusteum:

#627 - astrolabe, inv. no. K430/49, diameter 304 mm, dated by ZINNER to the sixteenth century, in fact this is a composite piece. The mater is fourteenth century, the replacement quatrefoil rete also early, and the replacement plate (for latitude 48°, Vienna?) from about 1600.

#628 - astrolabic plate for a single latitude (probably Vienna), inv. no. K 643/49, diameter 383 mm, dated by ZINNER to the sixteenth century, in fact from about 1500 and virtually identical with #4542 - Florence, MSS, inv. no. 1282 (unpublished). These instruments should now be compared with the Dorn pieces.

St. Paul im Lavanttal, Stiftsgymnasium:

Five astrolabes, all missing:

#629 - diameter 150 mm, signed by Georg Hartmann, Nuremberg, 1545. See ZINNER, *Instrumente*, p. 365.

#630 - diameter 237 mm, signed by Wilibaldus Menzel, dated 1566. See ZINNER, *Instrumente*, p. 443. No other works by this maker are recorded.

#631 - diameter 90 mm, unsigned and undated, dated by ZINNER to ca. 1500.

#632 – diameter 220 mm, unsigned and undated, dated by ZINNER to the sixteenth century.
#2075 – diameter unknown, signed by Laurentz van de Sype, undated. See ZINNER, *Instrumente*, p. 569. No other works by this maker are recorded.

Vienna, Uhrenmuseum:
#7519 – undated fifteenth-century Vienna sundial*, dimensions 55/43 mm.

Vienna, Technisches Museum für Industrie und Gewerbe:
#1100 – fourteenth-century unsigned Andalusian astrolabe in the tradition of Ibn Bāṣo (see CALVO, *"Ibn Bāṣo's Plate"*), with important sixteenth-century additions, probably added in Naples or Rome, inv. no. 15144, diameter 157 mm. See FIRNEIS, *"Moorish Astrolabe,"* on some aspects of this instrument. Three instruments signed by Ibn Bāṣo are preserved; this is an early copy of one of his instruments and lacks the quality of the originals.

Vienna, Kunsthistorisches Museum:
#7516 – Vienna sundial dated 1463*, inv. no. 122, dimensions 44/29 mm.
#4547 – astrological plate by Georg Hartmann dated 1541, inv. no. 9823, diameter 276 mm. See ZINNER, *Instrumente*, p. 365, and KLEMM, *Hartmann*, p. 28 (illustration).
#5505 – fourteenth- or fifteenth-century horary quadrant with movable cursor, inv. no. 733, radius 141 mm. The piece has been considered suspect because it is silver plated, probably since the nineteenth century. It is doubtless genuine, if incompetently made. It was until 1880 in Schloß Ambras near Innsbruck (inv. no. 1607).
#5510 – unsigned Viennese ivory quadrant dated 1438, inv. no. 166, radii 90/80 mm. See *Linz 1990 Catalogue*, II, p. 36 (no. 34) (brief account), with a colour plate of the back only.
#1159 – undated Persian astrolabe by ʿAbd al-ʿAlī and Muḥammad Bāqir, datable ca. 1710, inv. no. 9839, diameter 183 mm.

There are other important post-1550 European astrolabes and sundials in this collection. See *International Checklist*, p. 107 (sub VKM).

Bibliography and Bibliographical Abbreviations

Ameisenowa, "*Bylika Globe*":
Zofia Ameisenowa, *The Globe of Martin Bylika of Olkusz and Celestial Maps in the East and the West*, transl. by Andrzej Potocki (Warsaw, etc., 1959).

Amsterdam 1990 Catalogue:
Anthony J. Turner, (ed.), *time* (The Hague, 1990).

Billmeir Supplement Catalogue:
[Francis R. Maddison], *A Supplement to a Catalogue of Scientific Instruments in the Collection of J. A. Billmeir* etc. (Oxford, London, 1957).

Brieux - Maddison, *Répertoire*:
Alain Brieux - Francis R. Maddison, *Répertoire des facteurs d'astrolabes et de leurs œuvres*, I (Islam), to appear.

Brusa, "*Navicelle di Venezia*":
Giuseppe Brusa, "Le navicelle orarie di Venezia," *Annali dell'Istituto e Museo di Storia della Scienza di Firenze* 5 (1980), pp. 51-59.

Brussels 1984 Catalogue:
Georges Baptiste, et al., *La mesure du temps dans les collections belges* (Brussels, 1984).

Burczyk-Marona, "*Astrolabium Brożka*":
Danuta Burczyk-Marona, "Zegar słoneczny i astrolabium profesora Jana Brożka w zbiorach Muzeum Uniwersytetu Jagiellońskiego," *Opuscula Musealia* (*Zeszytu Naukowe Uniwersytetu Jagiellonskiego* 3) (1988).

Caldo, "*Astrolabi*":
Lorenzo Caldo, "Astrolabi del Museo Nazionale di Palermo," *Pubblicazioni dell'Osservatorio Astronomico di Palermo, Memorie* 65 (1936), also printed in: *Atti della Reale Accademia di Scienze, Lettre e Belle Arti di Palermo* 19/3 (1936), pp. 3-12, and 8 figs., reprinted in *Cœlum* 8 (June 1938), pp. 101-108, and 8 figs.

Calvo, "*Ibn Bāṣo's Plate*":
Emilia Calvo, "Ibn Bāṣo's Universal Plate and its Influence on European Astronomy," *Historia scientiarum* 18 (1992), pp. 61-70.

Dalton, "*Byzantine Astrolabe*":
O. M. Dalton, "The Byzantine Astrolabe at Brescia," *Proceedings of the British Academy* 12 (1926), pp. 133–146 and 3 pls.

Destombes, "*Astrolabe carolingien*":
Marcel Destombes, "Un astrolabe carolingien et l'origine de nos chiffres arabes," *Archives Internationales d'Histoire des Sciences* 58-59 (1962), pp. 1-45, reprinted in: G. Schilder - P. van der Krogt - S.

DE CLERCQ, (eds.), *Marcel Destombes (1905–1983): Selected Contributions to the History of Cartography and Scientific Instruments* (Utrecht, Paris, 1987), pp. 153–200.

Dict. Sci. Biog.:

Dictionary of Scientific Biography, 16 vols. (New York, 1970–80).

Enc. Islam:

Encyclopaedia of Islam, 2nd ed., 6 vols. to date (Leiden, 1960 to present).

FIRNEIS, *"Moorish Astrolabe"*:

MARIA G. FIRNEIS, "A Moorish Astrolabe from Granada," in: G. SWARUP – A. K. BAG – K. S. SHUKLA, (eds.), *History of Oriental Astronomy* (Cambridge, etc., 1987), pp. 227–232.

FRANSIS – NAQSHABANDI, *"Astrolabes"*:

B. FRANSĪS – N. AL-NAQSHBANDĪ, "Al-asṭurlābāt fī Dār al-āthār al-ʿarabiyya fī Baghdād" (in Arabic), *Sumer* (Baghdad) 13 (1957), pp. 9–33, and 3 plates.

GARCÍA FRANCO, *Astrolabios*:

S. GARCÍA FRANCO, *Catálogo crítico de astrolabios existentes en España* (Madrid, 1945).

GOLDSTEIN, *"Hebrew Astrolabe"*:

BERNARD R. GOLDSTEIN, "The Hebrew Astrolabe in the Adler Planetarium," *Journal of Near Eastern Studies* 35 (1976), pp. 251–260, reprinted in: IDEM, *Theory and Observation in Ancient and Medieval Astronomy* (London, 1985), XVIII.

GORDON, *"Technology"*:

ROBERT B. GORDON, "Sixteenth-Century Metalworking Technology used in the Manufacture of two German Astrolabes," *Annals of Science* 44 (1987), pp. 71–84.

GORDON, *"Metallography"*:

ROBERT B. GORDON, "Metallography of Brass in a 16th century astrolabe," *Journal of the Historical Metallurgy Society* 20 (1986), pp. 93–96.

GOUK, *Nuremberg Sundials*:

PENELOPE GOUK, *The Ivory Sundials of Nuremberg 1500–1700* (Cambridge, 1988).

Greenwich Astrolabe Booklet:

The Planispheric Astrolabe (Greenwich, 1976).

Greppin Catalogue:

La Collection Greppin – Instruments scientifiques (Zurich, 1975).

GUNTHER, *Astrolabes*:
ROBERT T. GUNTHER, *The Astrolabes of the World based upon the series of instruments in the Lewis Evans Collection ..., with notes on astrolabes in ... other public and private collections*, 2 vols. (Oxford, 1932, reprinted in 1 vol., London, 1976).

HAMANN, *J. von Gmunden*:
GÜNTHER HAMANN – HELMUTH GRÖSSING, (eds.), Der Weg der Naturwissenschaft von Johannes von Gmunden zu Johannes Kepler (*Veröffentlichungen der Kommission für Geschichte der Mathematik, Naturwissenschaften und Medizin* 46 = *Sb. Ak. Wien, phil.-hist. Klasse* 497) (Vienna, 1987).

HARTNER, *"Astrolabe"*:
WILLY HARTNER, "The Principle and Use of the Astrolabe," a chapter in: ARTHUR U. POPE, *A Survey of Persian Art*, III (London, New York, 1939), reprinted with the original plates in: *Astrolabica* 1 (1978), and the article "Asṭurlāb," in: *Enc. Islam*, I, pp. 722–728. Both reprinted in: IDEM, *Oriens–Occidens* (Hildesheim, 1968), pp. 287–318.

HILL, *Arabic Numerals*:
GEORGE F. HILL, *The Development of Arabic Numerals in Europe Exhibited in Sixty-Four Tables* (Oxford, 1915).

International Checklist:
See PRICE et al.

JANIN, *"Cadran solaire damascène"*:
LOUIS JANIN, "Le cadran solaire de la Mosquée Umayyade à Damas," *Centaurus* 16 (1971), pp. 285–298, reprinted in EDWARD S. KENNEDY – IMAD GHANEM, (eds.), *The Life and Work of Ibn al-Shāṭir* etc. (Aleppo, 1976), pp. 107–121.

JOHNEN, *Geschichte der Stenographie*:
CHRISTIAN JOHNEN, *Geschichte der Stenographie im Zusammenhang mit der allgemeinen Entwicklung der Schrift und der Schriftkürzung*, I: *Die Schriftkürzung und Kurzschrift im Altertum, Mittelalter und Reformationszeitalter* etc. (Berlin, 1911).

KENNEDY – KENNEDY, *Geographical Coordinates*:
EDWARD S. KENNEDY – MARY-HELEN KENNEDY, *Geographical Coordinates of Localities from Islamic Sources* (Frankfurt/M., 1987).

KING, *Catalogue*:
DAVID A. KING, *A Catalogue of Medieval Astronomical Instruments – Astrolabes, Quadrants and Sundials*, pts. 1–5: Eastern (Islamic and Indian) and 6–10: European, in preparation; See already IDEM, *"Report"*.

KING, *Studies* A–C:
DAVID A. KING, *Islamic Mathematical Astronomy* (London, 1986, 2nd rev. ed., Aldershot, 1993); *Islamic Astronomical Instruments* (London, 1987); *Astronomy in the Service of Islam* (Aldershot, 1993).
KING, "*Report*":
DAVID A. KING, "Medieval Astronomical Instruments: A Catalogue in Progress," *Bulletin of the Scientific Instrument Society* 31 (1991), pp. 3-7.
KING, "*Strumentazione*":
DAVID A. KING, "Strumentazione astronomica nel mondo medievale islamico," a chapter in: G. L'E. TURNER, *Strumenti*, pp. 154–189 and pp. 581–585 (bibliography). A shorter English version in: *Historia scientiarum* 18 (1992), pp. 5–23.
KING, "*Ciphers / Berselius Astrolabe*":
DAVID A. KING, "The Ciphers of the Monks and the Astrolabe of Berselius Reconsidered," in: SERGEI S. DEMIDOV - MENSO FOLKERTS - DAVID E. ROWE - CHRISTOPH J. SCRIBA, (eds.), *Amphora - Festschrift für Hans Wussing zu seinem 65. Geburtstag* (Basle, Boston/Mass., Berlin, 1992), pp. 375–38, and IDEM, "Rewriting History through Instruments: The Secrets of a Medieval Astrolabe from Picardy," in: ROBERT G. W. ANDERSON - JAMES A. BENNETT - WILL F. RYAN, (eds.), *Making Instruments Count. Studies on Historical Scientific Instruments presented to Gerard Turner* (Aldershot, 1993), pp. 42–62.
KING, *Ciphers*:
DAVID A. KING, *The Ciphers of the Monks*, in preparation.
KING – TURNER, "*Bessarion's Astrolabe*":
DAVID A. KING – GERARD L'E. TURNER, "The Astrolabe Dedicated to Cardinal Bessarion in 1462, Attributable to Regiomontanus," *Nuncius: Annali di Storia della Scienza* 9/1 (1994), pp. 165–206. An earlier version was published (with many printing errors) in *Venice 1994 Catalogue*, pp. 340–367.
KLEMM, *Hartmann*:
HANS GUNTHER KLEMM, Georg Hartmann aus Eggolsheim (1489 bis 1564). Leben und Werk eines fränkischen Mathematikers und Ingenieurs (*Wissenschaftliche und künstlerische Beiträge, Ehrenbürg-Gymnasium Forchheim* 8) (Forchheim, 1990).
KUNITZSCH, *Sternverzeichnisse*:
PAUL KUNITZSCH, *Typen von Sternverzeichnissen in astronomischen Handschriften des zehnten bis vierzehnten Jahrhunderts* (Wiesbaden, 1966).

Linton Catalogue:
Collection Leonard Linton et de divers amateurs: Instruments scientifiques, livres anciens (Paris, 1980) [a catalogue of the LEONARD LINTON collection offered for sale by E. LIBERT - A. CASTOR - A. BRIEUX in 1980].

Linz 1990 Catalogue:
WINFRIED SEIPEL, (ed.), Mensch und Kosmos. Oberösterreichische Landesausstellung 1990, 2 vols. (*Kataloge des OÖ. Landesmuseums, N. F.*, Nr. 33) (Linz, 1990).

London BM Catalogue:
F. A. B. WARD, *A Catalogue of European Scientific Instruments in the Department of Medieval and Later Antiquities of the British Museum* (London, 1981).

MAYER, *Islamic Astrolabists*:
LEO A. MAYER, *Islamic Astrolabists and Their Works* (Geneva, 1956); with a supplement also in: RICHARD ETTINGHAUSEN, (ed.), *Aus der Welt der islamischen Kunst* (Berlin, 1959), pp. 293–296.

MICHEL, *Astrolabe*:
HENRI MICHEL, *Traité de l'astrolabe* (Paris, 1947, reprinted Paris, 1976).

Milan MAA Catalogue:
TULLIO TOMBA - GUISEPPE BRUSA, *Museo d'Arti Applicate - Strumenti scientifici - orologi* (Milan, 1983).

Milan PA Catalogue:
TULLIO TOMBA, "Gli astrolabi della Collezione Settala nella Pinacoteca Ambrosiana," *Atti della Fondazione Giorgio Ronchi* 33/2 (1978), pp. 297-314 (also available as a separatum, Florence, n. d.), continued ibid. 33/4 (1978), pp. 3-14.

New York IM Catalogue:
ANONYMOUS, *The Industrial Museum of New York - Exhibit of Early Astronomical and Mathematical Instruments (Feb., 1930)* (New York/ N.Y., [1930]).

NORTH, *"Astrolabe"*:
JOHN D. NORTH, "The Astrolabe," *Scientific American* 230 (1974), pp. 96-106, reprinted with modifications in: IDEM, *Stars, Minds and Fate - Essays in Ancient and Medieval Cosmology* (London, Ronceverte/W. Virginia, 1989), pp. 211-220.

Nuremberg 1983 Catalogue:
JOHANNES WILLERS – KURT HOLZAMER, *Treasures of Astronomy: Arabic and German Instruments of the German National Museum* (Nuremberg, 1983).

Nuremberg 1992–93 Catalogue:
GERHARD BOTT, (ed.), *Focus Behaim-Globus*, 2 vols. (Nuremberg, 1992).

Paris IMA 1993–94 Exhibition Catalogue:
JEANNE MOULIÉRAC et al., (eds.), *Syrie. Mémoire et civilisation* (Paris, 1993).

POULLE, *"Instruments astronomiques"*:
EMMANUEL POULLE, "Les instruments astronomiques du moyen âge," *Museum of the History of Science, Oxford, Selected Off-prints* 7 (1969), reprinted from: *Le ruban rouge* 32 (mars 1967).

POULLE, *"Fabrication des astrolabes"*:
EMMANUEL POULLE, "La fabrication des astrolabes au moyen âge," *Techniques et Civilisations* no. 22 (vol. IV, no. 4) (1955), pp. 117–124.

PRICE, *"First Renaissance Instrument"*:
DEREK J. DE SOLLA PRICE, "The First Scientific Instrument of the Renaissance," *Physis: Rivista di Storia della Scienza* 1 (1959), pp. 26–30.

PRICE, *"Little Ship of Venice"*:
DEREK J. DE SOLLA PRICE, "The Little Ship of Venice – A Middle English Instrument Tract," *Journal of the History of Medicine and Applied Science* 15 (1960), pp. 399–407.

PRICE et al., *Checklist*:
DEREK J. DE SOLLA PRICE et al., *A Computerized Checklist of Astrolabes* (New Haven/Conn., 1973).

Rockford TM Catalogue:
BRUCE CHANDLER, (ed.), *The Time Museum – Catalogue of the Collection*, vol. 1: *Time Measuring Instruments*, Part 1: *Astrolabes – Astrolabe Related Instruments*, by ANTHONY J. TURNER (Rockford/Ill., 1985).

SAVAGE-SMITH, *Islamicate Globes*:
EMILIE SAVAGE-SMITH, *Islamicate Celestial Globes – Their History, Construction, and Use* (Washington/D.C., 1985).

SCHWARZINGER, *Sonnenuhren*:
KARL SCHWARZINGER, *Katalog der ortsfesten Sonnenuhren in Österreich* (Vienna, 1991).

Strasbourg 1987 Exhibition Catalogue:
FRANCIS DEBAUVAIS, *Catalogue des livres et objets scientifiques anciens exposés à la Bibliothèque Nationale et Universitaire de Strasbourg* (Strasbourg, 1987).

A. J. TURNER, *Instruments*:
ANTHONY J. TURNER, *Early Scientific Instruments – Europe 1400–1800* (London, 1987).

G. L'E. TURNER, *The Regiomontanus Astrolabe*:
GERARD L'E. TURNER, The Regiomontanus Astrolabe dated 1462 (Christie's Catalogue supplement to *Fine Scientific & Philosophical Instruments*, Thursday, September 28, 1989) (London, 1989), 12 pp.
G. L'E. TURNER, *Strumenti*:
GERARD L'E. TURNER, (ed.), Gli strumenti, = vol. 1 of: *Storia della scienza*, edited by PAOLO GALLUZZI (Turin, 1991).
Venice 1994 Catalogue:
GIANFRANCO FIACCADORI, (ed.), *Bessarione e l'Umanesimo* (Naples, 1994).
WARD, "*Sundials*":
F. A. B. WARD, "The Earliest String-Gnomon Sundials," *Antiquarian Horology* (June 1976), pp. 800-801.
Washington NMAH Catalogue:
SHARON GIBBS – GEORGE SALIBA, *Planispheric Astrolabes from the National Museum of American History* (Washington/D.C., 1984).
Washington 1992 Catalogue:
JAY A. LEVENSON, (ed.), *Circa 1492 - Art in the Age of Exploration* (Washington/D.C., New Haven/Conn., London, 1991). [A catalogue of an exhibition held at the National Gallery of Art, Washington/ D.C., in 1992.]
ZINNER, *Instrumente*:
ERNST ZINNER, *Deutsche und niederländische astronomische Instrumente des 11.–18. Jahrhunderts* (Munich, 1956).
ZINNER, *Sonnenuhren*:
ERNST ZINNER, *Alte Sonnenuhren an europäischen Gebäuden* (Wiesbaden, 1964).
ZINNER, "*Forschungen*":
ERNST ZINNER, "Forschungen zur Geschichte der Sternkunde. Vortrag beim Internationalen Symposium zur Geschichte der Astronomie in Hamburg 1964," *40. Jahresbericht der Naturforschenden Gesellschaft Bamberg* (1965), 7 pp. (Sonderdruck).
ZINNER, *Regiomontanus*:
ERNST ZINNER, *Leben und Wirken des Joh. Müller von Königsberg genannt Regiomontanus*, 2nd ed. (Osnabrück, 1968), translated by EZRA BROWN as: *Regiomontanus: His Life and Work* (Amsterdam, 1990).

II

THE EARLIEST KNOWN EUROPEAN ASTROLABE IN THE LIGHT OF OTHER EARLY ASTROLABES

SUMMARY — The forgotten treasures of the European Middle Ages, unknown to the vast majority of medievalists, are the numerous astronomical instruments from the 10th to the 15th century. When Marcel Destombes described the earliest known European astrolabe in 1962 he compared it with various other early Islamic and European astrolabes. The state of documentation of those instruments was at the time mainly catastrophic, and it is further to Destombes' credit that he included even a few pieces that were not then published. No-one since Destombes has compared his astrolabe with any others. However, circumstances have changed, the field has developed; indeed, it is progressing in leaps and bounds as each instrument is catalogued. This important instrument can now be compared with many more early astrolabes, all of which are now catalogued. (The catalogue has not yet been published but is available for consultation at the Institute for the History of Science in Frankfurt). This comparison confirms Destombes' dating and provenance of the piece but sets them on a firmer basis.

«Il y a un peu plus d'un an que je connais l'astrolabe de M. Destombes et que j'ai correspondu avec lui à ce sujet. Je considère cet instrument comme le document le plus intéressant qui nous soit parvenu du haut moyen âge, et j'ai attendu avec impatience le résultat des recherches de

* It is a pleasure to acknowledge my appreciation to Mme. Jeanne Mouliérac for her constant readiness to allow me unlimited access to the instruments of the Institute du Monde Arabe. My work has profited greatly from exchanges with Professor Paul Kunitzsch on star-names and Arabic and Latin orthography as well as with Burkhard Stautz on star-positions. To Silke Ackermann and Kurt Maier of Frankfurt my appreciation for their comments on various versions of this paper and suggestions for improvement. Research on the cataloguing project in Frankfurt has been generously funded by the Deutsche Forschungs-Gemeinschaft (1992-1996).

notre ami Destombes. Le manuscrit qu'il m'a envoyé en communication montre qu'il n'a pas perdu son temps: c'est un travail de valeur, très consciencieux et qui va servir de point de départ pour des recherches importantes» (Henri Michel, in a letter dated 9.5.1962 to Père Costabel, copy inspected by the author at the Institut du Monde Arabe in 1994).

INTRODUCTORY REMARKS

In 1962 Marcel Destombes published an astrolabe that had come into his possession in the previous year, labelling it as Carolingian, and asserting as provenance late-10th-century Catalonia.[1] He thereby sparked a controversy that has lasted until the present day.[2] One early outcome of this controversy was that this extremely important document from the Middle Ages – Henri Michel used the right word – did not go to any French national museum, but rather to the Institut du Monde Arabe in Paris, along with Destombes' other instruments. But even in its imposing new home the scepticism of the scholarly community prevailed, and one result of that was that the instrument did not feature in the catalogue of the Institut's collection of astronomical instruments published in 1989.[3] It has since been put on display and is now recognized by the Institut as their prize possession.[4] But the damage was done: «le document le plus intéressant qui nous soit parvenu du haut moyen âge» was, for example, virtually ignored in a book of over 600 pages on science in Carolingian times, published in 1993.[5]

---

[1] See M. DESTOMBES [1962] on International Instrument Checklist #3042. On the Checklist see *Appendix 3*.

[2] The only published criticism of Destombes' claims for the instrument is G. BEAUJOUAN [1972], pp. 658-662, now expanded in G. BEAUJOUAN [1995].

Some recent literature on the piece in which it is treated as genuine is W. STEVENS [1986], pp. 24-25 and note 25; [1981], p. 99 and note 54; A. BORST [1989], pp. 54-55; J. VERNET, J. SAMSO *ET AL.* (Eds.) [1992], pp. 192-193 (by J. Samsó, based on my notes); D.A. KING [1993b], pp. 30-35 (with an uninspired guess at a possible Sicilian connection); [1991b], p. 174 . See also note 4 below.

[3] See J. MOULIÉRAC [1989]. Actually it is perhaps just as well that it was not featured in that catalogue, for at the time no serious research had been done on the piece since the time of Destombes, and the ill-founded opinions of those who thought it was a fake would have had to have been mentioned.

[4] See the Institute's magazine «Qantara - Cultures en mouvement», XI, April-June 1994, p. 51.

[5] P.L. BUTZER, D. LOHRMANN (Eds.) [1993]. The instrument is mentioned in passing by P. KUNITZSCH [1993], pp. 205-220, esp. pp. 210-211; the provenance and date are given as «vielleicht» N.E. Spain, *ca.* 980. P. KUNITZSCH [1991-92], p. 3, accepts the date *ca.* 980 for the Latin inscriptions.

Now Destombes knew a great deal about astronomical instruments.[6] He was one of the very few scholars to have published a medieval instrument since the time of the pioneer Robert T. Gunther.[7] His opinions are therefore not to be dismissed lightly; indeed, they should be treated with the utmost respect. The controversy will surely continue, for now there is more at stake than the authenticity of the instrument, but at least it should now take account of the other early astrolabes that cast new light on this remarkable piece. The purpose of the present paper is to show that the Destombes astrolabe cannot be considered in isolation, as it has been by most people since 1962. Every single detail must be considered in the light of similar details, if such can be found, on earlier and later instruments.

It has been asserted, for example, that the Destombes astrolabe cannot predate the earliest known European texts on the astrolabe from the middle of the second half of the 10th century. We now know enough about medieval European astrolabes to realize that this is not the case. One only needs a few basic skills, and, if possible, some guidance, to construct an astrolabe, although only if one has actually seen such an instrument and had some guidance concerning the basic procedures. If this is all one has, the astrolabe may not be very good, but this is precisely the case with the Destombes astrolabe, which is crudely made when compared with any medieval Islamic astrolabe and any of the other earliest European astrolabes. The Destombes astrolabe is proof, and there are other examples, that some of the earliest European astrolabes were made

---

[6] See the various articles on instruments, Islamic and European, reprinted in G. SCHILDER, P. VAN DER KROGT, S. DE CLERQ (Eds.) [1987].

[7] Gunther's two monumental tomes dealing respectively with Islamic and European astrolabes remain the standard work. It is not Gunther's fault that its numerous errors, particularly in the attribution of various European instruments, have never been corrected. (For example, his 'Sicilian' #169 may be Italian, but the latitude underlying the single plate is 24° not 38° as Gunther thought; his Spanish #163 is French (which he recognized whilst his work was in press, see II, p. 309, note 1); his Italian #173 is German, and so on).

The articles and catalogues of various Milan collections by T. Tomba (see the bibliography) are the only detailed studies of medieval instruments to have appeared since Gunther (1932), apart from one description in S. García Franco of a French astrolabe (#476 – see his [1945], pp. 295-300 (no. 22)), accounts of various instruments featured in S.GIBBS, G. SALIBA [1984] (#186, #304, #2006); and my own descriptions of some preserved in Nuremberg (#558, #547, #548, #549, #640, #550 – see G. BOTT (Ed.) [1992] II, pp. 574-592) and of the remarkable 14th-century French astrolabe featuring monastic ciphers (#202 – see D.A. KING [1993b] and K. MAIER [1994]). Descriptions of medieval instruments (*e.g.* of #4512 – see B. CHANDLER (Ed.) [1985], pp. 124-127) which describe them in detail but ignore the star-names must be deemed deficient. The tabular representations of star-names in S.GIBBS, G. SALIBA [1984] are to be avoided; it is not only of interest to know which stars are represented, but also how their names are written. In short each instrument deserves a full description.

without reference to any written tradition. In some cases, they must have been copied from other instruments, the positions of the star-pointers getting more and more inaccurate and the star-names becoming more and more corrupt as the copying process continued. The procedures for engraving the astronomical markings on astrolabe plates could have been transmitted by word of mouth; one does not need much practice to master the basics. The existence of these early European pieces, poorly made and, from a practical point of view, quite useless for serious astronomical purposes – and such is the Destombes astrolabe – points to one of the basic problems of medieval European astronomy. They help explain why it took several centuries for Europeans to achieve the levels of sophistication reached by Ptolemy in the 2nd century or the Muslim astronomers of Baghdad in the 9th century.

Also it has been asserted that the instrument might be a modern fake. (The assertions that Destombes might have had the astrolabe made for himself are beneath contempt). Any faker who made this astrolabe knew more about the history of the astrolabe and about Latin epigraphy than all of the scholars who have ever looked at this piece – including all of the authors contributing to this volume – put together. For without doubt we are dealing with a genuine piece; the only questions are: How old is it? and: Where was it made? We are not competent to answer these questions definitively. Both of these questions can be answered roughly. In fact, they have been answered already.

There is no need to describe the instrument here, for it has been better described than most medieval astrolabes already, namely, by Marcel Destombes. It will be discussed in detail in the forthcoming catalogue of all medieval astronomical instruments.[8] I shall focus here on some of the most interesting aspects. The Destombes astrolabe is not an original invention. It is not a blank Islamic astrolabe made by a Muslim, with inscriptions added by a European;[9] no Islamic astrolabe of this crudity is known before the

---

[8] See D.A. King, [1991a].

[9] Paul Kunitzsch ([1991-92], p. 3), suggests that the astrolabe itself was made by a Muslim craftsman and that only the inscriptions were added by a Latin monk *ca.* 980, but more recently (with Elly Dekker [in print]) inclines to the view that it is a European production based upon an Islamic model.

Clearly there must have been a time when the first *Andalusian* made an astrolabe, which may have been in the 9th or 10th century, and surely this would not have been of the same quality as those being produced in the Islamic East at the time. Nevertheless the throne, the rete (but for the horizontal bar), and the star-pointers argue for non-Islamic influence. It is here, I suspect, that we see the initiative of the Christian monk, or even of his predecessor(s), for there is no proof that the Destombes astrolabe was the first ever made by a European. In fact, for a first attempt, the Destombes astrolabe is quite successful, which leads me to suspect that it was not the first.

17th century. Rather, it was copied by some European from another instrument. Therefore we have to look around for possible candidates, and, although no single piece comes into consideration, there is quite a lot of evidence to be gained from the other earliest surviving European astrolabes and from the earliest Greek and Islamic astrolabes. There may be other traditions, Roman or Sicilian, about which we currently know nothing.[10]

The only person ever to have compared any of the instruments with the Destombes astrolabe was Destombes himself, who considered some well-chosen early Eastern Islamic and early Western Islamic pieces, as well as an 11th-century illustration of a signed Western Islamic astrolabe reworked by a European. He also had limited information on one very early European astrolabe that he had not seen, and in addition he referred to the three different manuscript traditions (Eastern and Western Islamic and European).[11] No early astrolabes have ever been compared with the Destombes astrolabe by any other writer (although Dekker and Kunitzsch have compared a few other instruments for star-pointers).[12]

What is possibly the second earliest European astrolabe (#420) is incorrectly labelled as a fake in its present home (Greenwich).[13] This is

---

[10] See my remarks below on the problematic Italian tradition, which does not seem to be related to the Spanish tradition. Therefore Sicily comes into question as another possible location for transmission of the astrolabe to Europe. In D.A. KING [1993b], pp. 30-35, I hinted at a possible Sicilian provenance for the Destombes astrolabe (in J. VERNET, J. SAMSÓ *et al.* (Eds.) [1992], pp. 192-193, this opinion is also mentioned), but at the time of writing I was not aware that Prof. A. M. Mundó had already established beyond any shadow of a doubt that the inscriptions are in full accord with those attested elsewhere in Catalonian inscriptions of the 10th century.

[11] See M. DESTOMBES [1962], pp. 10-28, on #99, #101, #111 (recognized as possibly by al-Khujandī), #3713, #3702 and #1026 (all Eastern Islamic), and #3650 and #116 (Western Islamic, although the former has a replacement Eastern Islamic rete from *ca.* 1300), #161 (early European) and #4024 (illustration).

See J. SAMSÓ [1992], pp. 93-98, for a summary of the astrolabe treatises available in Arabic in Muslim Spain; A. VAN DE VYVER [1931]; W. BERGMANN [1985]; P. KUNITZSCH [1982]; *IDEM* [1966], on the earliest Latin treatises.

[12] At the Paris meeting on the Carolingian astrolabe (Institut du Monde Arabe, February, 1990) I suggested that all early astrolabes should be investigated and showed some examples to illustrate why this was necessary.

[13] The Greenwich piece, which is unpublished, has been radically cleaned. The star-positions are so inaccurate that it is not yet possible to understand them (information from B. Stautz).

Another piece that is worth mentioning is #4507, featured in M. MINIATI (Ed.) [1991], pp. 48-53. It has a very Renaissance-looking design on the lower part of the rete; also the equinox is at March 10, which does not accord with an early dating. This instrument may be a 16th-century piece but it is certainly modelled on another that was considerably older: the star-pointers correspond to no particular epoch (information from B. Stautz). However, the occurrence of various unusual star-names attested in the manuscript tradition but not on any other known instrument, confirm that the piece is also of considerable historical interest. In fact, the history of the astrolabe in Italy is so poorly documented that we must use such instruments, whether we like them or not.

another example, and there are more that will be mentioned below, of a tendency amongst certain colleagues to label any instrument that they do not understand as a fake. There is also a tendency to dismiss as dubious any instrument that does not come up to certain scientific standards. If an instrument could not be used to yield reasonable results, no-one, it is argued, could have used it. Likewise, if an instrument is incomplete, questions are raised about its authenticity. Yet amongst the more than 1000 surviving astrolabes currently known there are several that are incomplete. There is another regrettable tendency to dismiss any instrument that is not signed and dated. One reason why the earliest European instruments have been so neglected is that none of them is signed or dated: therefore they have no place in any repertory of makers. They are, however, of extreme historical importance and they present an additional challenge to the historian. Again, an instrument bearing two or more layers of inscriptions is often considered suspect. Yet these inscriptions can often be useful in determining the subsequent fate of the instrument.[14] These arguments against the authenticity of such instruments simply do not hold water, especially when one realizes that the vast majority of medieval astrolabes were not used for any scientific purpose anyway.[15] We do not dismiss manuscripts copied in more than one hand, or bearing marks of ownership; rather, we describe them and try to establish the provenance and history of the manuscript. We also do not look at individual manuscripts of a given work in isolation; rather, we compare them with other copies. Similar criteria now need to be applied to instruments, most of which can be grouped into families.

There is only one complete surviving Byzantine astrolabe with inscriptions in Greek (#2), two solitary plates with Greek inscriptions

---

[14] See, for example, D.A. KING [1993b], on the two layers of inscriptions on a 14th-century astrolabe from Picardy (#202), the first in Picard dialect of Old French (for details see K. MAIER [1994])and the second added by the Benedictine monk Berselius of Liège in 1522; S. ACKERMANN [1994], on a signed and dated 12th-century Persian astrolabe reworked in the 17th century (#3922); as well as P. SCHMIDL [1994], for a second layer of «13th-century markings» added to a genuine 17th-century Indo-Persian instrument by a modern faker in order to increase the value (#3923 = #3700 = #9002).

[15] This was first realized by Derek De Solla Price – see his [1980]. The popular notion that Islamic astrolabes were used for regulating the times of prayer, propounded by the 'Oxford school', is to be abandoned for lack of evidence. To be sure, many Western Islamic astrolabes are marked with curves for the times of prayer, but there were other more reliable means of time-keeping available in the major mosques, and even there it was not always the astronomers who regulated the times of prayer. Astronomy in Islam operated on two distinct levels, which we can label folk astronomy and mathematical astronomy – see D.A. KING [1986], [1987], [1993a]. The same held in ancient Greece and medieval Europe.

(#4509 and #109 = #3459), and one Eastern Islamic astrolabe (#3702) modelled after the very earliest Islamic astrolabes, which were based entirely on Byzantine instruments. The form of the standard Greek astrolabe has recently been reconstructed from the available evidence by Burkhard Stautz.[16] We should consider the possibility that the Destombes astrolabe was copied from a Greek or Roman instrument, but several arguments speak against this: above all, the calendrical scale, not known on Greek instruments, and then the alphanumerical notation, which is derived from the Western Islamic system rather than the Greek. But even clearer evidence is afforded are the azimuth curves for each 10° on the plates for latitudes [38;30°] and [41;30°] – corresponding to Sicily and Cordova, and to Rome and Barcelona / Saragossa, respectively. Azimuth curves were, according to our present information, first introduced on astrolabe plates by al-Khwārizmī in the early 9th century, and are found on some early Eastern Islamic astrolabes (#99 and #111) as well as on a 10th-century Western Islamic astrolabe (#110) as well as on the 11th-century ones (for example, #3650).[17] However, they are not mentioned in any known European treatises before the 12th century.[18]

So we are definitely dealing with an object that displays Islamic influence, an instrument copied from an early Western Islamic astrolabe, of a kind so primitive that no other example survives. Therefore we may anticipate that some of the Islamic features may be found only on other very early European instruments, and with this in mind let us continue to look at some of the most remarkable features of the instrument.

The Throne

The throne of the Destombes astrolabe is of a kind similar to those on at least three other early European astrolabes (#300, #558 and #169), but not on any known Islamic pieces. Since between the two lobes there is a depressed circular central part, there must originally have been a shackle with circular sides that would have been attached there (such a shackle is still found on #300 and #558, but is missing on #169).

---

[16] See B. Stautz [1994].

[17] Notes by al-Khwārizmī on finding the azimuth of the sun with an astrolabe are found in MS Istanbul Ayasofya 4830,13 (fols. 198v-200r) – see F. Sezgin [1978], vol. VI, p. 143. Other Islamic works on the construction of azimuth curves are known from the 9th and 10th centuries – see J. Samsó [1969], pp. 49-51, and J.L. Berggren [1991].

[18] P. Kunitzsch [1982], no. 44: *al-sumūt*.

THE RIM

The rim is brazed onto the back. It bears a scale divided into 5° intervals subdivided for each 1°. The 5°-divisions are characterized by the fact that they are not all radial. This unhappy feature also occurs on the scale on the back.

THE RETE

The rete originally bore no astronomical markings beyond the star-pointers and no inscriptions. The absurdly inaccurate divisions of the ecliptic scale were added by an incompetent. And the names of the signs were added by an incompetent, probably the same person. It seems beyond doubt that these were added during the period 1200-1400: a more exact dating is difficult because our knowledge of inscriptions on medieval instruments rests on very insecure foundations,[19] but fortunately it is irrelevant to a discussion of the original instrument. It should be mentioned, however, that even the engraving of the names Virgo and Pisces as VIRGVO and RISES (with an initial R!) is not without historical interest, because although the first form has not been found on any other instrument and is not etymologically correct, it possibly is to be understood as a false analogy to forms like *antiguo*. Likewise numerous variants on the standard Pisces, including PISES, are attested on other medieval instruments.[20] Another point to observe is that the CA of «CAPRICORNVS» is written in a different, older style than the rest of the names of the signs, though not in the hand of the original inscriptions of the back and the plates. The names of the signs were often abbreviated by their first or first two letters on astrolabes on which little room was available, such as Islamic astrolabes that fell into the hands of Europeans (see, for example, #3622 and #1099,[21] with the first two letters, and #118, with the first three).

This is not the only example of an astrolabe on which the rete was not completed. On one 14th-century French astrolabe (#4551), otherwise complete in every detail, the star-pointers have not been finished: a blob

---

[19] There is no corpus of inscriptions on metal objects with which we can compare the inscriptions on astrolabes.

[20] See, for example, 161 and 4509. See also D.A. KING [1993b], pp. 59-60, note 22.

[21] The latter has LV for July, indicating that the second engraver was an Italian. The piece was acquired in 1890 for the Germanisches Nationalmuseum in Nuremberg at a flea-market in Rome.

of metal around the position where the pointers should be indicates that the maker was interrupted in his work (and also shows us how star-pointers were made).

Another feature of the Destombes astrolabe that we can compare with other instruments is the horizontal bar on the rete, which is counter-changed at the centre and at each intersection with the ecliptic, thus:

[D]

Now the Byzantine astrolabe and all early Eastern Islamic astrolabes (Sections 1.1 and 1.2 in Appendix 2) have straight horizontal bars, thus:

[1]

a feature not found on any Andalusian astrolabes[22] but occasionally found on early European astrolabes (#300 and #169). The earliest Andalusian astrolabes introduce a single counter-change, thus (for #4024):

[2]

also found on some of the earliest European astrolabes (#161, #167, #420, #550 and #4553), or (for #110=#135):

[3]

[22] A straight bar appears on the rete of the Edinburgh astrolabe of the early-11th-century Andalusian Muḥammad ibn al-Ṣaffār (#3650), but it is a later replacement in the Eastern Islamic tradition, and B. Stautz dates it to *ca.* 1300.

But more complicated patterns are attested on various Andalusian astrolabes of the 11th century. One such (attested only on #2572) has three counter-changes:

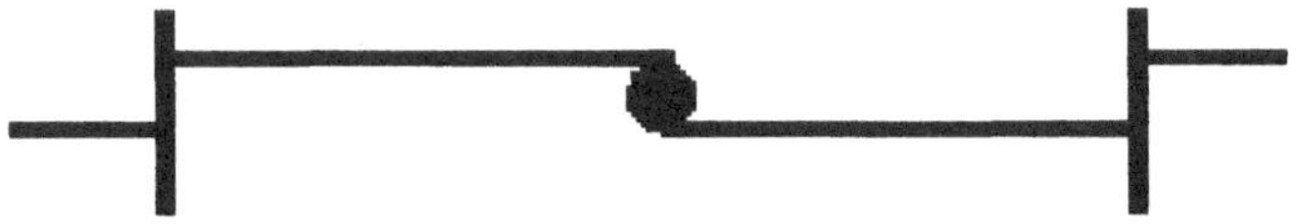

[4]

which is the same as that on the Destombes astrolabe when viewed upside-down. This is alas a problematic piece: the astrolabe was made by Muḥammad ibn al-Sahlī in 483 H [= 1090/91], but the rete bears inscriptions in Hebrew and is clearly a replacement.[23] A more common design (as in #116, #117, #118, #121, #2527, #1139 and #1099) has four counter-changes:

[5]

and is also found on later but still early European retes (#428, #4507, #191 and #558 (see further below on the star-pointers on this piece)). Other early European astrolabes have variations, such as (on #166):

[6]

or (as on #536):

[7]

[23] This piece has been studied in detail – see the bibliographical notes to #2572. In the Washington catalogue it is suggested that the rete was original; in the study by B. R. Goldstein and G. Saliba the rete is correctly recognized as a replacement. There is no trace of any original Arabic inscriptions, but the letters A, T, V, C and A are scratched on the spaces for the 1st, 2nd, 6th, 10th and 11th zodiacal signs. There are no European markings on the rest of the instrument, which fact constitutes yet another reason for supposing that the rete is not a reworked original.

or yet more complicated forms with as many as ten counter-changes (as on #2044).

I have not found a single rete with a horizontal bar having an identical form to the Destombes astrolabe, although, as noted above, one rete (#2572) with Hebrew inscriptions and probably of Andalusian provenance, has an identical pattern when viewed upside-down. On this the star-pointers display random errors, and an overall dating is not yet possible (the pointer at the beginning of Leo on the ecliptic near to that for Regulus is a clear indication of the fact that the maker was not well-informed). I would, however, not hesitate to date it between 1100 and 1300. In any case, the fact that the bar on the Destombes astrolabe has a limited number of counter-changes (actually three) is clear evidence on the one side that it was modelled after an early Andalusian astrolabe, not an early Eastern Islamic one, and on the other that it represents a very early European tradition.

The maker had no intention of marking the star-names because at least in the case of the upper star-pointers there is no room for any names. Other very early European astrolabes bore no star-names (for example #161 and #167). Various explanations are conceivable: the makers were incapable of writing the names of the stars in Latin; or they had no manuscripts with star-lists at their disposal; or they did not know which pointers belonged to which stars. This helps us to understand why the star-positions on the retes correspond only vaguely with the positions that they have in reality or that could have had, had they been based on, say, the table of the Andalusian astronomer, Maslama al-Majrīṭī (*ca.* 1000).[24] As Burkhard Stautz has shown, some very early European astrolabes (for example, #191, #428 and #300) do have star-pointers that are based on al-al-Majrīṭī's tables:[25] but the positions have not been updated, and so the retes are as useless from a practical point of view as those on the Destombes astrolabe. Stautz has also shown how other very early European astrolabes (for example, #420 and #4507) have star-positions that cannot be related to any of the star-lists so painstakingly documented by Professor Paul Kunitzsch. In the same way I can point to hundreds of star-names that are not attested in the manuscript tradition, all variant or further corruptions of names that are recorded there.

---

[24] The study by E. DEKKER, P. KUNITZSCH [in print] shows that the star-positions on the Carolingian astrolabe do not correspond to any available tables or illustrations.

[25] A computer-graphic display, such as that provided by B. Stautz, shows the problems of the positions of the star-pointers quite clearly.

The star-pointers on the Destombes astrolabe bear no resemblance to those on any other early Islamic or very early European astrolabe. The only Islamic astrolabes with star-pointers resembling those of the Destombes piece are a 14th-century Andalusian piece (#1161), an astrolabe made by a Maghribi *ca.* 1600 for use in the Yemen (#4170), and a 14th-century German quatrefoil astrolabe (#2072). This does not seem to be immediately significant. One early European astrolabe (#558) of uncertain provenance (the latitudes served run from 16° to 52° and there is no obvious means of establishing the provenance) has some star-pointers resembling those on the Destombes piece. The bases are perforated by two or three holes, but otherwise the resemblance is striking. For the time being I think we can assume that since the maker never intended to mark the names of the stars he allowed himself some artistic licence with the shape of the pointers. We do not find this shape elsewhere (especially for pointers on the upper part of the rete) simply because it is inconvenient for marking star-names.[26]

I leave the study of the stars represented on the Destombes to colleagues better qualified, such as Paul Kunitzsch and Elly Dekker, and Burkhard Stautz. The stars can be identified, but the selection is not typical of the clearly defined group of Andalusian astrolabes of the 11th century, which follow a distinct pattern. Rather they are to be compared with the stars represented on the earliest known Andalusian astrolabes such as #4024 and #110, as well as those on the other earliest European astrolabes (#161, #4504, #300, #166, #167, etc.). This is a task for the future.

The Original Inscriptions

All of the unusual forms in the inscriptions on the plates and back of the Destombes astrolabe are attested elsewhere in 10th-century Catalan inscriptions. This has been graphically demonstrated by Professor Anscari M. Mundó.[27] Henceforth there need be no further basic discussion of the inscriptions, only of details. In the future it might even be possible to localize the unusual forms to a particular region of Catalonia.

---

[26] On Islamic astrolabes, for example, the pointers in the upper part of the rete are long and thick enough that the star-names can be engraved upon them. For other parts of the rete shorter pointers can be used because the names can be written on various supporting bars and frames (though not on the ecliptic).

[27] A.M. Mundó [1995].

THE NUMERAL FORMS

The numeral forms on the Destombes astrolabe are to be explained as follows:[28]

1-5: A B C D E

Here the sequence is alphabetic (one might expect A, B, G, D, H in the light of the Arabic *alif*, *bā'*, *jīm*, *dāl*, *hā'*, but G and H are needed later on).

6-9: V Z H T

These correspond to the Arabic *wāw*, *zayn*, guttural *ḥā'* and plosive *ṭā'*, which are the appropriate Arabic letters for 6-9.

10-60: I K L M N O

Here again the order is alphabetic, corresponding to Arabic *yā'*, *kāf*, *lām*, *mīm* and *nūn,* and then avoiding representing the Western Arabic plosive *ṣād* with S, preferring to continue with O.

70-90: G F ∂

Here G is used for the guttural *'ayn*, which is attested elsewhere: this Arabic letter was rendered by a G in the earliest Latin texts on the astrolabe from late 10th century Catalonia.[29] The letter F corresponds to Arabic *fā'* for 80 and Roman Uncial ∂ is used for the Western Arabic plosive *ḍād* (90), the capital D having been used already.

It emerges, then, that the numeral forms used on the Destombes astrolabe constitute a highly ingenious attempt to render the Western Arabic alphanumerical forms in a consistent fashion. Other isolated attempts at devising new number systems are known from the Middle

---

[28] Letter from Professor Paul Kunitzsch dated 15.9.1993. Already M. DESTOMBES ([1962], p. 29) realized that the forms were derived from Western Arabic notation. See also his useful study listed as M. DESTOMBES [1960], which supplements R.A.K. IRANI [1955], and also P. KUNITZSCH [1991-92], especially pp. 3 and 14.

[29] P. KUNITZSCH [1982], no. 2: ALGAZAR for *al-'aṣr*; no. 16: HOTOALZAGAD for *khuṭūṭ al-sā'āt*; and no. 20: ALGHILAKA for *al-'ilāqa*.

It is not too much to hope that some manuscript may be identified one day featuring precisely the numerals on the Destombes astrolabe. The whole problem of the introduction of the Arabic numerals in Europe, a process that took half a millennium, demands reinvestigation. As an offshoot of such a study numerous hitherto undocumented numeral systems, variants of the standard forms, may be anticipated. A good example is the system of Greek numerals presented in the *Liber floridus* from St. Omer *ca.* 1120 (see A. DEROLEZ *et al.* (Eds.) [1968], pp. 172-173).

Ages,[30] the most remarkable being the system introduced by Cistercian monks in the late 13th century in what is now the border region between Belgium and France, which, in a slightly modified form, was actually used on an astrolabe in the 14th century.[31]

The Plates

The latitudes chosen for the plates are:

LV	36° mater
LT	39°
MAL	41;30°
ME	45°
MZL	47;30°

These are related to the climates of antiquity.[32] Since these are defined in terms of the length of longest daylight, the associated latitudes vary according to the obliquity of the ecliptic and thus with time. *Appendix 1* includes a table showing this dependence, with latitudes calculated for different values of the obliquity. It is immediately clear that we are dealing with the middle of the 4th climate, the boundary between the 4th and the 5th, and the middles of 5th (only approximately – see below), 6th and 7th climates. The only discrepancy is the latitude 41;30°, which should, according to this explanation, be 41°. Here it seems we are dealing with a more specific value than that for the middle of a climate, and it is appropriate to note that the latitudes of Saragossa and Rome in the Ptolemaic *Handy Tables* are 41;30° and 41;20°, whereas Ptolemy's own value for the

---

[30] Destombes ([1962], p. 29) noted the forms: lé, kef, lem, mim, non, zad, hain, fé and dat, on the scale of an astrolabe illustrated in MS Berne Bibliothèque de la Bourgeoisie lat. A 196, fol. 7v (late 10th or early 11th century, see A. VAN DE VYVER [1931], p. 266, pl. 2). See also the transliterated Arabic numerals in one of the earliest Latin treatises on the astrolabe, here dealing with the climates, their latitudes and maximum lengths of daylight, published in J.M. MILLÁS VALLICROSA [1931], pp. 290-292. An attempt to write numbers alphabetically and backwards is described in L. GILISSEN [1976].

[31] See D.A. KING [1993b], pp. 44-55, and the earlier writings of G. Beaujouan, J. Sesiano and others there cited.

[32] See the article *Iḳlīm* by A. MIQUEL in *Encyclopaedia of Islam*, and J.B. HARLEY, D. WOODWARD (Eds.) [1987], vol. I, pp. 182-183 (by O. A. W. Dilke), and vol. II, 1, pp. 96-102 (by G. R. Tibbetts). For a discussion of the climates and their latitudes by the leading scientist of medieval Islam see now A. DALLAL [1984], (the implication of P. D. A. Harvey (J.B. HARLEY, D. WOODWARD (Eds.) [1987], vol. I, p. 320, note 166) that al-Bīrūnī was not in control of his subject is based on misunderstanding).

latter is 41;40° (alas Barcelona is not listed).[33] The value used for the plate is thus most appropriate.

It was not without consequences that, almost by accident, the climates correspond to the latitudes of places of extreme importance in the history of astronomy on the one side and politics on the other. Thus the middle of the 4th climate corresponds to the latitude of Rhodes (for Greeks) and of Raqqa and Rayy (for Muslims); the middle of the 5th to Constantinople, Rome and Barcelona; the middle of the 6th to the Po Valley; and the middle of the 7th to Paris. What is intended on the Destombes astrolabe is probably that the 4th climate should serve, say, Tunis; the boundary between the 4th and 5th Southern Spain, particularly Cordova; the 5th Northern Spain, of which the eastern part was called Francia, at least by the maker of the Destombes astrolabe; the 6th and 7th vaguely serving southern central (for example, Lyons) and northern central France (for example, Paris). This does not mean that the astrolabe was necessarily ever used in any of these places, except for the one latitude particularly favoured by the maker, on which see below.

It is difficult to determine the value of the obliquity used when a set of latitudes is presented with values that are rounded to the nearest half-degree. And it is unwise to try to determine the date of a set of latitude values related to the climates, not least because tradition remained so supreme in the Middle Ages that Ptolemy's outdated value 23;51,20° was used for centuries in spite of the fact that better values were available (in Islamic astronomical instrumentation it was used, amongst other values, until the 19th century).

One should not expect that the recomputed latitudes shown in *Appendix 1* were those actually used by medieval astrolabists, but when, for example, we find the latitudes of the climates given in the illustrations of the earliest known Western Islamic astrolabe (#4024) as:

Climate	Daylight	Latitude	Errors in minutes
1	13^h	16;27°	[0] (on the mater)
2	13½	23;16	should be 23;49 !!
3	[14]	30;22	[+1]
4	14½	36; 6	should be 36;1 !!
5	15	40;56	[+3]
6	15½	45; 1	[0]
7	16	48;32	[+1]

[33] E. S. KENNEDY, M. H. KENNEDY [1987], pp. 305 and 286.

we can be sure that he was using the Ptolemaic value of the obliquity, for the errors in the minutes shown above are based on recomputation with that parameter.

We know from Greek texts, if not from the sole surviving Greek astrolabe (#2) (see below), that Greek astrolabes bore three plates, which together with the mater, provided seven surfaces for markings for each of the climates.[34] The purpose was to make the instrument universal; this trend is also reflected in certain Greek astronomical tables and was continued by the Muslims.[35] The first Islamic astrolabes (see, as an example, #3702) had the same arrangement for the climates. By the 10th century most Eastern Islamic astrolabes had plates for specific latitudes; all surviving Eastern Islamic astrolabes from that century have this feature. But the earliest Eastern astrolabes to reach Andalusia must have had plates for the climates because the tradition started afresh there. So, for example, the illustrations of what is probably the oldest known Western Islamic astrolabe (#4024) show that this is the case, but the earliest surviving instrument (#110 = #135) already has latitudes for specific localities; both of these are probably from the 10th century. The same feature, markings for specific latitudes, is found on all of the Western Islamic astrolabes from the 11th century. But then some of the earliest European astrolabes start afresh once again with the climates (for example, #161, #166, #167, #420, #303 and #589). The Destombes piece was not copied from an instrument serving all of the climates.

If one plate of an astrolabe bears additional astronomical markings or inscriptions distinguishing it from the others, this is a clear indication that the maker favoured that particular latitude, or indeed that he was actually working there. In the inscription «ROMA ET FRANCIA» the first name clearly refers first of all to Rome. But there is already a possible confusion here: any Islamic astrolabe that may have inspired the Destombes piece would surely not have had Rome marked on a plate. If it was in the Eastern Islamic tradition it might well have had *bilād al-Rūm*, «lands of the Byzantines» or Byzantium, on a plate that would serve Constantinople, and the Arabic word *Rūm* might then have been confused with the city of

---

[34] D.A. KING [1994a], pp. 152 and 168-169, stresses the importance of the climates in the study of astronomical instrumentation. See also D.A. KING, G.L'E. TURNER [1994], pp. 171, 195-196, on their influence on the astrolabes of the Vienna and Nuremberg schools in the 15th and 16th centuries.

[35] On universal solutions in Islamic astronomy see D.A. KING [1986], [1987], [1993a].

Rome.[36] If it was in the Western Islamic tradition it would probably have had some cities in Muslim Spain identified on the plate and then, on the plate for latitude 41°, perhaps *bilād al-Rūm wa-bilād al-Ifranj*, meaning the land of the Byzantines and the lands of the Franks, that is, those parts of Spain that had not been conquered by the Muslims. Julio Samsó has collected numerous references to *bilād al-Ifranj* as Catalonia in the Arabic sources,[37] and the Latin sources for the name Francia have been collected by C. H. Brühl,[38] albeit without reference to the Arabic tradition! Professor Beaujouan insists that Francia was not used for Catalonia in the 10th century. It is, however, used for Catalonia on the Destombes astrolabe, and the reason why is clear from the Arabic sources, if not from the Latin ones.

The earliest known Eastern Islamic astrolabes from before *ca.* 1000 do not generally have place-names indicated alongside the latitudes on the plates (the only exception is #99, which has Mecca, Miṣr = Fusṭāṭ, and Ḥarrān engraved on three plates). Yet the sole surviving Byzantine astrolabe (#2, independent of Islamic influence and dated 1062) does have places associated with the latitudes. The earliest known Western Islamic astrolabes from before *ca.* 1100 also do, unless they feature the climates (as with #4024). It appears, therefore, that this feature was first introduced on Islamic astrolabes in Andalusia (like the calendrical scale – see below). That the Destombes astrolabe belongs to a tradition and is not an isolated object in featuring Rome on a plate is indicated by the fact that several 11th-century Andalusian astrolabes have *Rūmā* (or *Rūmiyya*) *al-kubrā*, meaning Rome, at the end of a list of localities on the plate for *ca.* 41° (#121, #2527, and also #1069). The only Eastern Islamic astrolabe known to me bearing an inscription *al-Rūm* (= Constantinople) is a very late Egyptian piece (#1065), on which the plate for 41° is marked for Istanbul and that for 42° is marked for *al-Rūm*. It is not clear what the maker had in mind, for by his time Byzantium no longer existed (at least not in Asia Minor), its former capital was generally called *Qusṭanṭīniyya* even by Turks and Arabs, and Rome would have been spelled *Rūmā*. He was clearly following an old tradition.

---

[36] See the articles *Rūm* [Byzantium] and *Rūmā* [Rome] in *Encyclopaedia of Islam*. It should be pointed that no surviving early Eastern Islamic astrolabe bears a plate marked specifically for Byzantium and no surviving early Western Islamic astrolabe bears one mentioning *bilād al-ifranj*.

[37] J. SAMSÓ [1995].

[38] C.R. BRÜHL [1990], ch. 2, *Die geographisch-politische Terminologie des 9.-13. Jahrhunderts*, especially Section 1, *Francia und verwandte Begriffe*, pp. 83-130.

Finally, it should be remarked that the astronomical markings on the plates are crudely engraved with a heavy hand. They bear no resemblance to the careful markings of the earliest Eastern or Western Islamic astrolabes. They were engraved by an amateur, someone with little experience, yet not incompetent. As examples we may mention the section of the horizon above the inscription «ROMA ET FRANCIA» and the second hour-curve on the plate for latitude 36°, each of which has been engraved at least three times, as well as the hour-curves continued carelessly within the summer solstitial circle underneath the above inscription.

The Back

The fact that there is a calendar scale at all on the instrument is indicative that the astrolabe was copied from a Western Islamic astrolabe rather than an Eastern Islamic one. For in the Islamic East only rarely were such scales marked on astrolabes. We know that there were some in Central Asia about the year 1000;[39] otherwise the Eastern astrolabes displaying such scales are all later (#105, #4026, #140 and #4029, from Syria and the Yemen, 12th-14th century). For some reason yet to be understood, all Western Islamic astrolabes have a calendrical and solar scale on the back. Does this represent a lost Byzantine or a lost Roman tradition? Or were the first Eastern Islamic astrolabes to come to Andalusia by chance fitted with such scales? Or was the scale first developed by a Muslim astrolabist in Andalusia? The fact that the month names are Arabicized from the Latin forms indicated that these scales are a Western Islamic innovation, otherwise one would expect to find Semitic forms for the Syrian months or some other (Coptic or Byzantine) forms. In short, the origin of these scales has still to be explained. But the earliest known Andalusian astrolabes (#4024 and #110=#135), which in all other respects, even to the shape of the rete, resemble the earliest Eastern Islamic astrolabes (for example, #99 and #3501), do have a calendrical scale, and so does the Destombes astrolabe.

The dangers of dating instruments by the information that may be gained from calendrical scales are well-know to specialists. Here again

---

[39] al-Bīrūnī (E.S. Kennedy [1976], vol. I, p. 76, vol. II, pp. 28-29) relates a charming story about a muezzin who was upset because the Byzantine months were marked on an astrolabe where he would have preferred to see the months of the Muslim calendar. al-Bīrūnī tried to explain the matter to him, but the hapless muezzin solved his own problem by breaking the instrument.

tradition reigned supreme. But on the Destombes astrolabe we find, for example, that the equinoxes are at March 15 and September 17 and the solstices at June 18 and December 15. These we should compare with the information that may be gained from contemporaneous written calendars. The obvious place to begin is in the *Calendar of Cordova* dated 961 A.D.[40] Here we find the corresponding dates March 16 and September 18, and June 18 and December 15/17. The agreement is quite close, but this does not mean that the information on the Destombes astrolabe was derived from the *Calendar of Cordova*; more likely both were ultimately derived from a common source yet to be properly identified.[41]

The month-names on the Destombes astrolabe include various unusual forms ending with -AS and -ES/(E)S. The complete list is:

INRS FBRS MRTS APRILES MAIAS IVNIAS IVLIAS AGSTS SBTRS OCTBRS NBRS DCMBRS

The endings indicate genitive forms, rather than, as we should expect, the nominative singular of the nominal forms. (Western Islamic astrolabes from the 10th and 11th centuries use transliterations of more standard Latin forms).[42] But similar forms for the individual month-names are attested already in 3rd-century Latin inscriptions on mosaics from Tunisia,[43] and Destombes himself noted two manuscripts of Bede's *De*

---

[40] See CH. PELLAT [1961]. The agreement between these two sets of dates is somewhat closer than the relationship of the dates on the astrolabe with those presented by Gerbert (March 15, June 16, September 16, and December 15), as recorded by M. DESTOMBES [1962], p. 28, from MS Paris B.N. lat. 7412 (on which see the bibliography to #4024).

[41] M. VILADRICH [in print], verifies the connection to the solar tables of the *Mumtaḥan Zīj* compiled by the astronomers of the Caliph al-Ma'mūn (Baghdad, *ca.* 830), which work is actually mentioned in the *Calendar*. (The Mumtaḥan solar tables are contained in the unique Escorial manuscript of a recension of the *Zīj*, the related table showing the dates of the sun's entry into the signs is preserved in the Istanbul manuscript of the *Zīj* of Ḥabash (*ca.* 850) – see already M.-T. DEBARNOT [1986], p. 45.) But it is more likely that this simple calendrical information, which may indeed be derived from the *Zīj*, was transmitted in the form of an almanac, that is, in the folk astonomical tradition rather than the mathematical one. The utility of investigating calendrical information of this kind is well illustrated in O. NEUGEBAUER [1983].

[42] The endings -h on the Arabic forms māy•h, yūn•y•h and yūl•y•h (where • indicates an unspecified vowel) could indicate the Spanish Arabic pronunciation māyu, yūniyu and yūliyu, as in the name of the 11th-century Andalusian astronomer al-Zarqāllu, which in Arabic also ends with -h.

[43] The forms that occur on a Roman mosaic now in the Musée Archéologique, El-Djem, Tunisia are: IANVARIVS FEBRARIVS MARTIAS APRILES MAIAS IVNIVS IVLIVS AVGVSTAS SEPTEMBER OCTOBRES NOVEMBER DECEMBER. See M.R. SALZMAN [1990], figs. 59-70.

The form ARIAS for ARIES attested amongst later additions by a Catalan to the 11th-century Cordovan astrolabe #3622 is vernacular (information provided by Kurt Maier). Àrias and Àrias are found in the Catalan *Tractat d'astronomia* of Raymond Llull (Catalonia, ca 1300): see, for example, J. GAYÀ [1981], p. 218, line 32, and p. 219, line 37.

natura rerum in which they occur.[44] There are surely numerous other attestations of these variants in manuscripts unknown to us. In other words, these are indeed Latin, but not the kind one finds in Latin text-books. Why these forms were introduced instead of the nominative forms is more a matter for the linguists and the cultural historians: clearly they correspond to the forms of the standard Latin month-names in the typical Roman usage, for example, *ante diem quartum kalendas maias*, the 4th day before the Kalends of May. But they were doubtless copied from a calendar.[45] In short, the unusual forms of the month-names are used consistently on the Destombes astrolabe (though not on the Tunisian mosaics) and belong to a well-attested, if uncommon, tradition.

The forms for the zodiacal signs, on the other hand, are unremarkable:

ARIES TA^{V}RVS GEMINI CACER LEO VIRGO LIBRA SCORPIO SGTRS CPRS A^{q}VARIVS PISCES.

Of more historical interest are the labels on the shadow-scales, namely, VBRA RECTA and VBRA DEIECTA on the horizontal and vertical scales, respectively. The orthography VBRA for VMBRA is not, to my knowledge, attested elsewhere on an instrument, but the suppression of an M and an indication of its presence by a dash over the preceding letter is a frequent Latin abbreviation, so that VBRA is perfectly reasonable. RECTA is the standard Latin form for the standard Arabic term *mustawī*, meaning 'even' (in the sense of a plane) or of the common alternative *mabsūṭ*, which means 'laid out' (like a carpet). However, the term DEIECTA does not correspond happily to the standard Arabic terms for vertical shadows, *maʿkūs* or *mankūs*, both meaning 'reversed'; those terms were later usually rendered in Latin with VERSA.[46] DEIECTA, meaning

---

[44] MSS Munich Bayerische Nationalbibliothek lat. 210, fol. 137, and Marcianus lat. 3477, fol. 51, cited in M. DESTOMBES [1962], p. 3, note 1.

[45] In a 6th-century calendar from Carthage (reproduced in the article *Kalendaria* by H. Leclercq in *Dict. arch. chrét.*, VIII:1, cols. 644-645) we note, for example, the following extracts:

XIII kalend(as) maias	martyris Mappalici
XI kal. iunias	sanctorum Casti et Emili
XIII kal. iulias	sancti Gervasi et Protasi martyrum
idus iulias	sancti Catulini martyris
XV[I] kal. agustas	sanctorum Scilitanorum

The other month-names are drastically abbreviated.

[46] On the Arabic terms and their Latin equivalents see P. KUNITZSCH [1982], no. 61: *al-ẓill al-mabsūṭ*, and no. 62: *al-ẓill al-mankūs*.

'thrown downwards', could, in fact, easily be understood as referring to a horizontal shadow; perhaps there was some confusion with DERECTA a variant of DIRECTA.

The single shadow-square is a sign of an early Islamic tradition. As far as we know, this (and other less common) forms of shadow-scales were introduced in Baghdad in the early 9th century. They did not feature on Greek astrolabes (those on #2 are later additions), or on the earliest Eastern Islamic astrolabes (#99, #3501 and #3702). A single shadow square features on some of the earliest Eastern Islamic astrolabes (#1026, #1179, #100 and #4030).[47] The double shadow-square which later became standard is found for the first time on only one of the latest of the 10th-century Eastern Islamic astrolabes (#122).[48] The two earliest Western Islamic astrolabes (#4024 and #110) have but a single shadow-square; on the other hand, the 11th-century Western Islamic astrolabes (#116, #3650, #2527, #2572, #3622, *etc.*) have a double one.

Finally, we note that the astronomical markings on the back are reasonably accurate. Consider, for example, the divisions for the days on the calendrical scale. It is not a trivial undertaking to divide a circle into 365 divisions, and this the maker of the Destombes astrolabe has achieved with some measure of success.

The orthography VIRGVO on the rete of the Destombes astrolabe is probably indicative of a Catalan provenance for these later inscriptions. The same form is attested in the Catalan text (next to the Latin Virgo) of the *Tractat d'astronomia* of Ramon Llull - see, for example, J. GAYA [1981], especially p. 264, line 1599, and p. 267, line 1717 (information provided by Kurt Maier, Frankfurt). In addition the 'misspelled' form RISES (for PISES from PISCES) seems to be indicative of another Catalan tendency to write R for P or B: see the form OCTORER on another instrument of Catalan provenance, namely, #162, on which see D. A. KING, K. MAIER [in press].

---

[47] The most historically interesting piece in this regard is a 10th-century Eastern Islamic astrolabe featuring four different types of shadow-scales (#1179), all invented in the previous century.

[48] #101, another 10th-century Eastern Islamic piece, often paraded as the 'Astrolabe of Pope Sylvester', has a double shadow-square in the Andalusian style on the back, but this was added by a European, possibly in the 19th century.

THE ISLAMIC ASTROLABE ON WHICH THE DESTOMBES PIECE IS BASED

The evidence from Catalonian inscriptions gathered by Professor Anscari Mundó suggests that the provenance of this piece is Catalonia, although Professor Wesley Stevens, using *Capitalis quadrata* inscriptions from the larger trading and crafts area stretching between Catalonia and Northern Italy, warns that Southern France and Northern Italy should not be excluded. As to the question of the date, the late 10th century is a possibility, confirmed but not proven by the inscriptions. But it is also confirmed by other factors. First, the Destombes astrolabe was not copied from an astrolabe of the kind that we know as the earliest surviving Andalusian astrolabes (#4024 and #110 = #135). These appear to date from the 10th century; they certainly precede the 'standard' Andalusian astrolabes of the 11th, of which we have several examples (see *Appendix 2*, Section 1.3). How can we be sure that these two pieces are earlier? Because an illustration of one of these (#4024, alas available to us only in this form) has astrolabic markings for each of the seven climates, not attested on any other known Andalusian astrolabe, but common on very early European astrolabes (#161, #166, #167, #420, #303, #589). And the Destombes astrolabe precedes the other earliest European astrolabes (notably #420, which has a rete design that is a crude imitation of the earliest Eastern Islamic astrolabes – compare it, for example, with #99 and #3501). Other early European astrolabes based on the Western Islamic type (that is, as it had developed by the 11th century) are clearly later (#191, #428 and #3037). In other words, the Destombes piece is copied from a very early Western Islamic astrolabe whose rete-design would have been already out of date in the 11th century.

By the 12th and 13th centuries Europeans were able to make more sophisticated astrolabes. By the 13th century someone in Catalonia could make a perfectly respectable astrolabe with an original rete design (#416, with month-names in Catalan, dialect of Valencia, and a V-shaped frame within the ecliptic, later adopted on some English astrolabes in the Chaucer tradition). By the late 14th century (another date that has been proposed for the Destombes piece!) the Aragonese Petrus Raimundus in Barcelona was able to make a spectacular astrolabe with elegant design (#3053).

The possibility remains, in my opinion, that the Destombes astrolabe may have been made in the early 10th century, as opposed to the late 10th century. There were surely Islamic astrolabes available in Catalonia that early, and perhaps even one European who could make an attempt at copying one.

LATER EUROPEAN SCHOOLS OF ASTROLABE CONSTRUCTION

There are several classes of early European astrolabes, distinguished mainly by their rete design. There is the Destombes astrolabe, which is a copy of the earliest Andalusian astrolabes. Then there are the Eastern Islamic-type astrolabes (#420), modelled on the Abbasid astrolabes. Then there are the Hispano-Mauresque astrolabes (#191, #428 and #3037), clearly modelled on the 11th-century Andalusian astrolabes. Then there are the astrolabes with a quatrefoil as decoration (#162) or with several complete or partial quatrefoils as prominent components of the rete (#202, #290, #547 and #3203). At least the earliest of these are related in a way that is not yet clear to Eastern Islamic astrolabes that have a quatrefoil as decoration (#111 (dated 984/85), #3 (dated 984, but the replacement rete is from the 12th century (Stautz)), #5 (dated 1223/24), and a few later pieces) and Western Islamic astrolabes that exploit the quatrefoil in the structure of the rete (#154 from 13th-century Andalusia, misdated by Gunther to 1747). Likewise European astrolabe retes exhibiting zoomorphic features (especially #2041) appear to be derived from 10th-century Islamic predecessors (compare #111). The later (14th- and 15th-century) French astrolabes with an upper frame (#202, #337, and all of the Fusoris astrolabes) are related to a design known in 13th-century Syria but attested only on one Islamic instrument (#4050). It seems likely that this idea might have been picked up during the Crusades. The Y- and V-shaped frames within the ecliptic so characteristic of English astrolabes (#291, #301, #304, *etc.*) are found first on an early (13th-century?) Catalan astrolabe (#416).[49]

Many questions about the development of the astrolabe in Spain, remain unanswered, and these will continue to cloud our understanding of developments elsewhere. Also problematic is the development of the astrolabe in Italy. We have several examples of Italian instruments that do not fit into any pattern: not only the piece dubbed a fake (#4507) that we have already mentioned, but also two astrolabes associated with Urbino, one (#493) very medieval in appearance, and the other (#4506, dated 1462), with identical basic design, 'Renaissance' in style. It was these which inspired the earliest surviving astrolabe (from 1457) of the 15th-century Vienna school (#549); a simplified version was adopted on the instruments associated with Regiomontanus (#640, *etc.*) and later again by Georg Hartmann.

---

[49] This instrument has inscriptions in the Valencia dialect of Catalan: see K. MAIER [1994].

Here is material indeed for the art historians; but it is better that the retes be first interpreted and classified by a historian of astronomy because rete-styles changed over the centuries not only with local tastes but also as a result of the fact that the stars move with time, and new designs were developed to ensure that star-pointers did not become too long and ungainly. There is enough material available for a doctoral dissertation on rete design, which could build on the dissertation of Burkhard Stautz of Frankfurt University completed in 1995 and dealing *inter alia* with star-positions on retes.

AN UNRESOLVED PROBLEM OF PROVENANCE

The number 56.950 that appears by the side of a sketch of the instrument in Destombes' notes may the price in new French francs that he paid for it. This would have been appropriate for an important instrument in the early 1960s (the unique Islamic spherical astrolabe now in Oxford sold for £3,200 in 1962[50]). It is related that Marcel Destombes bought the instrument from a dealer named Gilbert Suc from Toulouse who was representing a Spanish collector. (No private Spanish collectors of astrolabes are currently known, so we may surmise that this Spaniard had but one astrolabe). It is a later addition to his notes, in the same blue ball-pen ink as the addition «je l'ai acheté le 13 mars (1961)». But there are other possibilities that cannot yet be excluded, for example, that Destombes first learned of the instrument in 1956 and his first dating of it was 950. The possibility that the figure was a museum inventory number has been considered and abandoned.[51] (On the mater there are traces of glue indicating that there was once a rectangular label attached below the horizon).[52]

---

[50] J. DE GRAEVE [1995].

[51] Upon returning to Frankfurt after the Congress, I found that one Andalusian astrolabe from the year 1067 (#117) bears an inventory number of 50762 from the Museo Arqueológico Nacional in Madrid. Enquiries at that museum by Professor Julio Samsó after the Saragossa Congress established that no object bearing the number 56950 had ever been housed there. And there is no trace of any number on the instrument now. There is, however, a 1 cm long section on the rim to the left of the throne where the metal is brighter than elsewhere, but no 5-digit number could have been painted there – the space is simply too small.

[52] The astrolabes in Spanish collections other than Barcelona have yet to be inspected for the new Frankfurt catalogue. Fortunately they were all catalogued by García Franco, but he did not give their inventory numbers.

Whatever the original provenance, in 1961 a miracle occurred. The instrument was bought by someone – the only person in the world at the time – who knew what it was. (Even Derek Price, probably the leading specialist on instruments at the time, was not sure). He showed it only to people who he thought might be able to form a sensible opinion about it (notably Henri Michel and Derek Price).

In short, Destombes did the right thing three times: buying the instrument, publishing it, and ensuring that on his death it went to a museum that would appreciate it.[53]

Concluding Remarks

There is now, in my opinion, nothing standing in the way of an attribution to late-10th-century Catalonia.[54] If this should be incorrect, then either the early 10th century or the early 11th century seem to me to be the only other plausible dates, the former being more likely than the latter. The field of the history of medieval astronomical instrumentation is progressing in leaps and bounds as each new instrument is catalogued, but various incorrect notions widespread in the secondary literature will be hard to kill.[55] The Carolingian astrolabe is the prime example of a historical instrument that has been been misunderstood by modern scholars, with the notable exception of Marcel Destombes, a man who knew in particular that instruments speak their own language, a language different

---

[53] The Institut du Monde Arabe is the only museum in Paris that has published a catalogue of its astronomical instruments. The documentation of instruments in other museums in Paris – the Bibliothèque Nationale, the Musée maritime, the Musée des arts et métiers, and the Observatoire de Paris – is pitiful, and the documentation of instruments in private collections in Paris non-existent (the collection in nearby Écouen has been catalogued but is an exception). All medieval instruments in these collections have been catalogued within the framework of the new Frankfurt catalogue.

[54] This is the opinion I expressed in Saragossa. It is apparently shared only by Professors Anscari M. Mundó and Julio Samsó of Barcelona. Professor Paul Kunitzsch of Munich writes (letter dated 3.12.1993) that judging solely from «the aspect of the (older) inscriptions, the *abjad* rendering (of the numerals) and the placing of the stars» (which aspects alone he has investigated), he sees «no reason why the instrument should not belong to the date proposed by Destombes».

[55] See, for example, the instrument (#640) described in D.J. De Solla Price [1959] and G. L'E. Turner [1989], with new information incorporated in D.A. King [1993b], pp. 47-51; D.A. King, G.L'E. Turner [1994]; D.A. King [1994a], pp. 175-176. Then there are two 14th-century English horary quadrants (#5521 and #5522) that have been declared fakes (D.J. De Solla Price [1982], p. 131), although clearly one bears later markings, as well various 15th-century Viennese sundials (#7511-#7520) likewise dismissed as fakes (*ibid.*, p. 132) – on the last-mentioned see now D.A. King [1994a], pp. 186-187.

from that of the manuscripts. For the time being the object remains a testimony to how little we still know about the introduction of the astrolabe to Europe. On the brighter side, we now know more about the other earliest surviving astrolabes, and consequently, perhaps the reader may concede, more about the Carolingian astrolabe.

Added in proof:

An astrolabe from ca. 1300 (#4556) that has only recently become available for study bears star-pointers some of which resemble those on the Destombes astrolabe. This early piece with a very unusual rete design as yet defies identification of its provenance. None of the names of the stars, months or zodiacal signs show any vernacular influence, and the plates serve the seven climates.

One thing still disturbs me about Destombes astrolabe, namely, that it does not *look* as if it was copied from an Islamic astrolabe. By 'Islamic atrolabe' I mean here the king of astrolabe that might have been brought from the Islamic East to Andalusia in the 9th or 10th century, or one that was copied by a Muslim craftsman in Andalusia from such an instrument. There are certain aspects of the piece which are definitely of 9th-century Baghdad origin, namely, the azimuth curves on two of the plates and the shadow-square on the back. And there is one aspect which betrays the influence of a Western Islamic astrolabe, namely, the alphanumerical number notation. But there are others which would not – as far as we know – have typified those instruments imported into Andalusia, namely, the general aspect of the rete, the shape of the star-pointers, and the name ROMA on one of the plates.

The origin of the calendrical/zodiacal sçale on the back of Western Islamic astrolabes is still clouded in obscurity. No Eastern Islamic astrolabes from before the 14th century are known that bear such a scale and the Western Islamic tradition of such scales, which appear on the Destombes astrolabe and on every known Western Islamic astrolabe, appears to be completely independent of Eastern Islamic influence. Are these scales, then, an Andalusian invention, or were they already available in Spain when Eastern Islamic astrolabes were introduced there, perhaps originally imported to Visigothic Spain from elsewhere? This question has been addressed by Julio Samsó ([1992], pp. 32-36), who has shown that tables in rectangular and circular from giving the positions of the sun throughout the year are attested from the 9th century in Latin-Visigothic manuscripts from Spain.

There is an explanation of the Destombes instrument at which we have hinted above, but which demands further consideration and investigation in sources that have not yet been identified, namely, that we are dealing with an astrolabe copied from an Islamic instrument based partly on an astrolabe imported from the Islamic East and partly on an 'indigenous' Latin – and here I mean ultimately Roman – tradition of the instrument. Of this latter tradition, if it ever existed, we currently know nothing, but that does not mean that it did not exist. This hypothesis certainly explains why ROMA should be featured on one of the plates of the Destombes astrolabe. A tradition of crude vertical sundials, unknown from Antiquity, existed in early medieval Europe (witness the 10th-century sundial from Canterbury). And how do we explain the existence of a curious pedagogical instrument (from 11th-century Regensburg) bearing an orthogonal projection of the celestial sphere associated with Aratus? These are both clearly survivals of a Roman tradition in early medieval European astronomy. This tradition which was eventually replaced in Europe by the Graeco-Arabic one, has been neglected by most historians of medieval European astronomy (with the notable exceptions of Bruce S. Eastwood and Stephen McCluskey in addition to Julio Samsó). There appear to have been astronomical (*i.e.* astrolabic) clocks in the Rome region in Antiquity: Varro (*ca.* 80 A.D.) mentions an *avarium* at Tusculum some 25 km from the capital, in which there was such a clock. And the monumental sundial of Augustus (*ca.* 9 B.C.) in Rome, excavated in the late 1970s, bears a calendrical/-zodiacal scale. In other words, it would not have been beyond the wits of the Romans to have made astrolabes, or to have devised a circular calendrical/zodiacal scale for the back of an astrolabe. One day it may be possible to show that the astrolabe, too, was known in the Roman Empire, and that the Destombes astrolabe is the sole surviving example of an astrolabe bearing some traces of the influence of such a Roman tradition.

Or is the Destombes astrolabe the only example? Why do certain early Italian astrolabes (and even some Renaissance pieces) demonstrate features not found on early astrolabes from other regions of Europe that were influenced by the Western Islamic tradition? There is a reason why Robert T. Gunther did not make any remarks preceding his section on Italian astrolabes: he simply did not know what to write. And, with the exception of the writings of Tullio Tomba, nothing of consequence has been written on medieval Italian astrolabes since the time of Gunther.

Plate A: The front of the Destombes astrolabe.
[Photo courtesy of the Institut de Monde Arabe, Paris.]

Plate B: The back of the Destombes astrolabe.
[Photo courtesy of the Institut de Monde Arabe, Paris.]

Fig. 1 – #3042 – Paris IMA. Details of the Destombes astrolabe. (a) The inscription on the plate for 41;30°. Note the poorly-engraved horizon above the inscription; no known Islamic astrolabe from before *ca.* 1600 has markings as crude as these. (b) Part of the scales on the back. Note the careless division of the calendrical scale. (c-d) The names of the signs on the rete are later additions. Here we see VIRGVO and RISES for Virgo and Pisces; the former is a medieval Catalan form attested elsewhere, and the latter a curious Catalan scribal variant (an R for a B or a P) with parallels attested on another medieval Catalan astrolabe (#162 – see Fig. 20).

Fig. 2 – #2 – Brescia MEC. The form of the rete of this sole surviving complete Byzantine astrolabe, dated 1062, is the form in which the instrument was transmitted to Muslim astronomers in the 8th century. The evidence is provided by #3702 (preserved in Baghdad – no photos are available, but see B. Stautz [1994]), which is identical (even down to the bird on the pointer for Vega) and is either an original 9th-century Eastern Islamic astrolabe or a copy of one.

Fig. 3 – #110 – London BM. The oldest surviving Andalusian astrolabe, made probably in the 10th century, or at the latest in the early 11th century. The star-pointers are mainly Abbasid in form, and a single counter-change has been introduced on the left-hand side of the equinoctial bar. The latitudes serve a mixture of localities in the Islamic East and West. Some of the Arabic inscriptions have been erased and replaced by a European. A second European erased those on the upper half of the rete; his orthography SAGIPT for Sagittarius is attested on other medieval astrolabes from Italy (#4554 and #4523, the latter dated 420).

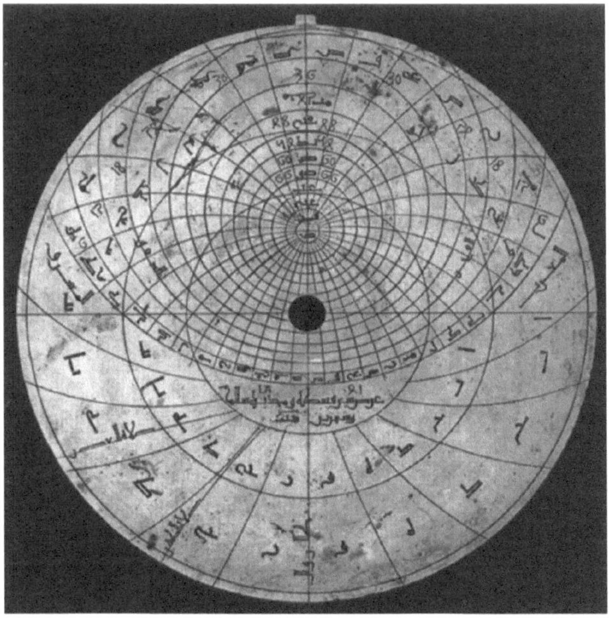

Fig. 4a – #110 – London BM. The plate for latitude 42° on the previous astrolabe. The locations mentioned in the Arabic inscription are Saragossa, Medinaceli and Santarem, but the European who added the Gothic inscriptions thought that the plate served 41°.

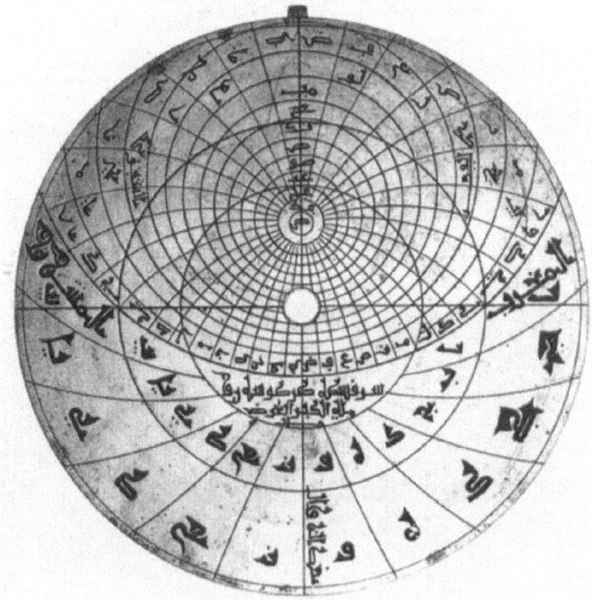

Fig. 4b – #2527 – Oxford MHS. The plate for latitude 42° on the astrolabe made by Muḥammad ibn Saʿīd al-Ṣabbān in Guadalajara in 474 H [=1081/82]. The localities specifically mentioned are Saragossa, Tortosa and Rome (*Rūmiyya al-kubrā*). This latitude-value of Rome is to be interpreted somewhat loosely. On the plate for latitude 41°30' on the astrolabe by Ibrāhīm al-Sahlī of Valencia, dated 478 H [=1086] (#121 – Kassel SKS), we find Saragossa, Rome (*Rūmā al-kubrā*) and Khwarizm, but the European who added the Gothic inscriptions to that plate interpreted the latitude as 41°.

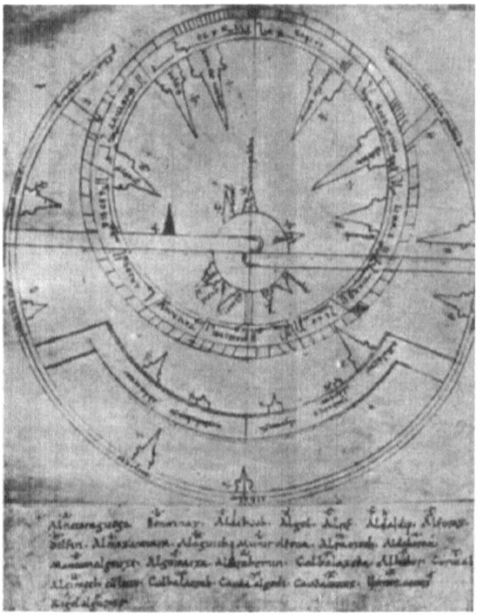

Fig. 5 – #4024 – Paris BN. An illustration of an early (10th-century?) Andalusian astrolabe by Khalaf ibn al-Mu'ādh in an 11th-century Latin manuscript. Most of the star-pointers are modified from the standard early Eastern Islamic type, and the equinoctial bar has a single counter-change at the centre. The plates (shown in other illustrations) are for the seven climates.

Fig. 6 – #116 – Berlin SBOA. The sole surviving complete astrolabe of Muḥammad ibn al-Ṣaffār, made in Cordova in 420 H [=1029]. (The Edinburgh piece by the same maker (#3650) has a replacement Eastern Islamic rete.) Note the Abbasid-style star-pointers in the upper part of the rete, and the spikey ones along the inside of the equinoctial bar, as well as the circular frame at the bottom. The latitudes represented on the plates are associated with a single locality on both of his instruments.

Fig. 7 – #3622 – Cracow JUM. An unsigned astrolabe made in Cordova in 446 H [= 1054/55], with later additional inscriptions (month-names, names of signs and place-names) in medieval Catalan, dialect of Valencia. Note the heart-shaped ring at the bottom of the rete, and the fact that the equinoctial bar is counter-changed only at the centre (all other surviving 11th-century Andalusian astrolabes have more counter-changes). The throne is also unusual (compare #2572), and can be seen behind various early European thrones.

Fig. 8 – #2572 – Washington NMAH. An astrolabe by Muḥammad ibn al-Sahlī dated 483 H [=1090/91], made in Valencia. The rete is a replacement, with some of the lower star-pointers resembling those in the Destombes piece. Also the counter-changes on the equinoctial bar are the reverse of those on the Destombes piece. The inscriptions on the rete are in a mixture of Hebrew and Judaeo-Arabic (Arabic written in Hebrew letters), further indication that the rete was added in Andalusia.

Fig. 9 – #420 – Greenwich NMM. Perhaps the oldest European astrolabe after the Destombes piece; it has also been dubbed a fake. But the maker was extremely well-informed about the early medieval star-names, and he engraved the plates for each of the seven climates. The design of the rete corresponds to the earliest Andalusian retes, that is, with Abbasid-style star-pointers and slight innovations, such as the circular frame at the bottom.

Fig. 10 – #166 – Oxford MHS. An early European astrolabe with at least two star pointers (inside the upper ecliptic) which resemble those on the Destombes astrolabe. The others have been either simplified or embellished. The throne is similar to that on the Destombes piece, and the photograph shows one possibility how the shackle and ring on the latter would have been attached. The plates serve the climates.

Fig. 11 – #161 – London BM. An early European astrolabe with a rete in the Abbasid style with star-pointers in the Andalusian tradition. Note that the equinoctial bar is counter-changed in the middle, an indication of early Andalusian influence. The mater and three plates are engraved for the seven climates in disguise: their latitudes are given as 16°, 23°, 30°, 36°, 41°, 45° and 48°.

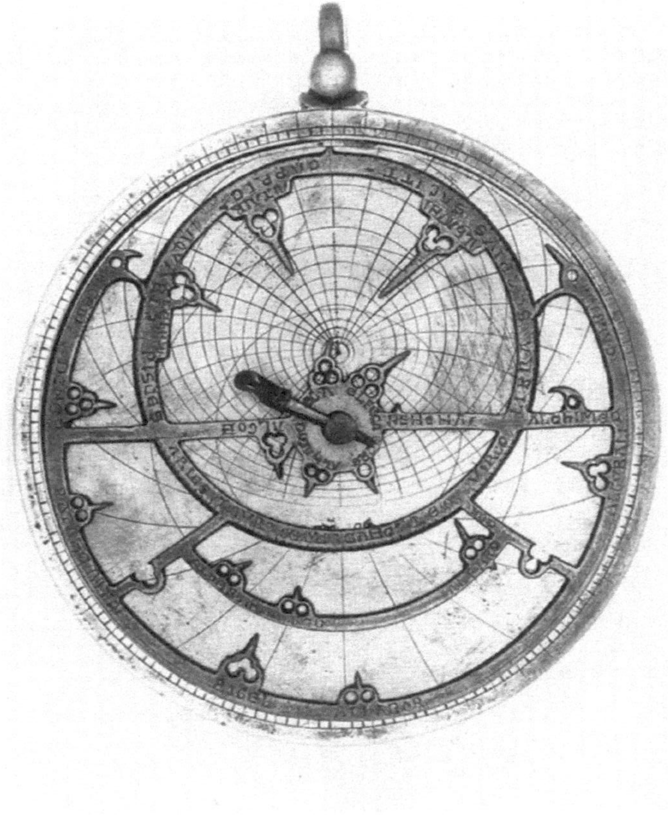

Fig. 12 – #300 – Oxford MHS. Another very early (12th- or 13th-century) European astrolabe, with a throne similar in shape to that on the Destombes piece, and with star-pointers likewise very similar, albeit pierced after the fashion of 11th-century Andalusian astrolabes. Note that the scales on both the front and the back are not labelled; the plates bear the latitudes engraved in Gothic numerals, but these are not necessarily original.

Fig. 13 – #621 – Munich DM. Note the throne and the simplified and reduced number of star-pointers on this early European astrolabe. This is a composite piece, with a back from an astrolabe with Hebrew inscriptions.

Fig. 14 – #416 – Greenwich NMM. A Catalan astrolabe from the 13th century. The V-shaped frame inside the rete explains the origin of the traditional Y- and V-forms on the English astrolabe retes associated with Chaucer. The Catalan provenance is apparent from the forms of the month-names on the back and some of the place-names on the plates, which are in the medieval dialect of Valencia.

Fig. 15 – # 3037 – Milan PC. An early European astrolabe with a rete clearly in the 11th-century Andalusian tradition. The plates serve latitudes between 36° and 48° and are not associated with any localities, which makes localizing the piece difficult.

Fig. 16 – #3053 – Boston, MFA. An astrolabe made in Barcelona ('Barchinone', with latitude 41° and longitude 39°!?) in 1375 by an Aragonese named Petrus Raimundus. It shows what an astrolabe made by a Spaniard some 300 years after the Destombes piece looks like.

Fig. 17 – #4551 – Rome OA. A 'Franco-Italian' astrolabe from the 14th century, complete but for some of the star-pointers. This piece provides us with a unique glimpse of the way in which star-pointers were cut out of a circular piece of metal. Its provenance is uncertain: Toledo, Paris, London and Rome are featured on the plates, as well as Carthage and Ephesus!

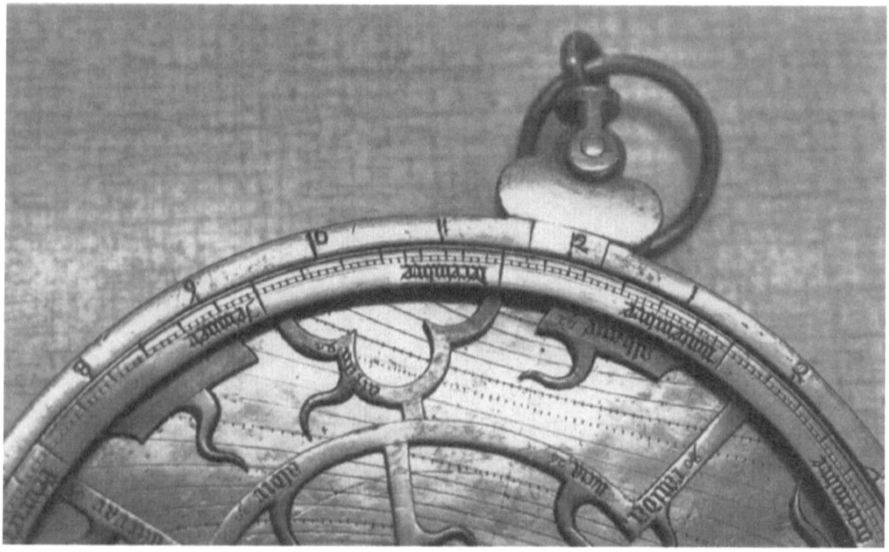

Fig. 18a – #198 – London SM. The throne on a remarkable geared astrolabe from N. France, datable to the 14th-century. This astrolabe comes from the same tradition as the Berselius astrolabe (#202), which can be localized to Picardy.

Fig. 18b – #190 – London VA. The throne on a 14th-century French quatrefoil astrolabe. (Courtesy of Prof. Gerard L'E. Turner.)

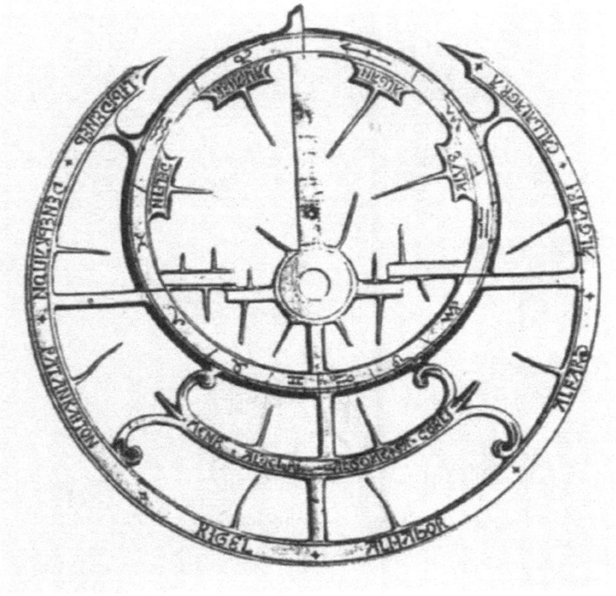

Fig. 19 – #4507 – Florence MNB. The anomalous rete of an astrolabe that has been thought to be an 19th-century fake. Whoever made this was very familiar with medieval astrolabes, too familiar to have lived in the 19th century. The star-positions are particularly problematic, but compare the pointers on the equinoctial bar with those on #116. Also, the unusual forms of the star-names and even the very number of star-pointers reflect a connection with a particular medieval star catalogue.

Fig. 20 – #162 – London SA. A Catalan astrolabe from *ca.* 1300 with a remarkable rete. The provenance has only recently been recognized: three star-names are in medieval Catalan. The rectangular frame inside the upper half is Islamic in origin (although no Islamic pieces with this survive); it inspired the distinctive forms of the astrolabes of Abū Bakr of Marrakesh *ca.* 1215 (see R.T. Gunther [1976], vol. I, p. 264), and it is also found, without the quatrefoil, on an astrolabe with Judaeo-Arabic inscriptions from *ca.* 1300 (#3915). The *miḥrābs* along the bottom of the rete are Andalusian (see #2527, which has one *miḥrāb* on the right). Likewise the distinctive quatrefoil is Islamic, being first attested on the astrolabe of al-Khujandī (#111, dated 984/85) in the same position. The bird's head representing Vega is found only on the Byzantine astrolabe of 1062 (#2) and on the earliest known Eastern Islamic astrolabe (#3702), on one if not on any other known Eastern Islamic astrolabe before *ca.* 1200 (#5 – see R.T. Gunther [1932], vol. I, pl. XXV after p. 118) or any known Western Islamic astrolabe.

Fig. 21 – #4556 – England, PC. This astrolabe came to light only in 1994. The general appearance of the piece suggests a dating as early as *ca.* 1300. The plates serve the climates, another indication of an early date. But the distinctive design of the rete bears no resemblance whatsoever to any other known rete. Nevertheless some of the star-pointers resemble those on the Destombes astrolabe. One of the plates has been reworked for latitude 43;45°, suggesting a location in Northen Italy.

APPENDIX 1

LIST OF SURVIVING GREEK AND EARLIEST ISLAMIC AND EARLIEST EUROPEAN ASTROLABES

Notes: This list is adapted from the table of contents of the new catalogue of medieval instruments (not restricted to astrolabes) now in preparation in Frankfurt. The list contains only the earliest instruments and therefore omits Islamic astrolabes after *ca.* 1100 and medieval European astrolabes of the (pan-European) quatrefoil and (French and Italian (?)) upper-frame and (English) Y-/V-shaped frame variety, as well as those of the clearly-defined Fusoris (Paris) and Regiomontanus (Vienna and Nuremberg) workshops. This list should not be quoted by the section numbers, which are not yet finalized (*e.g.* 1.2.1), but rather by the corresponding International Instrument Checklist Number (*e.g.* #3702). Numbers up to #4000 are from S.L. GIBBS, J.A. HENDERSON, D.J. DE SOLLA PRICE [1973]; numbers above that (to #9999) have been assigned by the present author.

The following abbreviations are used frequently (for the others the reader is referred to *Appendix 3* and to S.L. GIBBS, J.A. HENDERSON, D.J. DE SOLLA PRICE [1973]):

BM	British Museum (London)
BN	Bibliothèque Nationale (Paris)
GNM	Germanisches Nationalmuseum (Nuremberg)
IGN	Institut für Geschichte der Naturwissenschaften (Frankfurt)
IMA	Institut du Monde Arabe (Paris)
JUM	Jagiellonian University Museum (Cracow)
MHS	Museum of the History of Science (Oxford)
MSS	Museo di Storia della Scienza (Florence)
NMAH	National Museum of American History (Washington, D.C.)
NMM	National Maritime Museum (Greenwich)
PC	private collection
PLU	present location unknown
TM	Time Museum (Rockford, Ill.)

1. Early Eastern Astrolabes (to ca. 1500)

1.1. *Byzantine astrolabes*

1.1.1. An astrolabe dated 1062 (#2 – Brescia MEC)
1.1.2. Miscellaneous item
a) A single plate (#4509 – PC)
b) Markings on an Abbasid plate (#109 = #3549 – New York MMA)

1.2. *The earliest Eastern Islamic Astrolabes (9th-11th Centuries)*

1.2.1. A late (Ottoman?) astrolabe signed by Aḥmad ibn Kamāl, copied (or possibly remodelled) from an early Abbasid astrolabe (#3702 – Baghdad AM)
1.2.2. Three instruments by Khafīf, apprentice of ʻAlī ibn ʻĪsā
a) An undated astrolabe (#1026 – Oxford MHS)
b) A solitary rete (#2529 – Oxford MHS)
c) Parts of an astrolabe illustrated in a 16th-century Italian drawing (#4030 – Florence UGDS)
1.2.3. An undated astrolabe by Aḥmad ibn Khalaf (#99 – Paris BN)
1.2.4. Two instruments by (Muḥammad ibn ʻAbdallāh known as) Nasṭūlus
a) An astrolabe dated 315 H (#3501 – Kuwait MIA)
b) An undated mater (with various Mamluk Egyptian additions dated 714 H) (#1130 = #4023 – Cairo MIA)
1.2.5. An undated astrolabe by Muḥammad ibn Shaddād (al-Baladī) with a later Maghribi rete (#1179 – PLU, formerly Berlin PC)
1.2.6. An undated astrolabe by al-Muḥsin ibn Muḥammad al-Ṭabīb (#3522 = #3527 = #3904 = #3919 – Rockford TM)
1.2.7. An unsigned undated Abbasid astrolabe with later (19th-century?) additions by a European (the so-called ʻAstrolabe of Pope Sylvester II') (#101 = #9001 – Florence MSS)
1.2.8. An unsigned mater and plates (#4022 – London SM)
1.2.9. Two instruments by Ḥāmid ibn ʻAlī (al-Wāsiṭī)
a) A mater dated 343 H (#100 – PLU < Palermo MN – stolen ?)
b) A mater with an illegible date [3]?4 H (with a replacement rete and plates by [Muḥammad Ṣaffār]) (#3713 – Cairo MIA)
1.2.10. An astrolabe by Ḥāmid ibn Khiḍr al-Khujandī dated 374 H (#111 – Kuwait PC)
1.2.11. An astrolabe by Aḥmad and Muḥammad, sons of Ibrāhīm al-Iṣfahānī, dated 374 H and fitted with a replacement rete datable to *ca.* 1100 (#3 – Oxford MHS)
1.2.12. An astrolabe by Muḥammad ibn Abi l-Qāsim al-Iṣfahānī al-Ṣāliḥānī dated 496 H (#122 – Florence MSS)
1.2.13. A solitary Abbasid plate with ogival markings (#4021 – Copenhagen DS)

1.2.14. A badly-corroded Abbasid (?) astrolabe (#4020 – London AG)
1.2.15. Miscellaneous items
a) A solitary rete (#2529 – Oxford MHS)
b) A solitary plate (#109 = #3549 – New York MMA)
c) A single plate (#1026 – Oxford MHS)

1.3. *The Earliest Western Islamic Astrolabes (10th and 11th Centuries)*

1.3.1. An astrolabe by Khalaf ibn al-Muʿādh illustrated in a Latin manuscript (#4024 – Paris BN)
1.3.2. An unsigned, undated astrolabe with a rete in the early Andalusian 'Abbasid' style (and additional markings by a European) (#110 = #135 – London BM)
1.3.3. Some instruments by Muḥammad ibn al-Ṣaffār
a) A mater and plates dated 417 H (with a later Eastern Islamic rete) (#3650 – Edinburgh RSM)
b) An astrolabe dated 420 H (#116 – Berlin SBOA)
c) A solitary plate (#4025 – Palermo)
1.3.4. An unsigned Andalusian astrolabe dated 446 H, with later inscriptions in Catalan (#3622 – Cracow JUM)
1.3.5. Two instruments by Ibrāhīm ibn ʿAbd al-Karīm
a) A mater dated 458 H (#3714 – Fez MA)
b) An undated astrolabe (#1079 – Palermo MN – stolen)
1.3.6. Three astrolabes by Ibrāhīm ibn Saʿīd ibn Aṣbagh al-Anṣārī *thumma* al-Sahlī al-Mawāzīnī
a) Dated 459 H (#117 – Madrid MAN)
b) Dated 460 H (#118 – Oxford MHS)
c) Dated 463 H (#123 = #1167 – Rome OA – stolen)
1.3.7. An astrolabe by Ibrāhīm ibn al-Sahlī dated 478 H (#121 – Kassel SKS)
1.3.8. Two astrolabes by Muḥammad ibn Saʿīd al-Ṣabbān
a) Dated 474 H (#2527 – Oxford MHS)
b) Dated 496 H or 466 H (#1139 – Munich BNM – stolen (?))
1.3.9. An astrolabe by Muḥammad ibn al-Sahlī dated 483 H, with a replacement rete bearing Hebrew inscriptions (#2572 – Washington NMAH)
1.3.10. Some 11th-century Andalusian plates (in a composite Ottoman astrolabe) (#4040 – PC)
1.3.11. An astrolabe by Aḥmad ibn Muḥammad al-Naqqāsh dated 472 H (#1099 – Nuremberg GNM)
1.3.12. Miscellaneous items
a) A single plate (#2572 – Washington NMAH)
b) An isolated plate (#4025 – Palermo)
c) A single plate (#154 – Chicago AP)

2. Early European Astrolabes (to *ca.* 1500)

2.1. *Miscellaneous Early European Astrolabes (I)*

2.1.1. An astrolabe from 10th-century Catalonia (the 'Carolingian' astrolabe) (#3042 – Paris IMA)
2.1.2. An astrolabe for the seven climates (#161 – London BM)
2.1.3. An unsigned, undated (?) astrolabe (#4504 – PLU < PC Warren)
2.1.4. An English or French astrolabe for latitudes between 24° and 60°, including for [Paris] and [London] (#300 – Oxford MHS)
2.1.5. An Italian astrolabe with plates for the climates (#166 – Oxford MHS)
2.1.6. An Italian astrolabe with plates for the climates (#167 – London BM)
2.1.7. A French astrolabe with plates for the upper climates (#420 – Greenwich NMM)
2.1.8. A composite astrolabe with Hebrew inscriptions on the back of the mater (#621 – Munich DM)
2.1.9. An Italian (?) astrolabic plate for the second climate and a simpified rete combining northern and southern projections (#169 – Oxford MHS)
2.1.10. An Italian (?) astrolabe (#441 – London SM)
2.1.11. An Italian astrolabe slightly modified by Henricus de Hollandia in Paris (#548 – Nuremberg GNM)
2.1.12. An unsigned 15th-century (?) Dutch astrolabe in imitation of the style of the Nuremberg astrolabe 'signed' by Henricus de Hollandia (#566 – Prague NTM)
2.1.13. An astrolabe with an equatorial ring on the rete, dated 1375 and signed by Petrus Raimundus in Barcelona (#3053 – Boston MFA)
2.1.14. An early-15th-century astrolabe from the Low Countries (?) with a rete in an *art nouveau* style (#536 – Leiden MB)
2.1.15. An astrolabe with a latitude grid around the ecliptic on the rete and zodiacal figures on the back (#213 – PLU < PC Warren)
2.1.16. An English mater with plates for the climates (#303 – Oxford MC)
2.1.17. A mater and plates for the climates owned by Nicolaus of Cusa (#589 – Bernkastel-Kues NKS)
2.1.18. A ghost astrolabe dated 1374 (#3114 – non-existent < 'London PC Baldwin')
2.1.19. A diagram of a rete on an early European astrolabe in a 12th-century manuscript (#4553 – Vatican MS Reg. 598)

2.2. *Miscellaneous Early European Astrolabes (II)*

2.2.1. An Italian astrolabe (#3037 – Milan TT)
2.2.2. An astrolabe of uncertain provenance for latitudes between 16° and 52° (#558 – Nuremberg GNM)

2.2.3. A Catalan astrolabe with a V-shaped frame on the rete (#416 – Greenwich NMM)
2.2.4. A Spanish (?) astrolabe (for Saragossa and other latitudes, with later additions for Paris) (#191 – Oxford MHS)
2.2.5. A French astrolabe (#428 = #625 – Greenwich NMM)
2.2.6. A French astrolabe with a zoomorphic rete (#2041 – Oxford MHS)
2.2.7. An Italian astrolabe with a distinctive rectangular frame on the rete (#493 – Florence MSS)
2.2.8. A related astrolabe, signed "KP" and made in Urbino in 1462 (#4506 – Moulins MD – stolen)
2.2.9. A 15th-century South German astrolabe with an equatorial ring on the rete (#567 – Cracow JUM)
2.2.10. A German astrolabic plate with a tulip-shaped frame, dated 1468 and serving a latitude in N. Italy (#550 – Nuremberg GNM)
2.2.11. An Italian astrolabe with degenerate quatrefoil markings and a *shakkāziyya* grid (#2044 – Oxford MHS)
2.2.12. A late Italian astrolabe copied from an earlier type (#4507 – Florence MNB)
2.2.13. A mater with a universal projection and constellation figures on the front and an albion on the back (#4552 – PLU, stolen, formerly Rome OBS)
2.2.14. A ghost astrolabe dated 1395 (#645 – Seoul YDM)
2.2.15. Two "lost" astrolabes
a) (#625 – formerly Salzburg PS)
b) (#626 – formerly Salzburg PS)
2.2.16. Miscellaneous items
a) Some European plates in the astrolabe of al–Sahl al-Nīsābūrī (#4533 = #137 – Nuremberg GNM)

2.3. *Astrolabes with Quatrefoil Ornamentation (I)*

2.3.1. An astrolabe with a rectangular frame inside the ecliptic on the rete and plates for latitudes down to that of [Palestine] (#162 – London SA)

etc.

2.3.10. An astrolabe made for Ludolf de Scicte of Einbeck (#2072 – Cracow JUM)

etc.

APPENDIX 2

LIST OF INSTRUMENTS CITED IN THE TEXT

Note: The numbers denoted by the symbol # are those of the International Instrument Checklist (for medieval and Renaissance instruments) currently in preparation. Numbers from 1 to 3999 are those of S.L. GIBBS, J.A. HENDERSON, D.J. DE SOLLA PRICE [1973]. New numbers are being assigned as follows: 4001-4999 for additional astrolabes, 5001-5999 for quadrants, 6001-6999 for sundials, 7001-7999 for globes; 8001-8999 for miscellaneous; and 9001-9999 for fakes. The numbers preceded by ¶ refer to the new catalogue of medieval instruments currently in preparation in Frankfurt and are not yet definitive.

Instruments with numbers less than #337 are treated in R.T. GUNTHER [1932]. Other references are to literature in which the object in question has been catalogued or described in detail. For abbreviations in the accounts of the provenance see *Appendix 2*.

#2	Brescia, Museo dell'Età Cristiana, inv. no. 36 – Byzantine astrolabe dated 1062 (¶1.1.1) – see O.M. DALTON [1926].
#3	Oxford, MHS, inv. no. IC 3 – astrolabe by Aḥmad and Muḥammad, sons of Ibrāhīm al-Iṣfahānī, dated 374 H and fitted with a replacement rete datable to the 12th century (¶1.2.11).
#5	Oxford, MHS, inv. no. IC 5 – astrolabe with geared mechanism by Muḥammad ibn Abī Bakr al-Rāshidī al-Iṣfahānī, dated 618 H (¶1.4.7a).
#99	Paris, BN, inv. no. Ge. A.324 – astrolabe by Aḥmad ibn Khalaf (Baghdad, 10th century) (¶1.2.3).
#100	PLU, stolen from Palermo, Museo Nazionale – mater by Ḥāmid ibn 'Alī dated 343 H (¶1.2.9a) – see Mortillaro (1848), reprinted in F. SEZGIN *et al.* (Eds.) [1991], vol. I, pp. 162-191, and the more accurate account in L. CALDO [1936], pp. 6-9 (inadvertently omitted from F. SEZGIN *et al.* (Eds.) [1991]).
#101= #9001	Florence MSS, inv. no. 1113 – unsigned, undated Abbasid astrolabe with later (19th-century?) additions by a

European (the so-called 'Astrolabe of Pope Sylvester II') (¶1.2.7).

#105 London, BM, inv. no. 80 3-8 1 – unsigned Ayyubid astrolabe from Egypt (¶1.5.5).

#109=#3459 New York, Metropolitan Museum of Art, inv. no. 91.1.535 – astrolabe by the Yemeni Sultan al-Ashraf dated 690 H (containing an Abbasid plate with additional markings in Greek) (¶1.1.2a, 1.2.15b, 1.5.12) – see D.A. KING [1986], [1987], [1993a].

#110=#135 London, BM, inv. no. OA+371 – unsigned, undated astrolabe (10th century) with a rete in the early Andalusian 'Abbasid' style (and additional markings by a European) (¶1.3.2).

#111 Kuwait, PC – astrolabe by Ḥāmid ibn Khiḍr al-Khujandī dated 374 H (¶1.2.10).

#116 Berlin, Staatsbibliothek Preußischer Kulturbesitz, Orientabteilung, inv. no. 6567 (Sprenger 2050) – astrolabe by Muḥammad ibn Ṣaffār, dated 420 H (¶1.3.3b).

#117 Madrid, Museo Arqueológico Nacional, inv. no. 50762 – astrolabe by Ibrāhīm ibn Sa'īd al-Sahlī dated 459 H (¶1.3.6a) – see S. GARCÍA FRANCO [1945], pp. 229-235 (no. 12).

#118 Oxford MHS, inv. no. IC118 – astrolabe by Ibrāhīm ibn Sa'īd al-Sahlī dated 460 H (¶1.3.6b).

#121 Kassel, Staatliche Kunstsammlungen, inv. no. A38 – astrolabe by Ibrāhīm ibn al-Sahlī dated 478 H (¶1.3.7) – see the description by F. Woepcke (1858), reprinted in in F. SEZGIN ET AL. (Eds.) [1991], vol. II, pp. 1-36.

#122 Florence, MSS, inv. no. 1105 – astrolabe by Muḥammad ibn Abi l-Qāsim al-Iṣfahānī al-Ṣāliḥānī dated 496 H (¶1.2.12).

#135 See #110.

#140 Athens, Benaki Museum, inv. no. 13178 – astrolabe by Aḥmad ibn Abī Bakr known as Ibn al-Sarrāj, dated 729 H (¶1.5.13a) – see D.A. KING [1986], [1987], [1993a].

#154 Chicago, Adler Planetarium, inv. no. M-36 – astrolabe by Muḥammad ibn Yūsuf ibn Ḥātim, dated 638 H (¶1.6.3).

#161 London, BM, inv. no. 1961 12-1 1 – unsigned, undated astrolabe for the seven climates (¶6.1.2) – see F.A.B. WARD [1981], pp. 110 and 112 (no. 323).

#162 London, Society of Antiquaries – astrolabe with a rectangular frame inside the ecliptic on the rete and plates for latitudes down to that of [Palestine] (¶6.3.1), see D. A. KING, K. MAIER [in press].

#163 Oxford, MHS, inv. no. IC 163 – unsigned, undated French astrolabe (¶6.8.3).

#166 Oxford, MHS, inv. IC 166 – Italian astrolabe with plates for the climates (¶6.1.5).

#167 London, BM, inv. no. 67 7-5 22 – Italian astrolabe for the climates (¶6.1.6) – see F.A.B. WARD [1981], p. 114 (no. 329).

#169 Oxford, MHS, inv. no. IC 169 – unsigned, undated Italian (?) astrolabic plate for the second climate and a simplified rete combining northern and southern projections (¶6.1.9).

#173 Oxford, MHS, inv. no. IC 173 – unsigned, undated German astrolabe from *ca.* 1500 (¶7.2.5a).

#186 Washington, NMAH, inv. no. 316757 – unsigned, undated Italian astrolabe (¶6.7.10) – see S. GIBBS, G. SALIBA [1984], pp. 139-140 (no. 186).

#191 Oxford, MHS, inv. no. IC 191 – Spanish (?) astrolabe (for Saragossa and other latitudes, with later additions for Paris) (¶6.2.4).

#202 PC – late-14th-century Picard astrolabe bearing monastic ciphers, with an inscription by Berselius dated 1522 (¶6.4.6) – see D.A. KING [1993b].

#290 London, BM, Sloane 54 – an English quatrefoil astrolabe (¶6.3.4) – see F.A.B. WARD [1981], pp. 112 (no. 324).

#291 London, BM, inv. no. 1909 6-17 1 – Anglo-French astrolabe with plates for Montpellier, Oxford and elsewhere, dated 1326 (¶6.6.2) – see F.A.B. WARD [1981], pp. 112-113 (no. 325).

#300 Oxford, MHS, inv. no. IC 300 – English or French astrolabe for latitudes between 24° and 60°, including for [Paris] and [London] (¶6.1.4).

#301 Cambridge, Caius College – English astrolabe with markings for Oxford or London (¶6.6.1).

#303 Oxford, Merton College – English mater with plates for the climates (¶6.1.16).

#304 Washington, NMAH, inv. no. 316758 – unsigned, undated English astrolabe (¶6.8.7) – see S. GIBBS, G. SALIBA [1984], pp. 150-151 (no. 304).

#337 Greenwich, NMM, inv. no. A34/36-28C or A34/NA38-1661C – unsigned, undated French astrolabe bearing the name Thornoe (¶6.8.1).

#416 Greenwich, NMM, inv. no. A21/NA36-21c – Catalan astrolabe with a V-shaped frame on the rete (¶6.2.3).

#420	Greenwich, NMM, inv. no. 39.693.A43 – French astrolabe for the upper climates (¶6.1.7).
#428=#625	Greenwich, NMM, inv. no. A39/NA38-1661C – French astrolabe (¶1.6.5).
#476	Barcelona, Real Academia de Ciencias y Artes – unsigned, undated French astrolabe (¶6.8.2).
#493	Florence, MSS, inv. no. 1107 – Italian astrolabe with a distinctive rectangular frame on the rete (¶6.2.7).
#536	Leiden, Museum Boerhaave, inv. no. MBL 3102 (= A 1 (Obs.)) – a 15th-century (?) astrolabe from the Low Countries (?) with a rete in an *art nouveau* style (¶6.1.14)
#547	Nuremberg, GNM, inv. no. WI 21 – undated, unsigned Italian quatrefoil astrolabe with zoomorphic representations on the throne and back (¶6.3.5) – see G. BOTT (Ed.) [1992], vol. II, pp. 577-578 (no. 1.73).
#548	Nuremberg, GNM, inv. no. WI 6 – Italian quatrefoil astrolabe with additions by Henricus de Hollandia (¶6.1.11) – see G. BOTT (Ed.) [1992], vol. II, pp. 578-581 (no. 1.74).
#549	Nuremberg, GNM, inv. no. WI 129 – astrolabe dated 1457, probably by Peuerbach (¶6.9.1a) – see G. BOTT (Ed.) [1992], vol. II, pp. 582-586 (no. 1.75).
#550	Nuremberg, GNM, inv. no. WI 5 – unsigned German astrolabic plate serving a latitude in N. Italy, dated 1468 (¶6.7) – see G. BOTT (Ed.) [1992], vol. II, pp. 589-592 (no. 1.77).
#558	Nuremberg, GNM, inv. no. WI 282 – astrolabe of uncertain provenance with plates for latitudes from 16° to 52° (¶6.1.14) – see G. BOTT (Ed.) [1992], vol. II, pp. 574-576 (no. 1.72).
#589	Bernkastel-Kues, Nicolaus Kusanus-Stift, inv. no. 3 – mater and plates for the climates owned by Nikolaus of Cusa (¶6.1.17) – see J. HARTMANN [1919].
#625	See #428.
#640	PC – astrolabe dedicated by Ioannus (Regiomontanus) to Cardinal Bessarion in 1462 (¶6.9.3) – see D.A. KING, G. L'E. TURNER [1994].
#1026	Oxford, MHS, inv. no. 57-84/155 – undated astrolabe (10th century) by Khafīf (¶1.2.2a)
#1065	Antwerp, Nationaal Scheepvaartmuseum, inv. no. A.S. 54.11.36 – late, unsigned Egyptian astrolabe (¶2.2.3)
#1069	Madrid, Instituto de Valencia de Don Juan – astrolabe by [Abū Bakr ibn Yusuf (of Marrakesh) *ca.* 1200], reworked

by a European (¶1.6.1f) – see García Franco [1945], pp. 277-282 (no. 19).

#1099 Nuremberg, Germanisches Museum, inv. no. WI 353 – astrolabe by Aḥmad ibn Muḥammad al-Naqqāsh dated 472 H (¶1.3.11) – see G. Bott (Ed.) [1992], vol. II, pp. 568-570 (no. 1.70).

#1139 Munich, Bayerisches Nationalmuseum, inv. no. 33/243 (missing, presumed stolen, photos are available) – astrolabe by Muḥammad ibn Sa'īd al-Ṣabbān dated 496 H or 466 H (¶1.3.8b).

#1161 Stockholm, Statens Sjöhistoriska Museum, inv. no. S.1565 – astrolabe by Aḥmad ibn 'Alī al-Sharafī dated 729 H (¶1.6.11).

#1179 PLU, formerly Berlin PC – undated astrolabe (10th century) by Muḥammad ibn Shaddād (al-Baladī) with a later Maghribi rete (¶1.2.5) – see the description by Dorn (1865), reprinted in F. Sezgin *et al.* (Eds.) [1991], vol. I, pp. 345-498, especially pp. 461-464.

#2006 Washington, D.C., NMAH, inv. no. 318198 – unsigned, undated English astrolabe (¶6.7.10) – see S. Gibbs, G. Saliba [1984], pp. 153-154 (no. 2006).

#2041 Oxford, MHS, inv. no. 57-84/173 – unsigned, undated French astrolabe with a zoomorphic rete (¶6.2.6).

#2072 Cracow, JUM, inv. no. 41/V 4043 – astrolabe made for Ludolf de Scicte of Einbeck (¶6.3.10).

#2527 Oxford, MHS, inv. no. Billmeir 157 – astrolabe by Muḥammad ibn Sa'īd al-Ṣabbān dated 474 H (¶1.3.8a) – see F.R. Maddison [1957], pp. 18-19 (no. 157).

#2572 Washington, D.C., NMAH, inv. no. 318178 – astrolabe by Muḥammad ibn al-Sahlī dated 483 H with a replacement rete bearing Hebrew inscriptions (¶1.3.9) – see S. Gibbs, G. Saliba [1984], pp. 174-177 (no. 2752) (with illustrations of the front and back) (and my review in «Isis», LXXVII, 1986, pp. 711-713), and B.R. Goldstein, G. Saliba [1983] (with more illustrations), particularly on the Hebrew inscriptions on the rete.

#3037 Milan, PC (Tullio Tomba) – unsigned, undated Italian astrolabe (¶6.2.1).

#3042 Paris, IMA, inv. no. AI 86-31 – unsigned, undated astrolabe from Catalonia, late or early 10th century (¶1.6.1).

#3053 Boston, Mass., Museum of Fine Arts, inv. no. 88.654 – astrolabe with an equatorial ring on the rete, dated 1375 and signed by Petrus Raimundus in Barcelona (¶6.1.13).

#3203 Turin, Museo Palazzo Madama di Torino, inv. no. 3401/922B – unsigned, undated Italian quatrefoil astrolabe (¶6.3.8).

#3459 See #109.

#3501 Kuwait, Museum of Islamic Archaeology, inv. no. LNS 36M – astrolabe by (Muḥammad ibn ʿAbdallāh known as) Nasṭūlus, dated 315 H (¶1.2.4a).

#3622 Cracow, JUM, inv. no. 4037 – 35/V. – unsigned Andalusian astrolabe dated 446 H, with later inscriptions in Catalan (¶1.3.4) – see, most recently, D. BURCZYK-MARONA [1991], and K. MAIER [1994].

#3650 Edinburgh, Royal Scottish Museum, inv. no. T1959-62 – mater and plates Muḥammad ibn al-Ṣaffār dated 417, H, fitted with a later Eastern Islamic rete (¶1.3.3a).

#3700 See #3923.

#3702 Baghdad, Archaeological Museum, inv. no. 9723 – Late (Ottoman ?) astrolabe signed by Aḥmad ibn Kamāl, copied (or possibly remodelled) from an early Abbasid astrolabe (¶1.2.1) – see B. FRANSĪS, N. NAQSHABANDĪ [1957] and B. STAUTZ [1994].

#3713 Cairo, Museum of Islamic Art, inv. no. 15352 – mater by Ḥāmid ibn ʿAlī with an illegible date [3]?4 H (with a replacement rete and plates by [Muḥammad Ṣaffār]) (¶1.2.9b).

#3922 Frankfurt, IGN, inv. no. 1 – astrolabe by Muḥammad ibn Aḥmad ibn ʿAlī ibn Muḥammad dated 588 H, reworked by Muḥammad Khalīl in the 17th century (¶1.4.5) – see S. ACKERMANN [1994].

#3923=#3700=#9002 Frankfurt, IGN, inv. no. 2 – astrolabe by ʿAbd al-Qādir Muḥibb (17th century) with 20th-century fake markings to make it look like an astrolabe by Naṣīr al-Dīn al-Ṭūsī (mid 13th century) – see SCHMIDL [1994].

#4024 Paris, BN, manuscript lat. 7412, fols. 19v-23v – astrolabe by Khalaf ibn al-Muʿādh illustrated in a Latin manuscript (¶1.3.1) – see A. VAN DE VYVER [1931]; M. DESTOMBES [1962], pp. 23 and 41-43; E. POULLE [1964], p. 900 and 4 plates; P. KUNITZSCH [1959], pp. 90-92; and E. DEKKER, P. KUNITZSCH [in print], source 4.

#4026 Rockford, TM, inv. no. 3529 – undated astrolabic plate by ʿAlī al-Wadāʿī (¶1.5.1a) – see B. CHANDLER (Ed.) [1985], pp. 168-173 (no. 22), and D.A. KING [1986], [1987], [1993a].

#4029 Paris, IMA, inv. no. AI 86-15 – unsigned, undated

Rasulid Yemeni astrolabe (¶1.5.13) – see J. MOULIÉRAC [1989], pp. 89-90 (no. 5).

#4030 Florence, Florence, Uffizi Galleries, Gabinetto dei Disegni e Stampi, U1454A recto and verso – drawing by Antonio da Sangallo the Younger (1520) of an astrolabe by Khafīf (¶1.2.2c) – see G. SALIBA [1991].

#4050 Istanbul, Maritime Museum, inv. no. 264 – astrolabe by ʻAbd al-Raḥmān ibn Sinān al-Baʻlabakkī, dated 619 H [= 1222/23], diameter 560 mm. (!) (¶1.5.4) – see S. CLUZAN, J. MOULIÉRAC (Eds.) [1993], p. 480 (this instrument was not available for the exhibition), and a more detailed desciption in my contribution to a forthcoming issue of *Erdem* (Ankara) dedicated to the memory of Professor Aydın Sayışlı.

#4170 Cambridge (Mass.), Harvard University, Collection of Historical Scientific Instruments, inv. no. 7320 – astrolabe made by a Maghribi *ca.* 1600 for use in the Yemen (¶1.6) – unpublished, will be featured in a forthcoming catalogue edited by Will Andrewes.

#4506 Moulins, Musée Départemental, stolen – astrolabe, signed 'KP' and made in Urbino in 1462 (¶6.2.8).

#4507 Florence, Museo Nazionale del Bargello, inv. no. 1156 – late Italian astrolabe copied from an earlier type (¶6.2.12) – see M. MINIATI (Ed.) [1991], pp. 48-53 (no. 22).

#4509 PC – 'cosmopolitan' Italian astrolabe with a replacement plate from a Byzantine astrolabe and additional markings in Hebrew (¶1.1.2a, 6.4.15).

#4512 Rockford (Ill.), TM, inv. no. 3392 – undated French astrolabe (¶6.7.2) – see B. CHANDLER (Ed.) [1985], pp. 124-127 (no. 16).

#4523 Private collection (Germany) – Northern Italian astrolabe dated 1420 – discussed in detail in a forthcoming study by Burkhard Stautz.

#4551 Rome, Osservatorio Astronomico – 14th-century French astrolabe with an unfinished rete (¶6.7.7)

#4553 A diagram of a rete on an early European astrolabe in MS Vatican Regin. 598, fol. 120r, copied in the 12th-century (¶6.1.19). See, for example, J.M. MILLÀS VALLICROSA [1931], plate X; W. BERGMANN [1985], pp. 101, 104, etc.; and E. DEKKER, P. KUNITZSCH [in print], source 5.

#4554 Milan, Pinacoteca Ambrosiana – Early medieval Italian astrolabe – see T. TOMBA [1978], pp. 9-10 and fig. 3 on p. 6 and fig. 4 on p. 11 (pagination of the reprint).

#4556	England, private collection (note: this piece became available for study only in 1994) – Medieval European astrolabe of uncertain provenance, datable *ca.* 1300 (?) – see *Christie's London 29.9.94 Catalogue*, pp. 34-39 (lot 136).
#5521	Dorchester, Museum – English quadrant dated 1398 (¶8.4.1a) – see F.A.B. WARD [1981], pp. 55-56 (*ad* no. 146).
#5522	London, BM, inv. no. 60 5-19 1 – English quadrant dated 1399 (¶8.4.1b) – see F.A.B. WARD [1981], pp. 55-56 (no. 146).
#7511-#7520	Viennese sundials from 1450-80 (¶9.5) – see D.A. KING [1994], *Appendix A*.
#7511	Innsbruck, TLMF, inv. no. U5, dated 1451.
#7512	London, BM, inv. no. 93 6-16 9, dated 1453.
#7513	Graz, LMJ, inv. no. 4525, dated 1455.
#7514	Ecouen, MR, inv. no. ? (formerly Cluny 18546), dated 1455.
#7515	Munich, BNM, inv. no. Phys. 22, dated 1456.
#7516	Vienna, KHM, inv. no. 122, dated 1463.
#7517	Pécs (H), PC, dated 1465.
#7518	Nuremberg, GNM, inv. no. WI 7, datable *ca.* 465.
#7519	Vienna, UM, undated.
#7520	Berlin, MKG, inv. no. ?, undated?
#9001	See #101.
#9002	See #3923.

BIBLIOGRAPHY

S. ACKERMANN [1994], *Mutabor: Die Umarbeitung eines mittelalterlichen Astrolabs im 17. Jahrhundert*, in A. VON GOTSTEDTER [1994], pp. 193-209.

G. BEAUJOUAN [1972], *L'enseignement du* quadrivium, in *La scuola nell'Occidente latino dell'alto medievo*, («Settimane di studio del Centro italiano di studi sull'alto medioevo», XIX), Spoleto, 1972, pp. 639-668; reprinted in IDEM, *Par raison de nombres*, Aldershot, Variorum, 1991, vol. I

G. BEAUJOUAN [1995], *L'authenticité de l'astrolabe dit 'carolingien' à la lumière d'analyses et recherches récentes* – see pp. 439-450 of this volume (I have consulted a preliminary version of 7 pages distributed by the author at Saragossa)

J. L. BERGGREN [1991], *Medieval Islamic Methods for Drawing Azimuth Circles on the Astrolabe*, «Centaurus», XXXIV, 1991, pp. 309-344

W. BERGMANN [1985], *Innovationen im Quadrivium des 10. und 11. Jahrhunderts – Studien zur Einführung von Astrolab und Abakus im lateinischen Mittelalter*, («Sudhoffs Archiv – Zeitschrift für Wissenschaftsgeschichte», XXVI), Wiesbaden, Franz Steiner Verlag, 1985

A. BORST [1989], *Astrolab und Klosterreform an der Jahrtausendwende*, («Sitzungsberichte der Heidelberger Akademie der Wissenschaften, Philosophisch-historische Klasse»), Heidelberg, Carl Winter, 1989

G. BOTT (Ed.) [1992], *Focus Behaim-Globus*, 2 vols., Nuremberg, Germanisches Nationalmuseum, 1992

C. R. BRÜHL [1990], *Deutschland - Frankreich - Die Geburt zweier Völker*, Köln-Wien, Böhlau, 1990

D. BURCZYK-MARONA [1991], *Problem datowania dawnych instrumentów naukowych - Uwagi na marginesie artykulu A. Bunsch o astrolabium arabskim w zbiorach Muzeum UJ*, «Opuscula musealia», V, 1991, pp. 95-116

P. L. BUTZER, D. LOHRMANN (Eds.) [1993], *Science in Western and Eastern Civilization in Carolingian Times*, Basle-Boston-Berlin, Birkhäuser, 1993

F. CABROL, H. LECLERCQ (Eds.) [1913-1953], *Dictionnaire d'archéologie chrétienne et de liturgie*, 30 vols., Paris, Letouzey et Ané, 1913-53

L. CALDO [1936], *Astrolabi del Museo Nazionale di Palermo*, «Memorie dell'Osservatorio astronomico di Palermo», LXV, 1936; also printed in «Atti della Reale Accademia di Scienze, Lettere e Belle Arti di Palermo», XIX, 1936, pp. 3-12; reprinted in «Coelum», VIII, 1938, pp. 101-108

J. CASSULERAS, J. SAMSÓ (eds.), *From Baghdad to Barcelona, Studies in the Islamic Exact Sciences in Honour of Professor Juan Vernet*, Barcelona, in press.

B. CHANDLER (Ed.) [1985], *The Time Museum - Catalogue of the Collection*, vol. I, *Time Measuring Instruments*, part. I, *Astrolabes - Astrolabe Related Instruments*, Rockford (Ill.), The Time Museum, 1985

S. CLUZAN, J. MOULIÉRAC, E. DELMONT (Eds.) [1993], *Syrie, Mémoire et Civilisation*, Paris, Flammarion-Institut du Monde Arabe, 1993

A. DALLAL [1984], *Al-Bīrūnī on the Climates*, «Archives Internationales d'Histoire des Sciences», XXXIV, 1984, pp. 3-18

O. M. DALTON [1926], *The Byzantine Astrolabe at Brescia*, «Proceedings of the British Academy», XII, 1926, pp. 133-146

M.-T. DEBARNOT [1986], *The Zīj of Ḥabash al-Ḥāsib: A Survey of MS Istanbul Yeni Cami*

784/2, in D. A. KING, G. SALIBA (Eds.), *From Deferent to Equant: Studies in the History of Science in the Ancient and Medieval Near East in Honor of E. S. Kennedy*, «Annals of the New York Academy of Sciences», D, 1986, pp. 35-69

J. DE GRAEVE [1995], *La provenance de l'astrolabe carolingien de Marcel Destombes*, paper presented at the Saragossa Congress, Session 20 (delivered by Robert Halleux as *La correspondence entre M. Destombes et H. Michel*)– see pp. 209-215 of this volume

E. DEKKER, P. KUNITZSCH [in print], *The Stars on the Rete of the Carolingian Astrolabe*, in J. CASSULERAS, J. SAMSÓ [in print]

A. DEROLEZ *ET AL.* (Eds.) [1968], *Lamberti* [...] *Liber Floridus - Codex autographus Bibliothecae Universitatis Gandavensis*, Ghent, Story-Scientia, 1968

D.J. DE SOLLA PRICE [1959], *The First Scientific Instrument of the Renaissance*, «Physis», I, 1959, pp. 26-30

D.J. DE SOLLA PRICE [1980], *Philosophical Mechanism and Mechanical Philosophy - Some Notes Towards a Philosophy of Scientific Instruments*, «Annali dell'Istituto e Museo di Storia della Scienza di Firenze», V, 1980, pp. 75-85

D.J. DE SOLLA PRICE [1982], *Instruments in the British Museum*, «Journal of the History of Astronomy», XIII, 1982, pp. 129-132

M. DESTOMBES [1960], *Les chiffres coufiques des instruments astronomiques arabes*, «Physis», II, 1960, pp. 197-210

M. DESTOMBES [1962], *Un astrolabe carolingien et l'origine de nos chiffres arabes*, «Archives Internationales d'Histoire des Sciences», XV, 1962, pp. 3-45

Encyclopaedia of Islam, 8 vols. to date, Leiden, E. J. Brill, 1960 to present

B. FRANSIS, N. NAQSHABANDI [1957], *Al-asṭurlābāt fī Dār al-Āthār al- 'Arabiyya fī Baghdād* (in Arabic), *Sumer* (Baghdad) XIII, 1957, pp. 9-33 and 3 pls.

S. GARCÍA FRANCO [1945], *Catálogo crítico de los astrolabios existentes en España*, Madrid, Instituto Histórico de Marina, 1945

J. GAYA [1981], *Ramon Llull - Tractat d'astronomia,* in J. VERNET (ed.), *Textos y estudios sobre astronomia española en el siglo XIII*, Barcelona, Universidad Autonoma de Barcelona, 1981, pp. 205-323.

S.L. GIBBS, J.A. HENDERSON, D.J. DE SOLLA PRICE [1973], *A Computerized Checklist of Astrolabes*, New Haven, Yale University, 1973

S. GIBBS, G. SALIBA [1984], *Planispheric Astrolabes from the National Museum of American History*, Washington (D.C.), Smithsonian Institution, 1984

L. GILISSEN [1976], *Curieux foliolage d'un manuscrit de droit civil, la Somme d'Azzon (Bruxelles 9251 et 9252)*, in I. FORCHIELLI, A. M. STICKLER (Eds.), *Studia Gratiana post octava decreti saecularia*, («Collectanea historiae iuris canonici», XIX, *Mélanges G(érard) Fransen*, I), Roma, 1976 , pp. 303-311

B.R. GOLDSTEIN, G. SALIBA [1983], *A Hispano-Arabic Astrolabe with Hebrew Star Names*, «Annali dell'Istituto e Museo di Storia della Scienza di Firenze», VIII, 1983, pp. 19-28; reprinted in B.R. GOLDSTEIN, *Theory and Observation in Ancient and Medieval Astronomy*, London, Variorum Reprints, 1985

A. VON GOTSTEDTER (Ed.) [1994], *Ad radices - Festband zum 50jährigen Bestehen des Frankfurter Instituts für Geschichte der Naturwissenschaften,* Stuttgart, Franz Steiner, 1994

R.T. GUNTHER [1932], *The Astrolabes of the World* , 2 vols., Oxford, Oxford University Press, 1932, reprint in 1 vol., London, Hollandm 1976.

J.B. HARLEY, D. WOODWARD (Eds.), *The History of Cartography*, vol. I, *Cartography in Prehistoric, Ancient, and Medieval Europe and the Mediterranean*, vol. II, part I,

Cartography in the Traditional Islamic and South Asian Societies, Chicago (Ill.)-London, The University of Chicago Press, 1987 and 1992

J. HARTMANN [1919], *Die astronomischen Instrumente des Kartinals Nikolaus Cusanus*, «*Abhandlungen der Königlichen Gesellschaft der Wissenshaften zu Göttingen*», *Math.-Phys. Klasse*, N.F. Band X, Nr. 6, Berlin, Wiedmannsche Buchhandlung, 1919.

R.A.K. IRANI [1955], *Arabic Numeral Forms*, «Centaurus», IV, 1955, pp. 1-12; reprinted in E.S. KENNEDY *ET AL.* [1983], pp. 710-721

E.S. KENNEDY [1976], *The Exhaustive Treatise on Shadows by Abū al-Rayḥān [...] al-Bīrūnī*, vol. I, *Translation*, vol. II, *Commentary*, Aleppo, Institute for the History of Arabic Science, 1976

E.S. KENNEDY, M.H. KENNEDY [1987], *Geographical Coordinates of Localities from Islamic Sources*, Frankfurt am Main, Institut für Geschichte der Arabisch-Islamischen Wissenschaften, 1987

E.S. KENNEDY *ET AL.* [1983], *Studies in the Islamic Exact Sciences*, Beirut, American University of Beirut Press, 1983

D.A. KING [1986], *Islamic Mathematical Astronomy*, London, Variorum, 1986

D.A. KING [1987], *Islamic Astronomical Instruments*, London, Variorum, 1987

D.A. KING [1991a], *Medieval Astronomical Instruments: A Catalogue in Preparation*, «Bulletin of the Scientific Instrument Society», XXXI, 1991, pp. 3-7

D.A. KING [1991b], *Strumentazione astronomica nel mondo medievale islamico*, in G. L'E. TURNER (Ed.), *Gli strumenti*, Torino, Giulio Einaudi, 1991, pp. 154-189 and 581-585; an English summary is in *Some Remarks on Islamic Astronomical Instruments*, «Scientiarum Historia», XVIII, 1992, pp. 5-23

D.A. KING [1993a], *Astronomy in the Service of Islam*, Aldershot, Variorum, 1993

D.A. KING [1993b], *Rewriting History through Instruments: The Secrets of a Medieval Astrolabe from Picardy*, in R.G. ANDERSON, J. A. BENNETT, W. F. RYAN (Eds.), *Making Instruments Count - Essays on Historical Scientific Instruments presented to Gerard L'Estrange Turner*, Aldershot, Variorum, 1993, pp. 42-62

D.A. KING [1993c], *Some Medieval Astronomical Instruments and Their Secrets*, in R. MAZZOLINI (Ed.), *Non-Verbal Sources in Science before 1900*, Firenze, Leo S. Olschki, 1993, pp. 29-52

D.A. KING [1993d], *1992 – A Good Year for Medieval Astronomical Instruments*, «Bulletin of the Scientific Instrument Society», XXXVI, 1993, pp. 17-18

D.A. KING [1994a], *Astronomical Instruments between East and West*, in H. KÜHNEL (Ed.), *Kommunikation zwischen Orient und Okzident* («*Sitzungsberichte der Österreichischen Akademie der Wissenschaften, Phil.-Hist. Klasse*», XVII), Wien, Österreichische Akademie der Wissenschaften, 1994, pp. 143-198; summarized in «Medium aevum quotidianum», XXVII, 1992, pp. 125-130

D.A. KING [in press], *Der Frankfurter Katalog mittelalterlicher astronomischer Instrumente*, in G. ENDREß (Ed.), *Symposium Graeco-Arabicum III*, Leiden, E. J. Brill, in press.

D.A. KING, *Aspects of Applied Science in Mosques and Monasteries*, paper prepared for the Symposium *Science and Theology in Medieval Islam, Judaism, and Christendom*, University of Wisconsin at Madison, April 15-17, 1993. A German translation, *Aspekte angewandter Wissenschaften in Moscheen und Klöstern*, is in «Berichte zur Wissenschaftsgeschichte» (Organ der Gesellschaft für Wissenschaftsgeschichte) XVIII, 1995, pp. 85-95 and 137-149

D.A. KING, K. MAIER [in press], *The Medieval Catalonian Astrolabe of the Society of Antiquaries, London*, in J. CASSULERAS, J. SAMSÓ [in press].

D.A. KING, G.L'E. TURNER [1994], *The Astrolabe Presented to Cardinal Bessarion in 1462, Attributable to Regiomontanus*, «Nuncius», IX, 1994, 1, pp. 165-206

P. KUNITZSCH [1959], *Arabische Sternnamen in Europa*, Wiesbaden, Otto Harrassowitz, 1959

P. KUNITZSCH [1961], *Untersuchungen zur Sternnomenklatur der Araber,* Wiesbaden, Otto Harrassowitz, 1961

P. KUNITZSCH [1966], *Typen von Sternverzeichnissen in astronomischen Handschriften des zehnten bis vierzehnten Jahrhunderts*, Wiesbaden, Otto Harrassowitz, 1966

P. KUNITZSCH [1982], *Glossar der arabischen Fachausdrücke in der mittelalterlichen europäischen Astrolabliteratur*, «Nachrichten der Akademie der Wissenschaften in Göttingen, Phil.-Hist. Klasse», XI, 1982

P. KUNITZSCH [1991-92], *Letters in Geometrical Diagrams: Greek - Arabic - Latin*, «Zeitschrift für Geschichte der arabisch-islamischen Wissenschaften», VII, 1991-92, pp. 1-20

P. KUNITZSCH [1993], *Arabische Astronomie im 8. bis 10. Jahrhundert*, in P.L. BUTZER, D. LOHRMANN (Eds.) [1993], pp. 205-220

F. R. MADDISON [1957], *A Supplement to a Catalogue of Scientific Instruments in the Collection of J. A. Billmeir, Esq., C.B.E.*, Oxford, Museum of the History of Science, London, Frank Partridge and Sons, 1957

K. MAIER [1994], *Bemerkungen zu romanischen Monatsnamen auf mittelalterlichen Astrolabien*, in A. VON GOTSTEDTER (Ed.) [1994], pp. 237-254

J. M. MILLÀS VALLICROSA [1931], *Assaig d'història de les idees físiques i matemàtiques a la Catalunya medieval*, Barcelona (Estudis Universitaris Catalans), 1931

M. MINIATI (a cura di) [1991], *Strumenti scientifici della Collezione Carrand*, Firenze, Museo Nazionale del Bargello, 1991

J. MOULIÉRAC [1989], *La Collection Marcel Destombes*, in A. J. TURNER (Ed.), *Astrolabica No. 5 - Etudes 1987-1989*, Paris, Institut du Monde Arabe- Société internationale de l'astrolabe, 1989, pp. 77-126

A. M. MUNDÓ [1995], *Analyse paléographique de l'écriture de l'astrolabe carolingien*, paper presented at the Saragossa Congress, Session 20, printed in this volume, pp. 303-321

Museo Poldi Pezzoli - Orologi - Oreficerie, Milano, Electa, 1981

O. NEUGEBAUER [1983], *Astronomical and Calendrical Data in the Très Riches Heures* [1974], reprinted in IDEM, *Astronomy and History - Selected Essays*, New York, Springer, 1983, pp. 507-520

CH. PELLAT [1961], *Le calendrier de Cordoue publié par R. Dozy - Nouvelle édition accompagnée d'une traduction française annotée*, Leiden, E. J. Brill, 1961

E. POULLE [1956], *Peut-on dater les astrolabes médiévaux?,* «Revue d'histoire des sciences», IX, 1956, pp. 301-322

E. POULLE [1964], *Le traité d'astrolabe de Raymond de Marseille*, «Studi medievali», s.3, V, 1964, pp. 866-900

G. SALIBA [1991], *A Sixteenth-Century Drawing of an Astrolabe Made by Khafīf Ghulām 'Alī b. 'Īsā (c. 850 A.D.)*, «Nuncius», VI, 1991, pp. 109-119

M.R. SALZMAN [1990], *On Roman Time - The Codex-Calendar of 354 and the Rhythms of Urban Life in Late Antiquity*, Berkeley-Los Angeles (Ca.)-Oxford, University of California Press, 1990

J. SAMSÓ [1969], *Estudios sobre Abū Naṣr Manṣūr b. 'Alī b. 'Irāq*, Barcelona, Diputación Provincial de Barcelona - Asociación para la Historia de la Ciencia Española, 1969

J. SAMSÓ [1992], *Las ciencias de los antiguos en al-Andalus*, Madrid, Mapfre, 1992

J. SAMSÓ [1995], *Roma et Francia (= Ifranja) in M. Destombes' Carolingian Astrolabe*, paper presented at the Saragossa Congress, Session 20 - see pp. 239-251 of this volume

G. SCHILDER, P. VAN DER KROGT, S. DE CLERCQ (Eds.) [1987], *M. Destombes (1905-1983): Selected Contributions to the History of Cartography and Scientific Instruments*, Utrecht, Hes Publishers, Paris, A. G. Nizet, 1987

P. SCHMIDL [1994], *Ein Astrolab aus dem 17. Jahrhundert, prachtvoll und verfälscht*, in A. VON GOTSTEDTER (Ed.) [1994], pp. 291-314

F. SEZGIN [1978], *Geschichte des arabischen Schrifttums*, vol. VI, *Astronomie bis ca. 430 H.*, Leiden, E. J. Brill, 1978

F. SEZGIN *ET AL.* (Eds.) [1991], *Arabische Instrumente in orientalistischen Studien*, 6 voll., Frankfurt am Main, Institut für Geschichte der Arabisch-Islamischen Wissenschaften, 1991

B. STAUTZ [1994], *Die früheste bekannte Formgebung des Astrolabs*, in A. VON GOTSTEDTER (Ed.) [1994], pp. 315-328

W. STEVENS [1981], *Scientific Instruction in Early Insular Schools*, in M. HERREN (Ed.), *Insular Latin Studies*, Toronto, Pontifical Institute of Mediaeval Studies, 1981, pp. 83-111

W. STEVENS [1986], *Bede's Scientific Achievement* («The Jarrow Lecture», 1985), Jarrow-upon-Tyne, Parish of St. Paul's Church, 1986

T. TOMBA [1966], *Due astrolabi latini del XIV secolo conservati a Milano*, «Physis», VIII, 1966, pp. 295-306

T. TOMBA [1968], *Nuove osservazioni di due astrolabi latini del XIV secolo conservati a Milano*, «Physis», X, 1968, pp. 119-122

T. TOMBA [1970], *Un astrolabio del XIV secolo di probabile origine italiana*, «Physis», XII, 1970, pp. 82-87

T. TOMBA [1978], *Gli astrolabi della Collezione Settala nella Pinacoteca Ambrosiana*, «Atti della Fondazione Giorgio Ronchi», XXXIII, 1978, 2, pp. 297-314

T. TOMBA, G. BRUSA [1983], *Museo d'Arti Applicate - Strumenti scientifici - Orologi* , Milano, Electa, 1983

G. L'E. TURNER [1989], *The Regiomontanus Astrolabe dated 1462*, catalogue supplement to *Fine Scientific and Philosophical Instruments* (Thursday, September 28, 1989), London, Christie's South Kensington, 1989

A. VAN DE VYVER [1931], *Les premières traductions latines (X^e^-XIe s.) de traités arabes sur l'astrolabe*, in *Actes du 1er Congrès International de Géographie Historique*, voll. 2, Brussels, 1931, pp. 266-290 and 3 plates; reprinted in F. SEZGIN *ET AL.* (Eds.) [1991], vol. VI, pp. 377-405

J. VERNET, J. SAMSÓ, *ET AL.* (Eds.) [1992], *El legado científico andalusí*, Madrid, Ministerio de Cultura, 1992

M. VILADRICH, *The* Mumtaḥan *Tradition in al-Andalus. Analysis of the Data from the* Calendar of Cordova *Related to the Entrance of the Sun in the Zodiacal Signs*, in J. CASSULERAS, J. SAMSÓ [in press].

F.A.B. WARD [1981], *A Catalogue of European Scientific Instruments in the Department of Medieval and Later Antiquities of the British Museum*, London, British Museum Publications, 1981.

III

REWRITING HISTORY THROUGH INSTRUMENTS: THE SECRETS OF A MEDIEVAL ASTROLABE FROM PICARDY

INTRODUCTORY REMARKS

Inevitably what one learns from the best teachers is what should have been obvious anyway. I had already *seen* several hundred medieval instruments and recorded their astronomical features and inscriptions when Gerard Turner introduced me to new approaches. So it is fitting that I dedicate this paper to Gerard because he taught me how to *look* at instruments. Now I travel burdened with a magnifying glass, microscope and camera. Gerard thus caused me much extra work, but I am nevertheless extremely grateful to him, not only because the instruments are worth the trouble, but also because looking at them *à la Gérard*, or still better *avec Gérard*, is a lot of fun.

EARLY MEDIEVAL EUROPEAN INSTRUMENTS—THEIR NEGLECT AND THEIR POTENTIAL

The vast majority of medieval European astronomical instruments are unpublished, in the sense that they are not documented pictorially and their various inscriptions have been either ignored in the literature or never studied at all. It is not an exaggeration to assert that the present state of documentation of medieval instruments is primitive. R. T. Gunther's uneven treatment of those few medieval European astrolabes known to him[1] has been supplemented by a few studies (of which the best are by S. García Franco and Tullio Tomba for instruments in Spain and Italy respectively),[2] but his misidentifications of many provenances remain uncorrected. Gunther listed only about thirty European astrolabes that clearly predate 1450; the number now known is about 120. The state of docu-

---

[1] R. T. Gunther, *The Astrolabes of the World*, 2 vols., Oxford, 1932 (reprinted in 1 vol., London, 1976). Numbers introduced by # refer to astrolabes listed in D. J. de Solla Price *et al.*, *A Computerized Checklist of Astrolabes*, New Haven, Conn., 1973. The American symbol # for 'number' is retained here and in my forthcoming catalogue of medieval instruments (see n. 3 below). A new version of the *Checklist* entitled *World Astrolabes Inventory* has been prepared by Leonard Linton of Point Lookout, Long Island, and is currently being improved and updated.

[2] See S. García Franco, *Catálogo crítico de astrolabios existentes en España*, Madrid, 1945, and T. Tomba, 'Due astrolabi latini del XIV secolo conservati a Milano', *Physis. Rivista di storia della scienza*, 8, 1966, pp. 295–306; *idem*, 'Nuove osservazioni sui due astrolabi latini del XIV secolo conservati a Milano', ibid., 10, 1968, pp. 119–22; *idem*, 'Un astrolabio del XIV secolo di probabile origine italiana', ibid., 12, 1970, pp. 82–7, and the same author's contributions to catalogues of various collections and exhibitions in Italy. Catalogue entries by other scholars for miscellaneous museums could be cited, but these persistently ignore star names and are hence inadequate.

mentation of medieval quadrants and sundials is actually far worse than that for astrolabes. Now, however, some momentum has been achieved for a systematic investigation of these important historical sources.[3]

Most of these medieval instruments are unsigned and undated. But the fact that they thus have no place in repertories of makers and their works is no reason to ignore them altogether; rather they merit extra attention and present a special challenge to the historian. For there are indeed clues that can help to identify their provenance: the design of the rete and throne, the star names used on the rete, regional forms for month names and names of the zodiacal signs, and the latitudes and localities featured on plates. These are often more reliable than styles of calligraphic engraving or decoration, which are still insufficiently recorded. In particular, star names are of the utmost importance since they help to relate the instruments to the manuscript tradition so painstakingly documented by Professor Paul Kunitzsch of Munich.[4] Virtually none of the available data relating to star names on medieval instruments has been exploited yet, and the same holds for geographical and calendrical data. Star positions can be used for dating only if one has control over the star names and access to the star catalogues in the relevant manuscript traditions.[5] But metal analysis is more of a *bête noire*: investigation of a few pieces has thus far revealed only how little we know about medieval metalwork. A systematic analysis could reveal information few of us could dream of. With so many factors involved we are often still a long way from being able to date undated or dubiously dated pieces.

Most medieval instruments have a single layer of inscriptions. But a few bear later additions which add to the historical importance of the piece and can help to locate the instrument in some later cultural milieu. Different layers of inscriptions, if they can be understood, add to the 'personality' of the instrument.

Numerals on medieval European instruments are also important. In general we find the Gothic forms of the Hindu-Arabic numerals. But a few early instruments

---

[3] See my reports 'Medieval Astronomical Instruments: A Catalogue in Progress', *Bulletin of the Scientific Instrument Society*, no. 31, Dec., 1991, pp. 3–7, and '1992—A Good Year for Medieval Scientific Instruments', ibid., no. 36, March, 1993, pp. 17–18. Funding for the second stage of this project (1992–94) has been obtained from the Deutsche Forschungsgemeinschaft.

[4] See, for example, P. Kunitzsch, *Typen von Sternverzeichnissen in astronomischen Handschriften des zehnten bis vierzehnten Jahrhunderts*, Wiesbaden, 1966.

[5] An unhappy procedure for dating astrolabe retes, which ignores both star names and the irregular motions of astrolabe stars, as well as the rest of the instruments, is described in R. K. E. Torode, 'A Mathematical System for Identifying Stars of an Astrolabe and Finding its Age', *Astrolabica 5. Études 1987–1989*, Paris, 1989, pp. 53–76, and *idem*, 'A Study of Astrolabes', *Journal of the British Astronomical Association*, 102, 1, 1992, pp. 25–30. The former is reviewed in detail by E. Dekker, 'Of Astrolabes and Dates and Dead Ends', *Annals of Science*, 49, 1992, pp. 175–84. Some of the problems confronting us are outlined already in E. Poulle, 'Peut-on dater les astrolabes médiévaux?', *Revue de l'histoire des sciences et de leurs applications*, 9, 1956, pp. 301–22, but the facilities available to us now were inconceivable in the 50s—see n. 18 below.

have no numerals at all on the scales, perhaps because only the Roman numerals were acceptable to their engravers and it was simply too inconvenient to use these. Until now the study of the development of numeral forms has been conducted almost exclusively from manuscripts;[6] instruments, however, offer an equally important historical source, as yet untapped.

THE PICARD ASTROLABE

Now an astrolabe recently auctioned at Christie's in London—the so-called 'Astrolabe of Berselius'—is a perfect example of what instruments can reveal upon close inspection. Its numerals are expressed in a remarkable notation, in which each number is represented by a single cipher.[7] Further, the instrument bears three layers of inscriptions. The first layer includes the ciphers, which were clearly engraved by the maker of the instrument. The special attention afforded certain of the plates suggests a geographical provenance in Northern France or Flanders. The second layer is inscriptions in Gothic script, probably, but not necessarily, by the same person who engraved the ciphers; these can be shown to pinpoint the provenance to Picardy. The third layer consists of a dated dedication that apparently postdates the construction of the instrument by about 150 years. The persons mentioned in the dedication are now identified, and the question arises: why was the instrument given as a present? To this we can now offer some suggestions. But before we take a closer look at the astrolabe it is necessary to present an overview of the ciphers.

ON CIPHER NUMERALS

While ciphers are fairly well documented in various scholarly articles that are difficult of access, their elegance and utility have escaped appreciation. Major contributions to our knowledge of the ciphers were made by the late German palaeographer Bernhard Bischoff, who identified close to twenty manuscripts in which they occur, the French palaeographer Guy Beaujouan, who discovered an

---

[6] The most useful study is G. F. Hill, *The Development of Arabic Numerals in Europe Exhibited in Sixty-Four Tables*, Oxford, 1915. Dr Charles Burnett of the Warburg Institute, London, is currently undertaking a new study of the Hindu-Arabic numerals in early European manuscripts.

[7] A detailed account of the ciphers from a cultural historical perspective and a full description of the Picard astrolabe is contained in my forthcoming book entitled *On Ciphers and Monks, Astrolabes and Virgins—Forgotten Episodes in the History of the Middle Ages and Renaissance.* (Berselius also owned a statue of the Virgin and Child—see n. 25 below—and in this book I have discussed not only the development of rete styles, quatrefoils and the like, but also surveyed all known European instruments exhibiting religious symbolism, including representations of the Virgin.) A brief summary intended for historians of mathematics is in my paper 'The Ciphers of the Monks and the Astrolabe of Berselius Reconsidered,' in S. S. Demidov, M. Folkerts, D. A. Rowe and C. J. Scriba, eds, *Amphora. Festschrift für Hans Wussing zum 65. Geburtstag*, Basel, etc., 1992, pp. 375–88. It should be mentioned that the appellation 'cipher' is entirely my own usage; in the manuscripts they are referred to as numbers or figures.

additional manuscript from Normandy, and the Swiss historian of mathematics Jacques Sesiano, who sorted out the various traditions attested in these manuscripts (and a few more).[8] The ciphers were used to represent letters of the alphabet as well as numbers. In both of their manifestations the ciphers are as important in manuscripts from the thirteenth to the fifteenth century as are the runes in manuscripts from the eighth to the twelfth century.[9]

The simplest variety of numerical ciphers was introduced in England in the early thirteenth century, a time when the 'Hindu-Arabic' numerals were still a novelty and used only in limited circles. Any number up to 99 can be represented by a cipher [see Fig. 1] by applying appendages for 1–9 and 10–90 to the left and right side of a vertical stem, which itself has no numerical value. It was the monk John of Basingstoke (d. 1252) who introduced these ciphers, and his biographer, the contemporary chronicler Matthew Paris, stated that he had learned of them in Athens. They do indeed bear strong physical resemblance to the basic signs of a fourth-century-BC syllabic shorthand described in a tablet found on the Acropolis at the end of the last century,[10] but there is as yet no evidence of any *numerical* ciphers of this kind having been used in ancient Greece or thirteenth-century Athens. On the other hand, the ciphers introduced by John of Basingstoke *in the early thirteenth century* have the same shape as those used for an alphabetical shorthand in a *late-12th-century* English *ars notaria*, in which the Tironean notes of Antiquity were revived in an ingenious, modified form.[11]

---

[8] On cipher numerals see B. Bischoff, 'Die sogenannten <<griechischen>> und <<chaldäischen>> Zahlzeichen des abendländischen Mittelalters' in *idem*, *Mittelalterliche Studien. Ausgewählte Aufsätze zur Schriftkunde und Literaturgeschichte*, 3 vols, Stuttgart, 1966–81, I, pp. 67–73 (with numerous additions to the original 1944 version); *idem*, 'Übersicht über die nichtdiplomatischen Geheimschriften des Mittelalters,' in *idem*, *Studien*, III, pp. 120–48, especially pp. 139–40, and 146 n. 25 (incorporating additions to the original 1954 version); J. Sesiano, 'Un Système artificiel de numération au moyen âge' in M. Folkerts and U. Lindgren, eds, *Mathemata. Festschrift für Helmuth Gericke* (Reihe 'Boethius', Bd. 12), Wiesbaden, 1985, pp. 165–96; and G. Beaujouan, 'Les Soi-disant Chiffres grecs ou chaldéens (XIIe–XVIe siècles)', *Revue d'histoire des sciences*, 3, 1950, pp. 170–74, and *idem*, 'The Place of Nicolas Chuquet in a Typology of 15th-century French Arithmetics' in C. Hay, ed., *Mathematics from Manuscript to Print, 1300–1600*, Oxford, 1988, pp. 73–88, both reprinted in *idem*, *Par raison de nombres. L'art du calcul et les savoirs scientifiques médiévaux*, Aldershot, 1990, XII and XIII.

[9] See, for example, R. Derolez, *Runica Manuscripta. The English Tradition*, Bruges, 1954.

[10] Published in Th. Gomperz, 'Ueber ein bisher unbekanntes griechisches Schriftsystem aus der Mitte des vierten vorchristlichen Jahrhunderts ... ', *Sitzungsberichte der philosophisch-historischen Classe der Kaiserlichen Akademie der Wissenschaften* (Vienna), 107, 1884, pp. 339–95, and 1 plate. See Chr. Johnen, *Geschichte der Stenographie im Zusammenhang mit der allgemeinen Entwicklung der Schrift und der Schriftkürzung*, I: *Die Schriftkürzung und Kurzschrift im Altertum, Mittelalter und Reformationszeitalter ...* , Berlin, 1911, pp. 105–17, for different interpretations by various classicists, and also A. Mentz, *Geschichte der Kurzschrift*, 2nd edn, Wolfenbüttel, 1949, pp. 10–11. There is apparently no material of consequence in English on the Acropolis text.

[11] Published in V. Rose, 'Ars notaria. Tironische Noten und Stenographie im 12. Jahrhundert',

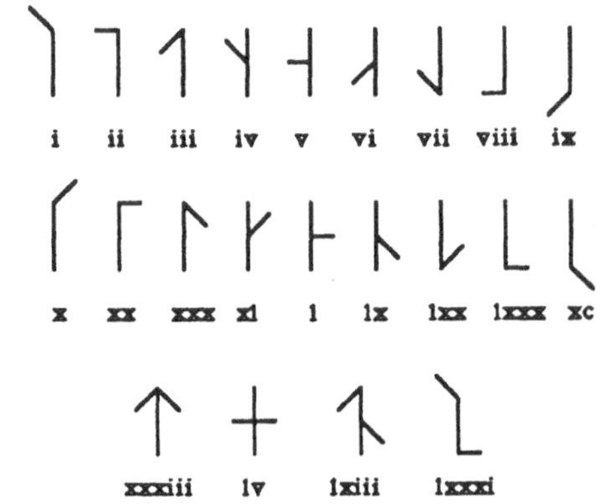

Fig. 1. The ciphers of John of Basingstoke.

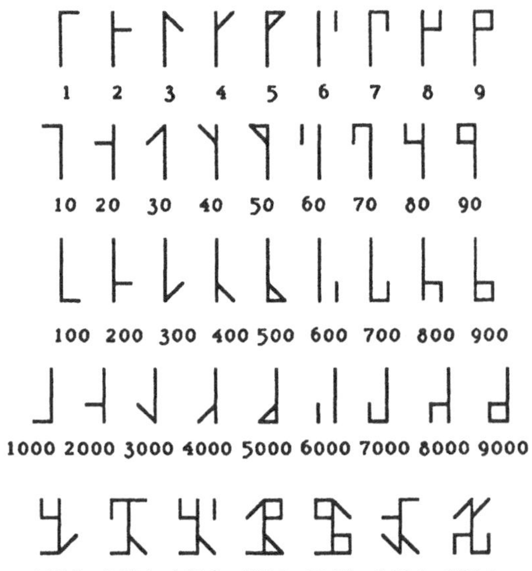

Fig. 2. The ciphers on the Picard astrolabe and in the Norman treatise on arithmetic.

Hermes. Zeitschrift für classische Philologie, 8, 1874 (reprinted Vaduz, 1963), pp. 303–26. See also A. Mentz, 'Zwei Stenographiesysteme des späteren Mittelalters', *Korrespondenzblatt des Königlichen Stenographischen Landesamts zu Dresden,* 57, 1912, pp. 163–79 and 227–53, also published separately, Dresden, 1912; Johnen, *Geschichte der Stenographie* (see n. 10 above), pp. 247–63; and Mentz, *Geschichte der Kurzschrift* (see n. 10 above), pp. 31–4; and the brief note by M. B. Parkes, 'Tachygraphy in the Middle Ages: Writing Techniques Employed for *reportationes* of Lectures and Sermons' in *idem, Scribes, Scripts and Readers,* London and Rio Grande (Ohio), 1991, pp. 19–34, esp. 32–3 and n. 30.

By the end of the thirteenth century a more useful set of ciphers, originating in Cistercian monasteries in Hainault, had been devised to represent any number up to 9999. There were two main varieties, the first with horizontal stems being the more common, and the second with vertical stems, attested only in Northern France from the fourteenth and fifteenth centuries. (We also have evidence of various other experimental varieties, with different appendages applied in alternative ways, but these were not as successful.) Now the appendages for the units and tens are on either side of one half of the stem, and those for the hundreds and thousands are on the other half—see Fig. 2 for the ciphers used in Northern France.

Most of the twenty-odd manuscripts in which numeral ciphers of one sort or another occur are of religious content and date from the thirteenth to the fifteenth century. Their provenance spans England to Italy, Normandy to Sweden, a wide geographical distribution reflecting to some extent the expansion of the Cistercian order. In these sources the ciphers are used to represent numbers for foliation, divisions or subdivisions of texts, for numbering marginal notes and items in lists or the lines of a staff in musical notation, referring to columns of text in separate indexes, and writing arguments in Easter tables and year-numbers in dates [see Fig. 3],[12] providing an excellent alternative to the cumbersome Roman numerals and the alien Hindu-Arabic numerals. They were *not* intended for calculation (although, in fact, they are sufficiently versatile that they could be); for this purpose the abacus and finger arithmetic were available anyway. They were intended simply as a means of representing numbers.

Vertical numeral ciphers have been found so far only in a Norman treatise on arithmetic (MS Paris, Bibliothèque nationale, fr. 1339), written in Old French and dating from the mid-fifteenth-century [see Figs. 2 and 4]. Here they are introduced as an alternative to the Hindu-Arabic numerals. The treatise itself is anonymous, but in the same manuscript we find the treatise on the construction of the astrolabe compiled by Jean Fusoris of Paris (c. 1365–1436).[13] Unfortunately we have no information on the anonymous author's source for the ciphers, which he calls 'algorisme grec'. Certainly he did not devise them himself, for in form they are virtually identical to the most common variety of Cistercian ciphers, and besides identical ciphers feature on the Picard astrolabe, which was surely made several decades before this treatise was compiled. Nor can we judge

[12] They have been overlooked by those scholars who have written on Cistercian aids to scholarship. See, for example, R. H. Rouse, 'Cistercian Aids to Study in the Thirteenth Century', *Studies in Cistercian History*, 2, 1976, pp. 123–34.

[13] On Fusoris and his treatise, as well as for preliminary remarks on some of the astrolabes of his school see E. Poulle, *Un Constructeur d'instruments astronomiques au 15e siècle—Jean Fusoris*, Paris, 1963, and also Poulle's article 'Fusoris' in *Dictionary of Scientific Biography*, 16 vols, New York, 1970–80, 15, pp. 162–4. King, *Catalogue* (mentioned above), Section 6.8, forthcoming, is to include descriptions of about thirty 'Fusoris' instruments and copies thereof.

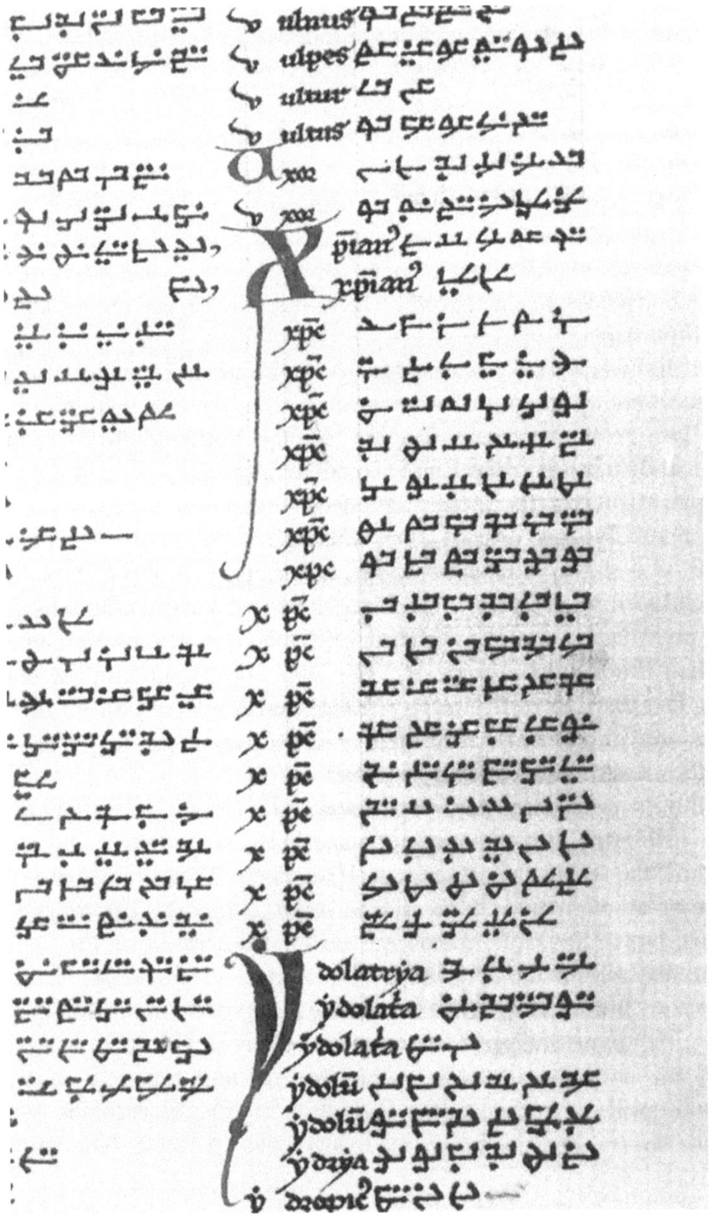

Fig. 3. Horizontal ciphers used in an index to the sermons of Jehan Halgrin of Abbeville in a thirteenth-century manuscript from the Cistercian monastery of Aulne-sur-Sambre near Charleroi. The numbers refer to the columns of text, two to a page, in which key words occur, the columns themselves being numbered in Roman notation. (The appendages are different from those shown in Fig. 2.) From MS Brussels B.R. II.1051, courtesy of the Bibliothèque royale.

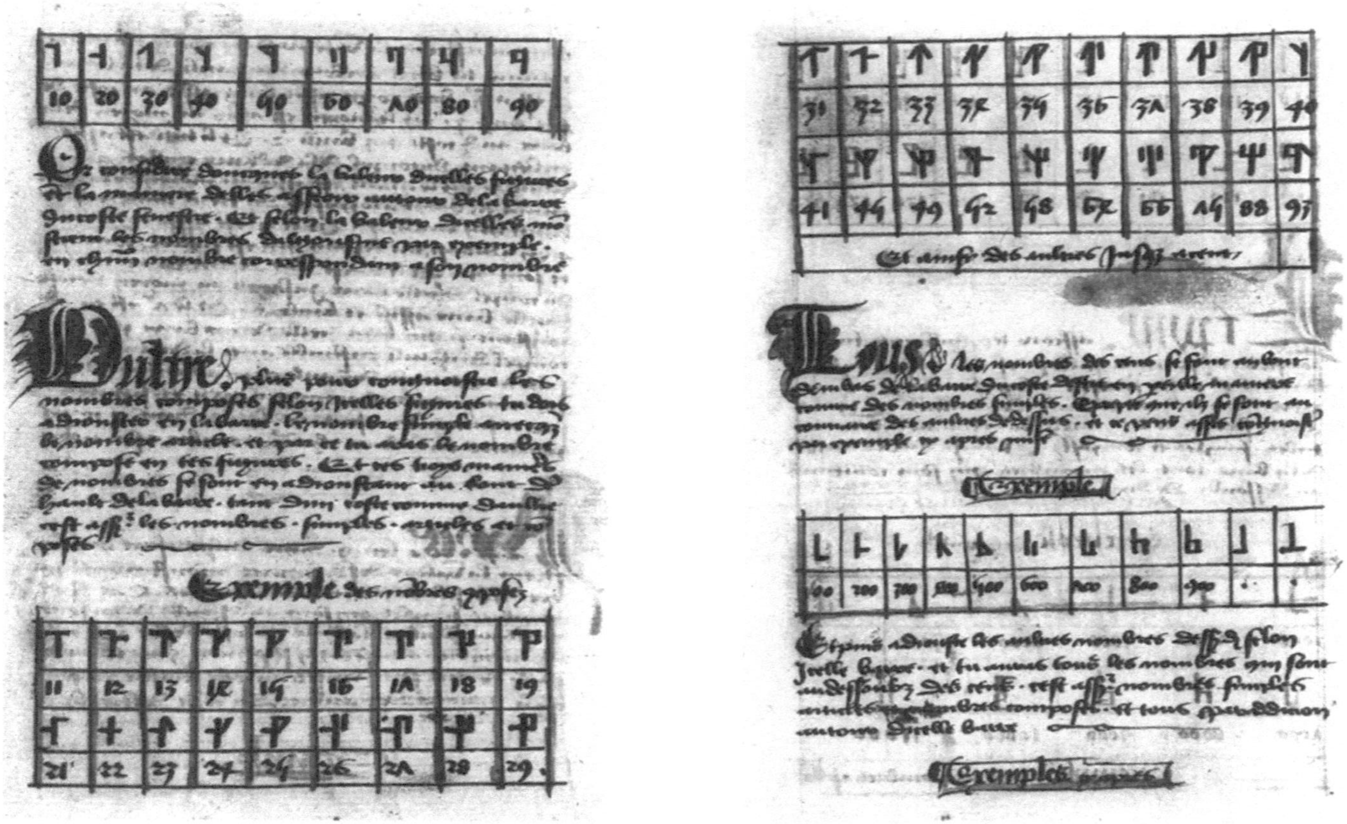

Fig. 4. An extract from the Norman treatise on arithmetic featuring the ciphers. From MS Paris BN fr. 1339, courtesy of the Bibliothèque nationale.

Fig. 5. The front of the Picard astrolabe. Note the ciphers 1–12 and again 1–12 around the scale on the rim. Photograph courtesy of Christie's of London and the owner.

Fig. 6. The back of the Picard astrolabe. Note the ciphers on the four altitude scales, the solar and calendrical scales, the scales of the shadow squares, and at the ends of the curves on the universal horary quadrant. Photograph courtesy of Christie's of London and the owner.

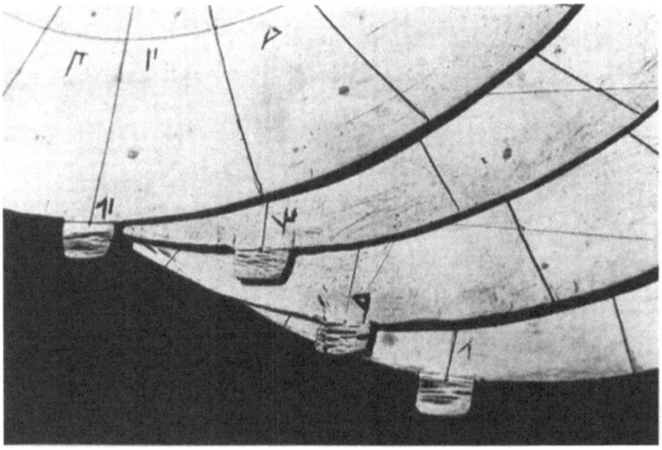

(a)

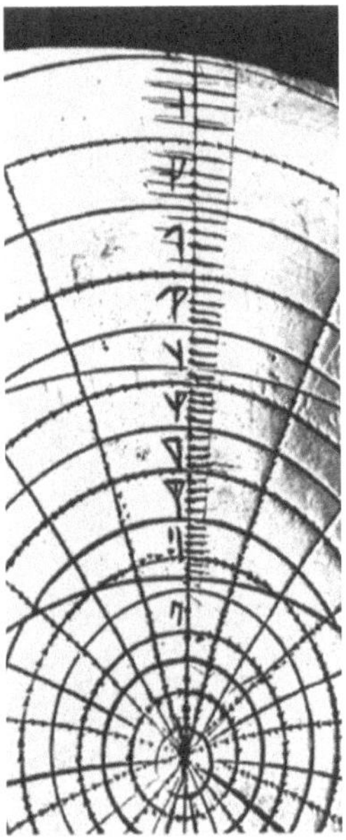

(c)

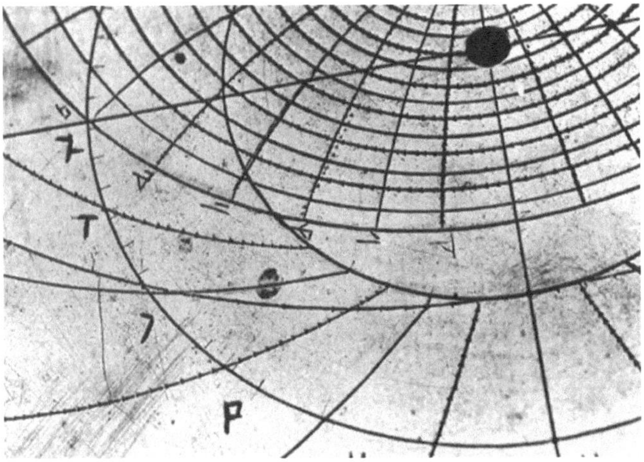

(b)

Fig. 7. Various details of the Picard astrolabe (photographs by the author): (a) Ciphers engraved with varying force to indicate the latitudes served by the plates. (b) Ciphers indicating the arguments for the altitude and azimuth circles on the plate for 51°. Note the twilight curve tangential to the equinoctial circle. (c) An unusual scale indicating individual degrees, as found on some of the plates. (d) The incorrect divisions of the shadow scales, and the construction marks used to achieve them. (e) Scales on the back of the ecliptic ring of the rete. (f) Chisel marks on two plates.

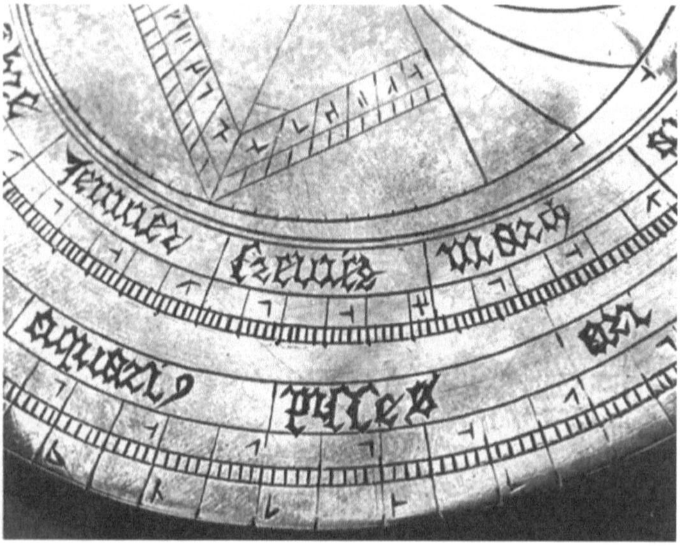
(d)

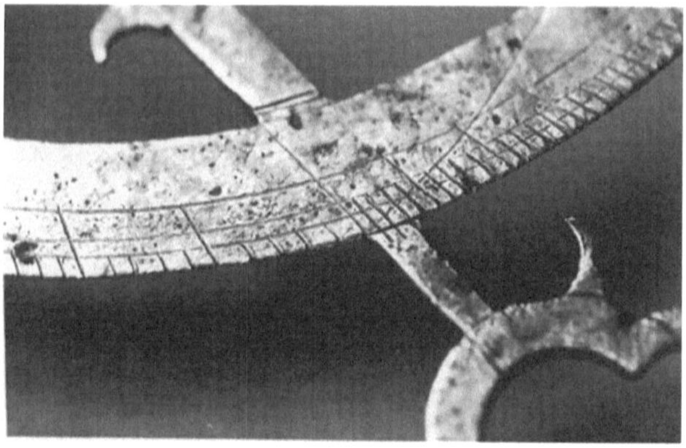
(e)

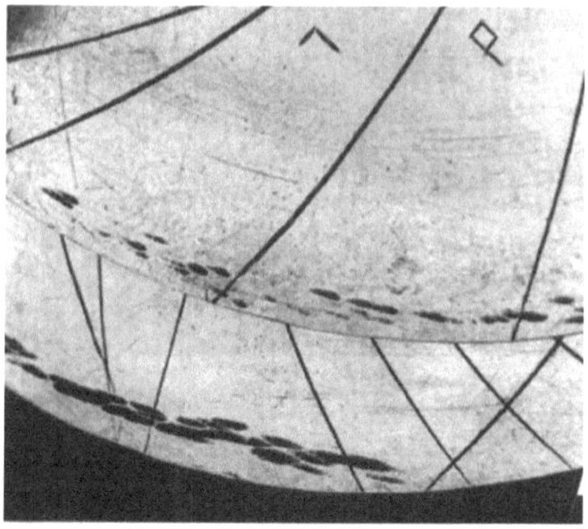
(f)

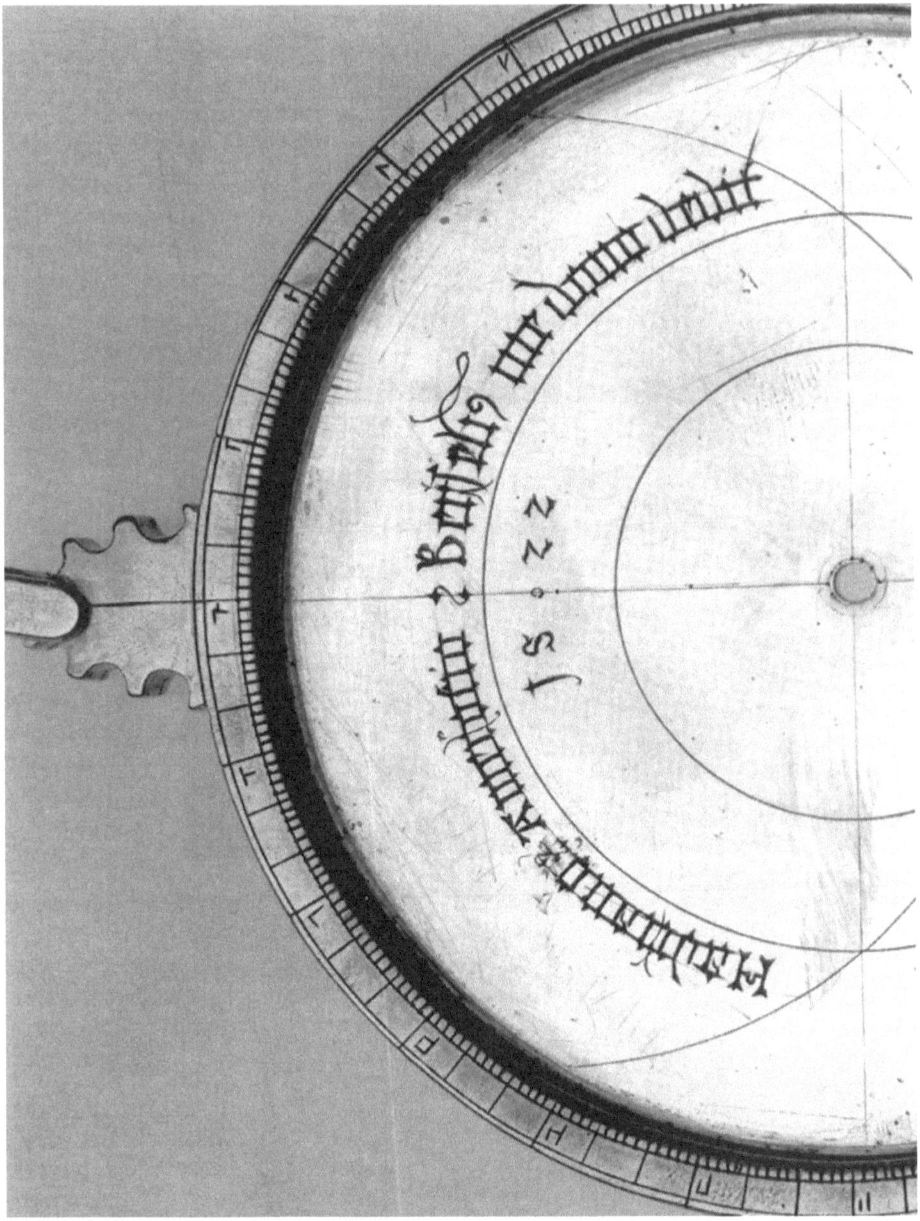

Fig. 8. The dedication on the mater.

the immediate success of the treatise. But certainly within a few more decades the ciphers had been forgotten as far as the purpose for which they were intended was concerned.

The history of the ciphers enters a second phase in the sixteenth century. The celebrated scholar Henricus Agrippa of Nettesheim wrote that he had seen them in two very old books on magic and astrology. He presented what I call the Northern French ciphers after the numerals of the Hebrews and Chaldeans, and the position of his treatment of them led his contemporaries to associate them with the Chaldeans. As 'Chaldean' or 'astrologer's' ciphers they were consigned for posterity to the junk-heap of the pseudo-sciences: they were misrepresented in numerous early printed works and histories of mathematics from the sixteenth century (starting with Noviomagus) to the twentieth. There is, in fact, but scant evidence that these ciphers were used by magicians or astrologers before the time of Agrippa or thereafter. Some ingenious developments for representing numbers larger than 9999 were proposed by the author of the Norman treatise and by the sixteenth-century Italian mathematician Cardano. The development of printing in the sixteenth century sounded their death knell: even in the treatises of Agrippa and Cardano one cipher is printed in reverse and another upside-down.

Contrary to Agrippa's assertion as to where he had encountered the ciphers, the manuscripts available to us indicate that the numeral ciphers were used mainly at the hands of monks, with two possible exceptions: the Norman treatise on arithmetic and the Picard astrolabe. However, I am at present inclined to think that the astrolabe was made in a monastic environment. Certainly it ended up in a monastery, for its later owner, Berselius, was a Benedictine monk.

In addition to the sources identified and studied by the colleagues mentioned above I cannot claim to have found more than an Easter table and some musical notation using the ciphers. The same basic ciphers can, however, be found in earlier Greek and Latin shorthands and in the earliest Renaissance shorthands and coded scripts.[14] I do not doubt that the ciphers, used either as numerals or as letters of the alphabet, are found in many other medieval manuscripts of religious content, but a search for these is a task for someone younger than myself. The Picard astrolabe and the Norman treatise provide definitive proof that the ciphers were actually used outside the scriptoria of the Middle Ages.

THE CIPHERS AND OTHER UNUSUAL FEATURES OF THE PICARD ASTROLABE

Made of brass, 116 mm in diameter and 6 mm thick, this intriguing astrolabe[15]

---

[14] See Mentz, *Geschichte der Kurzschrift* (see n. 10 above), pp. 36–7, on Renaissance English and German shorthands and coded scripts featuring the John of Basingstoke and the Northern French ciphers, respectively.

[15] In the time of Gunther the Picard astrolabe was in the Collection of Mr Claude Fry. It was auctioned at Sotheby's of London in 1957 (not sold!), and in 1958, and was until recently in a private collection in Paris. During those thirty or so years apparently no attempt was made to identify

was already a collector's item by 1522 when the monk Berselius added a dedication to his friend Amerotius. [See Figs 5 and 6 for the front and back.]

The ciphers, which constitute the first layer of inscriptions, are competently and correctly executed. But the rather awkward Gothic inscriptions on the rete and the back are not marked by the same level of competence, and this is my sole reason for suspecting that two different individuals might have been involved. The Gothic script has been dated by Guy Beaujouan to c. 1380, but it is probably contemporary with a Northern French astrolabe with gear mechanism now in the Science Museum, London, which has been dated by others to c.1300.[16] It still has to be determined whether these two astrolabes were made in the early or the late fourteenth century, that is, before or after the Black Death.

Unusual features abound on the Picard astrolabe. The throne is unique, resembling a conifer in cross-section. On the scale on the rim the hours are labelled in ciphers 1–12 and again 1–12. The rete, too, is remarkable in that it bears a half-quatrefoil on the upper solstitial bar as well as an upper bar concentric with the ecliptic.[17] There are also half-quatrefoil frames at the bottom of the solstitial bar and at each end of the equinoctial bar; the Science Museum

---

Berselius or Amerotius or to investigate the medieval French inscriptions. The astrolabe was auctioned on 26 September 1991 by Christie's of London, and is again in a private collection.

It is mentioned in passing in Gunther, *Astrolabes* (see n. 1 above), II, p. 349 (no. 202), where attention is drawn to the numeral forms and the dedication. It features in London Sotheby's catalogues, 14 March 1957 (lot 147), 27 June 1957 (lot 136), and 31 July 1958 (lot 193), as well as S. Guye and H. Michel, *Mesures du temps et de l'espace. Horloges, montres et instruments anciens*, Fribourg (CH), 1970, pl. 209 (illustrations of the mater, rete and plates) and 218 (back) with captions on pp. 224 and 231. A brief description by the present writer—based solely on photographs—is in London Christie's 29 September 1991 catalogue, pp. 39–42 (lot no. 82), where the front is depicted in reverse, and a complete description is in King, *Catalogue*, Section 6.4.6, forthcoming, amongst quatrefoil astrolabes (Sections 6.3–6.4).

The ciphers on the astrolabe are mentioned in various works cited in n. 8 above, namely, Sesiano, 'Système', p. 18; Beaŭjouan, 'Chiffres', pp. 172 and 174, and 'Chuquet', p. 76.

[16] Gunther, *Astrolabes* (see n. 1 above), II, p. 347 (no. 198), and pls. LXXX (front with gear mechanism) and LXXXI (back, plate for England, and gear mechanism separately), and J. North, 'Opus quarundam rotarum mirabilium', *Physis*, 8, 1966, pp. 337–72, especially pp. 370–72, reprinted in *idem, Stars, Minds and Fate—Essays in Ancient and Medieval Cosmology*, London, 1989, no. 11, pp. 135–69, especially pp. 167–9, on the gear mechanism. The main features of this instrument do not concern us here, save to note that it has apparently never before been recognized as the missing link between the astrolabe and the earliest European astrolabic clock: this astrolabe, like those clocks, is based on a *southern* stereographic projection. It should act a warning to all those colleagues who believe that truth is only to be found in manuscripts that one of the earliest known European instruments represents an important tradition apparently not attested in any known manuscripts. On the astrolabic clocks see J. D. North, 'Monasticism and the First Mechanical Clocks' in J. T. Fraser and N. Lawrence, eds, *The Study of Time*, New York, 1975, pp. 381–93, reprinted as no. 12 in North, *Essays*, pp. 171–86, especially p. 181–2.

[17] The only similar rete is on a French astrolabe (#461) now in the Bijloke Museum in Ghent: see G. Baptiste *et al.*, *La Mesure du temps dans les collections belges*, Brussels, 1984, pp. 38–9 (no. 9), where the front is depicted in reverse.

astrolabe has half-quatrefoils on both perpendicular bars. On the ecliptic scale each 10° is labelled in ciphers. The twenty-six star pointers are in the wavy 'Fusoris' tradition but, like those on the geared astrolabe, clearly predate it. On both astrolabes many of the star names have been omitted. Whoever labelled the star pointers on the Picard astrolabe was using defective sources. The star positions on the rete confirm the need in the fourteenth century for new star tables.[18] Artistically it combines two traditions, the quatrefoil design being widely used in France, Italy and England in the Middle Ages, and the upper frame featuring on early Italian and French astrolabes and culminating in the instruments of Jean Fusoris.

There are four plates and the latitudes are indicated in ciphers just above the pegs at the bottom [see Fig. 7a]. The latitudes served are as follows:

23°, 30°, 36°, 41°, 45°, 48°, 50°, and 51°.

The mater is engraved with an incomplete set of astrolabic markings for an unspecified latitude, actually about [51°]. The choice of latitudes results from a development of the ancient notion of the seven geographical climates, defined in terms of the length of longest daylight, and was intended to make the instrument 'universal'. Thus latitude 23° serves the second climate (and no city of consequence); 30° serves the third climate (and also Cairo), 36° the fourth (and also Rhodes (for Greeks), Raqqa and Rayy (for Muslims), 41° the fifth (and also Barcelona and Rome), 45° the sixth (and also South Central France and the Po Valley), 48° the seventh (and also Paris). The higher latitudes 50° and 51° are for 'points north', which would include Picardy, the Low Countries and the coast of Kent.

The plates for 50° and 51° have the altitude circles and the seasonal hours numbered in cipher script, and on the latter the azimuth curves are labelled in ciphers [see Fig. 7b]. Clearly the maker favoured latitudes 50° and 51°, and this can be taken as pointing to a Northern French or Low Countries provenance for the instrument.

Another unusual feature occurs on some of the plates: a degree-scale along the meridian, which could be used for measuring meridian altitudes or celestial declinations [see Fig. 7c]. Indeed, such scales are not attested on any other medieval astrolabe currently known to me. The same can be said of the curves for twilight, which on each plate are drawn according to an approximate method so that they are tangential to the circle corresponding to the summer solstice. Now it happens that for the latitude of Paris, assuming an angle of solar depression of 18°, the twilight curve will have this property, and it was clearly a plate

---

[18] My doctoral student Burkhard Stautz has undertaken analyses of various families of medieval astrolabe retes using computer graphics, comparing them with reconstructions using contemporaneous (and often already outdated) star catalogues. Some preliminary findings were presented at the 1992 Annual Meeting of the Société internationale de l'astrolabe in Frankfurt. Judging by what has been achieved thus far, this research will contribute greatly to our understanding of medieval astrolabes.

with such markings that inspired those on the Picard astrolabe. For latitudes around 50° the error is negligible, but for low latitudes it is substantial.

On the back all arguments on the various scales and the curves of universal horary quadrant are expressed in ciphers. The equinox is at March 12, which corresponds to the early fourteenth century, but it would be unwise to use this alone to date the instrument. The scales on the double shadow square are incorrectly drawn, their divisions constructed by joining the centre to equi-spaced points (still visible) inside the calendrical scale [see Fig. 7d]; this feature is attested on the geared astrolabe from the Science Museum but not on any other medieval astrolabe currently known to me. It was clearly favoured by astrolabe-makers in Northern France in the fourteenth century.

The back of the rete bears various kinds of construction marks [see Fig. 7e], but more unusual is the fact that the edges of some of the plates show evidence of rather heavy-handed hammering [see Fig. 7f]. Several sets of construction marks are visible on the plates and on the diameter of the universal horary quadrant. These cannot be discussed here, not simply for lack of space, but also since no such marks have ever been studied on any medieval astrolabes.

THE GOTHIC INSCRIPTIONS

The various star names applied to the pointers and the names of the zodiacal signs on the rete and to the names of the months and signs on the back are all in the same hand, which is neither consistent nor steady and in fact gives the impression of advancing years. I have already alluded to the possible monastic origin of the astrolabe; certainly an elderly monk would have stood a better chance of being forgiven for incorrectly identifying a few stars than a professional astrolabe-maker.

The star names featured on the rete can be divided into two groups,[19] those

---

[19] (1) written more like 'fooois'—for 'finis fluvii', 'the end of the river', θ Eridani. Several medieval astrolabes feature this star, usually labelled 'finis fluxus' or simply 'finis'. (2) for 'septime'. This is an unusual astrolabe star (see note 20) and is attested, as I far as I know, only on two other astrolabes, both French—the geared astrolabe (#198) has 'septime', and another French astrolabe dating from about 1425 (#163) has 'sept(ime)' (as well as various other mysterious star names). (3) for Aldebaran, standard for α Tauri. (4) standard (if sometimes spelled 'rigel'), referring to the foot (Arabic *rijl*) of Orion, β Orionis. (5) for 'algomeisa', standard for Procyon, α Canis Minoris. (6) written 'leoo' (with the 'e' upside-down)—for Cor Leonis, standard for Regulus, α Leonis. (7) standard (if incorrect from Arabic *al-fakka*) for α Coronae Borealis, usually written 'elfeta' on Fusoris astrolabes. (8) standard, from Arabic (*al-nasr*) *al-tā'ir*, the flying (eagle), α Aquilae. (9) standard, from Arabic *al-'abūr* for the star Sirius, α Canis Maioris. The name is an error for '(ras)alhaue' from Arabic (*ra's*) *al-ḥawwā'*, referring to α Ophiuchi. (10) standard although other renderings are common, from Arabic *al-fard*, 'the solitary' star α Hydrae, called 'ydra' on some Fusoris astrolabes (no doubt following French pronunciation) and 'alfard' on others. The name is an error for 'alferas' or 'alferaz' from Arabic *al-faras* 'horse', for Pegasus. (11) for Corvus, the raven (the standard abbreviation for -us is here printed as (us)). The star γ Corvi was a popular astrolabe star. The name has here been applied to the pointer to the left of the right-hand support for the equatorial frame,

correctly associated with pointers:

(1) finis (?) (2) sept (3) alde (4) Rigil
(5) algon (6) cor Leon (?) (7) feta (8) altair

and others applied to the wrong pointers:

(9) alhabor (10) alfart (11) cor(vus)

Both groups can be compared with star names featuring on contemporary French astrolabes, in particular with the Science Museum piece. These star names, with one exception (sept), are found in the star catalogue associated with Messahalla, which derives originally from the eleventh-century astronomer Maslama al-Majrīṭī (Kunitzsch's Typ VIII).[20] But the engraver of the Picard astrolabe was merely consulting a defective star catalogue or copying the star names from another instrument based on such a catalogue.[21]

The names of the signs found on the rete and on the back are as follows, unusual forms being printed bold:

Aries taur(us) **gm̄y** **cance** **leon** uirgo
libra scorpius sagitarius capricorn(us) **Acarius** Pisses
aries taur(us) **geminis** cancer **leoni** uirgo
libra scorpius **sagitari**(us) **caṗcorn**(us) aquari(us) **pisses**

During the Middle Ages the names of the signs were written in various ways, including numerous abbreviations. It is not unusual for different forms to be used on the same instrument, as on the Picard astrolabe; sometimes the forms can cast light on the provenance of a given instrument. These particular ones have been compared with those found on various other instruments all predating 1450: we can see some influence from northern Spain, and some accord with other early French astrolabes.[22]

---

rather than the one below the half-quatrefoil at the right-hand end of the equinoctial bar. This might be judged the least 'serious' of the three errors.

[20] Kunitzsch, *Sternverzeichnisse* (see n. 4 above), especially pp. 51–8 (Typ VIII) and pp. 73–85 (Typ XII). The star name 'septime' is not attested in the medieval star catalogues that have been studied, although in the Toledan star table (Typ XII) there is a 'septentrionale cornu'. 'Septime' may refer to the Pleiades, whose name in some popular European traditions (e.g. German *Siebengestrirn*) may imply that seven stars were visible.

[21] In the star catalogue mentioned above (Typ VIII) the star 'alfard' is mistakenly associated with the constellation Equus (for Pegasus): ibid., p. 55 (Typ VIII, no. 19), and the notes on p. 57.

[22] The forms 'geminis' and 'acarius' are attested together only on an early Catalan astrolabe (#416), although an Andalusian astrolabe with later Spanish inscriptions (#1148) has 'geminis' and 'aqari(us)', an English zoomorphic astrolabe (#2006) 'gemıııs̄' and a French astrolabe (#163) 'akar-ius'. The astrolabe of Petrus Raimundus of Aragon (#3057), made in Barcelona in 1375 has 'acari(us)', and Catalan additions to an eleventh-century Islamic astrolabe from Cordova (#3622) have 'acari'.'Geminy' is found only on two other French pieces (#337 and #461). 'Leon(i)' is not attested on any other known instrument. Sagittarius was misspelled with one t as often in the Middle Ages as it is nowadays. The same abbreviation of Capricornus, with a dot over the p indicating that some letters have been omitted, is attested on a French astrolabe (#337) and a Dutch one (#566), as well

Particularly revealing are the month names found on the back of the astrolabe, namely:

Jenuier freuier march auril may Juing
Juillet ~~a~~auoust septe(m)bre octe(m)bre noue(m)bre dece(m)bre

(The first *a* on the eighth month appears to have been crossed out and the first *u* and the *o* have been written together.) The distinctive forms for the second, third and sixth months, as well as the more common variant for the tenth month, are attested in the medieval dialect of Picardy.[23] This linguistic evidence accords with the geographical information provided by the latitudes favoured on the plates. It should be realized that the astrolabe presents a complete set of month names and is itself a valuable new linguistic source. The only other medieval astrolabe with old French names for the months is the geared astrolabe in the Science Museum, which has a plate serving 'engletiere' and 'hollande'. The double horary quadrant on the back, however, serves a latitude in Northern France, and the month names are a mixture of what is attested in either Northern or Eastern French (in other words, they have yet to be associated with a specific region).

THE DEDICATION

Dedications in which the object speaks in the first person are attested on other instruments.[24] Here the inscription on the mater, engraved in bold Gothic script

---

as various other English and German astrolabes, whereas the geared French astrolabe (#198) has 'cap̄n(us)'. The double s in 'Pisses' is found on three French astrolabes (#337, #461, and #536), and one Fusoris astrolabe (#194), as well as on an Italian astrolabe (#479) and at least two early English astrolabes (#301 and #304, the former actually with Piss(us)!). These names predate the Fusoris astrolabes, which had their own distinctive forms (for example, 'Caper' for 'Capricorn').

[23] See, for example, W. von Wartburg, *Französisches etymologisches Wörterbuch. Eine Darstellung des galloromanischen Sprachschatzes*, 23 vols, Bonn, etc., 1928–69 (reprinted Basle, 1969–83), especially III, p. 441b (frevrier / frevier), and VII, p. 390a (march), and C. Th. Gossen, *Grammaire de l'ancien picard*, Paris, 1970, especially pp. 44, 94 and 114, and also 166 (jenvier), 159 (fevrier), 163 and 167 (march), 169 (octembre). While von Wartburg does not list *juing* as Picard, this name is indeed attested in medieval Picard; see, for example, (Abbé) Hénocque, *Histoire de l'abbaye et de la ville de Saint-Riquier*, 3 vols, Amiens, 1880–1888, II, p. 135. The Picard connection was first recognized by my assistant Kurt Maier, who is currently engaged on a study of all vernacular month-names on medieval instruments. Besides the two French astrolabes mentioned in this study, other medieval astrolabes (some Islamic with later European additions) exist on which the month names indicate Spanish-Jewish, South Catalan and Italian provenance.

[24] See, for example, the astrolabe in Nuremberg associated with, but not made by, Henricus de Hollandia (#548, now catalogued in Gerhard Bott, ed., *Focus Behaim Globus*, 2 vols, Nuremberg, 1992, II, pp. 578–81), and the astrolabe dedicated by Regiomontanus to his patron Bessarion. This is #640 and was first described in D. J. de Solla Price, 'The First Scientific Instrument of the Renaissance', *Physis*, 1, 1959, pp. 26–30. Price was not aware that there are at least ten other fifteenth-century German astrolabes from the same school; the rete on one (#452) was engraved by the same person as the engraver of the Regiomontanus piece. Ironically, they are all listed in his *Checklist* (#256, #452, #516, #530, #545, #549, #592, #620, and #643 or #644), and it is a sad reflection on

with humanistic flourishes [see Fig. 8], reads: 'Hadriano Amerotio Berseli(us) me dono dedit 1522' (that is, 'Berselius gave me as a gift to Hadrianus Amerotius in 1522'). Berselius and Amerotius were leading figures in the humanist movement in the Low Countries in the early sixteenth century. Paschasius Berselius was a Benedictine monk from Liège, an artist and poet, as well as a writer.[25] Of his writings other than his poetry, alas, just a part of his correspondence with Erasmus has survived; his library was apparently destroyed when the French dissolved his monastery at the end of the eighteenth century. Berselius was also involved with the French humanist Budé, as well as with the Spanish humanist Vives during the latter's sojourn in Louvain.

Hadrianus Amerotius, a native of Soissons on the southern border of Picardy, was the second Professor of Greek at the newly-founded College of the Three Languages (Latin, Greek and Hebrew) at Louvain.[26] The backgrounds of these two men and the relations between them I have discussed in detail elsewhere.[27] Why did Berselius give Amerotius the astrolabe? Amerotius mentioned Berselius as his friend in the introduction of his Greek primer published in 1520, and at least for this, Berselius was in his debt. Certainly the year 1522, when Berselius gave Amerotius the astrolabe, was of significance to both men: Amerotius received his doctor's degree, their mutual friend Adrian of Utrecht was proclaimed pope; and about the same time Berselius left Louvain to return to his monastery in Liège. Perhaps Berselius thought that a Picard astrolabe would be a particularly appropriate gift for his friend, himself a Picard?

Berselius was certainly a man of passion; this is evident from his exchanges of letters and gifts with Erasmus, and from the fact that he commissioned an exquisite statue of the Virgin and Child on whose pedestal he engraved some of his poetry.[28] This piece, however, he clearly had no intention of giving away: one

---

the state of the field that none of them had ever been studied until recently. For details see *Focus Behaim Globus*, II, pp. 581–9 (the illustrations 1.76.3–4 on p. 588 are of #452); King, *Catalogue*, Section 6.9 (fifteenth-century German astrolabes), forthcoming; and my joint paper with Gerard Turner entitled 'The Astrolabe Dedicated by Regiomontanus to Cardinal Bessarion in 1462' to appear in *Nuncius*.

[25] The most detailed account of his life is in J. Ceyssens, 'Berselius et la Statue de la Vierge de Dalhem', *La Vie Wallonne—Revue mensuelle illustrée*, 5, 1924–5, pp. 341–55 and 381–7, and *idem*, 'Encore Berselius', ibid., pp. 446–50. On his relations with Erasmus see also P. S. Allen and H. M. Allen, *Opvs epistolarvm Des. Erasmi Roterodami*, Oxford, 1922, III, pp. 95–6. See also n. 28 below.

[26] On Amerotius see, for example, H. De Vocht, *History of the Foundation and Rise of the Collegium Trilingue Lovaniense 1517–1550*, 4 pts, Louvain, 1951–4, pt. IV, pp. 252–68.

[27] See n. 7 above.

[28] On Berselius's statue of the Virgin and Child, made for him by the Swabian sculptor Daniel Mauch but painted and inscribed by himself, see in addition to the references cited in n. 25, for example, R. Didier *et al.*, *Les Sculptures médiévales allemandes dans les collections belges / Duitse middeleeuwse beeldhouwwerken in Belgische verzamelingen*, Brussels, 1977, pp. 220–25 (no. 102), with several illustrations in black and white as well as colour illustrations of the front and back on the front and back covers.

of the inscriptions (this time in bold capital letters) reads 'Sum Berselii'. The astrolabe he had probably acquired from a fellow monk, and he was happy to give it to a colleague and valued friend.

CONCLUDING REMARKS

The appearance of the cipher numerals on the Picard astrolabe renders it unique amongst surviving medieval instruments and an invaluable source for new historical ideas. In fact the ciphers are a most convenient way of representing numbers, and because of their compactness they are particularly suited to marking numbers on astronomical instruments. However, since the ciphers on the Picard astrolabe are engraved almost casually and there is an associated textual tradition attested by the unique Norman manuscript, it would seem likely that other instruments marked with ciphers were also made. These, like the vast majority of instruments constructed in the Middle Ages, have simply not survived the ravages of time.

Not only the ciphers, but also the fact that the astrolabe came into the possession of a monk, suggest a monastic origin for the piece. Thus it is doubly unique amongst surviving medieval instruments. The quatrefoil decoration on the rete connects it with about thirty other surviving examples made in France, Italy, Germany and England before the introduction of the distinctive rete designs of fifteenth-century French (Fusoris-type) and German (Regiomontanus-type) astrolabes. The Picard additions associate it with a very small sub-group of instruments bearing vernacular month names and local variants of the names for the signs. The still later dedication establishes its association with two of the leading personalities in the exciting intellectual milieu of early sixteenth-century Louvain.

The astrolabe is speaking to us, and we can only receive its message if we devote some time and effort to interpreting each of its components and engravings. One just has to consult previous literature on the ciphers to realize that a most ingenious medieval technique of numeration has never received due attention or appreciation in modern times, and that it is only by returning to the original sources and investigating their provenance and content that one can understand how the ciphers were used. Likewise previous writings on the Picard astrolabe demonstrate how little can be written about instruments on the basis of photographs alone. But above all it is only by using Gerard Turner's approach to looking at instruments and also by listening to them that their secrets will reveal themselves.

IV

THE MEDIEVAL CATALAN ASTROLABE OF THE SOCIETY OF ANTIQUARIES, LONDON

David A. King and Kurt Maier

Introduction

The Society of Antiquaries in London possesses an exquisite medieval European astrolabe,[1] donated by the Reverend I. G. Lloyd in 1893.[2] It is made of brass and is 100 mm in diameter and 6 mm thick. Its existence was announced by the Society's Secretary C. H. Read in the *Proceedings* of the Society that same year.[3] Read presented a careful description of the piece but was somewhat condescending about its intrinsic value for the Society and for posterity because it was Spanish rather than English. Robert T. Gunther listed it as Spanish in his monumental work on astrolabes published close to 40 years later, taking over much of Read's description and interpretation.[4] He also reproduced Read's lithograph of the back of the instrument.

No further research was conducted on the piece until 1993,[5] when the first author had the privilege of examining it in detail. The first discovery was that the back had additional markings not shown on the published lithograph. The second (which could have been made from Gun-

---

[1] The best introduction to the astrolabe in English is still *Greenwich Astrolabe Booklet*. For Catalan readers there is the excellent Chabàs & Bosch, *L'astrolabi pla*. See also G. L'E. Turner on other popular accounts of the instrument in *Archives internationales d'histoire des sciences* in 1996.

[2] Inventory number Cat. 559 ("Cat." stands for Catalogue). The piece bears the International Instrument Checklist Number #162. (Numbers between 1 and 9999 are being assigned to all medieval instruments in the new Frankfurt-based project to catalogue all medieval scientific instruments (on which see note 7 below); numbers less than 330 are for astrolabes treated in Gunther, *Astrolabes of the World*.) For bibliographical information on the various instruments mentioned in this paper see Appendix 2.

[3] Read, "Astrolabe", reproduced in Appendix 1.

[4] Gunther, *Astrolabes of the World*, II, pp. 306-309 (no. 162).

[5] Here one might compare the so-called "Berselius astrolabe" (#202), whose existence has been known since Gunther briefly mentioned it in 1932, but which bears monastic ciphers, whose history was not investigated before *ca.* 1991.

ther's description) was that at least one of the star-names (*corbus*) was in a vernacular form of Latin. More recently the second author, who has identified the provenance of various other medieval astrolabes by the vernacular forms of the month-names used on them,[6] discovered other vernacular star-names on this astrolabe and, using them, identified the provenance of the piece.

We continued our research, drawing on the photographic archives of the Frankfurt-based project to catalogue all surviving medieval Islamic and European instruments,[7] and exploiting the vast corpora of data on medieval star-names and stellar coordinates,[8] and of geographical coordinates,[9] compiled respectively by Paul Kunitzsch and Edward and Mary Helen Kennedy. The result is the present paper, of which the main conclusion about the provenance, determined from the *vernacular star-names* - this is the first time this has ever been done - and confirmed by the latitudes represented on the plates, was presented at a lecture in the Society's rooms on November 16th, 1994. The title of the lecture was "Making Instruments Talk - Some Medieval Astronomical Instruments and Their Secrets",[10] appropriate perhaps not least since the secrets of this astrolabe have been kept for a century since it was acquired by the Society.

---

[6] Maier, "Romanische Monatsnamen". The instruments in question are: #198 (N. French, precise provenance not yet identified from the month-names although the horary quadrants on the back are for latitude *ca.* 51°); #202 (Picard), #416 (Catalan), and #3622 (Islamic, with later Catalan inscriptions); #1148 (Islamic, with later inscriptions from N. Spain).

[7] On this see King, "Instrument Catalogue", and on some preliminary results see *idem*, "Astronomical Instruments".

[8] Kunitzsch, *Arabische Sternnamen in Europa* and *Typen von Sternverzeichnissen*.

[9] Kennedy & Kennedy, *Islamic Geographical Coordinates*. There is no comparable compilation for medieval European sources, although Sezgin *et al.*, eds., *European Geographical Tables*, is a useful collection of some of the materials. The latitudes used on astronomical instruments do not necessarily correspond to those found in the medieval geographical tables and for comparative purposes it is actually more useful to refer to other instruments - see, for example, notes 60 below.

[10] Now published as King, "Making Instruments Talk".

Description of the astrolabe

The instrument is a fine Gothic-type astrolabe, carefully executed and with inscriptions in a clear Gothic hand.[11] The basic design of the rete is Islamic, and the name "Hispano-Mauresque" is sometimes applied to this design. The state of our present knowledge of medieval European instruments is such that it is not always possible from the design alone to determine the provenance. In this case we are dealing with an instrument whose elegant rete is without parallel in design amongst all known European astrolabes. We shall return to various aspects of this basic description in the commentaries.

The throne is high with foliate decoration resulting in three pairs of double small lobes and one larger lobe as well as one hole on each side. The suspensory apparatus is of the standard Ω-shaped type with bowl-shaped "feet" and is fitted with a circular ring - both appear to be original. The back is brazed onto the rim. The scale of the rim is labelled for each 10°-interval in Gothic numerals and subdivided for each 5° and then each 1°. The outside of the rim is marked for each hour and hours 1-10 are labelled in the same hand as the main markings.

The rete is of particular interest since in addition to a series of five contiguous *miḥrābs*, with trefoil-type decoration at the top of each, between the circumferential frame and the lower ecliptic there is a rectangular frame inside the southern ecliptic. The equinoctial bar is counter-changed between the circumferential frame and the ecliptic and again between the ecliptic and the central disc, that is twice, on each side. Furthermore the axis from the central disc to the winter solstice intersects this rectangle in a quatrefoil. Notice the circular protrusions at the lower corners of the rectangle and the small frames connecting the ecliptic to the circumferential frame. The scale of the ecliptic is divided and labelled for each 10° of each sign and subdivided for each 5°. The names of the signs are written (here and elsewhere inscriptions in bold are of particular historical interest):

---

[11] Our studies of scripts on medieval instruments are still in their infancy. On the Gothic forms of the Arabic numerals, albeit mainly from manuscripts, the best source remains Hill, *Numerals*.

ARIES TAUR^9 GEMINI CANCER LEO UIRGO
LIBRA **SCORPIUS** **SAGITARIUS** CAPRICORN^9 AQUARIUS **PISCIS**

The pointer for the star CAENT (= Vega) is in the form of a cock's head. There are curved pointers with one or two lobes at the base, some with a silver button in the middle of the base. They serve 20 stars, named as follows:[12]

			K nos.
1.	CAYTOS	ζ Cet	8 - 68
2.	ALDABERAN	α Tau	18 - 16
3.	ALGOL	β Per - see below	14 - 24
4.	**MENCER**	α Ori - see below	22 - 72
5.	ALABOR	α CMa	23 - 26
6.	ALGOMIC	α CMi	25 - 93
7.	BENNAZ*	η UMa	40 - 71
8.	COR LEONIS	α Leo	30 - 109
9.	**CORB^9**	ghost - see below	(32) - ?
10.	MAN^9 URSI^9	ι UMa	28 - ?
11.	AhZEL	α Vir	39 - 66
12.	ELRAMECh	α Boo	41 - 46
13.	COR SCORPIUS	α Sco	48 - 110
14.	**ELAYYE**	α Oph	51 - 162
15.	**CAENT**	α Lyr	53 - 198
16.	CAUDA CAPRICORN^9	δ Cap	59 - 82
17.	**UOLANT**	α Aql	54 - 52
18.	ALREDAF	α Cyg	56 - 61/81
19.	ALFARAS	β Peg	62/63 - 44
20.	CAUDA CAYTOS	ι Cet	4 - 85

* The letter Z is written "backwards".

The positions of the stars are reasonably accurate (see further below) with two exceptions that seem to result from confusing two pairs of sets of coordinates. No. 3 *algol* is in a position that would be more appropriate for *alhaioc*, α Aur, and no. 4 *mencer* is in a position that would be more

[12] The order is that of increasing right ascension from the vernal equinox, and the stars are arranged in the four quadrants of the ecliptic. Modern identifications are also given, as well as the numbers in Paul Kunitzsch's lists of "Astrolabsterne" and individual *Arabic* star-names in *idem*, *Arabische Sternnamen in Europa*, pp. 59-96 and 97-224 (references here are separated by -). Gunther had incorrectly BENNAS, ALREDAh, and ELAVVE.

approriate for *algueze*, α Ori (sometimes known as *mankeb*, the Arabic name being *mankib al-jawzā'*).

The mater and two plates serve the following five latitudes:

M	32;30°
1a	38;30
2a	**39;40**
2b	41
1b	42 .

There is room for two additional plates.[13] All have altitude circles for each 5° and azimuth circles for each 10°. The set for 32;30° also has curves for the hours since sunset, which is quite unusual on an early European astrolabe (such curves are found already in the 10th century on Western Islamic astrolabes (see, for example, #111). None of these are marked with arguments. All latitude-dependent markings except for those on the mater have unlabelled curves for twilight at *ca.* 18° below the horizon. The seasonal hour curves are labelled in Gothic numerals.

The back bears four altitude scales divided for each 10°, subdivided for each 5° and then for each 1° and labelled for each 10°. These scales serve as a solar longitude scale starting at 90° and running counterclockwise. Inside the scale of names of the signs

ARIES TAURUS GEMINI CANCER LEO UIRGO
LIBRA **SCORPIUS** **SAGITARI⁹** CAPRICORN⁹ AQUARIUS **PISCIS** ,

is a decorative foliated band, and within this a *concentric* calendrical scale

IANUARIUS FEBRUARI⁹ **MARCI⁹** APRILIS
MADIUS IUNIUS IULIUS AUGUST⁹
SEPTEMBER **OCTORER** NOUEMBER DECEMBER .

The equinox is at March 12. Associated with each month is a number:

---

[13] Read did not notice this; Gunther thought that only one plate was missing.

3 6 5 1 3 6 1 4 7 2 5 7,

the so-called "solar regulars". Inside the central space are three circular scales showing the numbers 1 to 28 of the solar cycle, the so-called concurrents for those years, and which years are leap (each fourth year is marked with a B for *annus bisextus* or *bisextilis*), thus :

1	2	3	4	5	6	7	8	9	10	11	12	13	14	15	16	17	18	19	20	21	22	23	24	25	26	27	28
			B				B				B				B				B				B				B
1	2	3	4	6	7	1	2	4	5	6	7	2	3	4	5	7	1	2	3	5	6	7	1	3	4	5	6

The solar regulars serve to determine the day of the week (which recurs on the same date every 28 years). The year 9 B.C. is a leap-year beginning on a Monday. So is, for example, the year 1392 because $1392+9 = 1401 \equiv 1$ (modulo 28). May 1 of that year is a Wednesday because $31 + 29 + 31 + 30 + 1 = 122 \equiv 3$ (modulo 7). Alternatively one can take the solar regular for May 1, which is 3, and add it to the concurrent, which is 1, and obtain 4, which corresponds to a Wednesday (now counting Sunday as day 1).[14]

Below the horizontal diameter is a double shadow-box (bases 12/12) with scales divided and labelled for each 3 units and subdivided for each 1 unit. They bear standard inscriptions UMBRA RECTA and UMBRA UERSA. There are some later markings in the upper left quadrant, not all visible in the lithograph published by Read and Gunther (see below), although the equinoctial quarter-circle is indeed visible there and looks very curious in isolation. It invites closer investigation but nothing more is visible on the photographs provided to the first author by the Society some years ago.

The alidade is counterchanged and bears no markings. Both it and the pin appear to be original; the simple wedge is surely a replacement and is made to serve the two surviving plates.

Other markings: A quarter circle and vertical radius have been engraved in the upper left quadrant of the central space on the back. There are traces of markings for an horary quadrant but things seems to

---

[14] For assistance in fathoming the mysteries of calendrics we are, as always, grateful to Silke Ackermann, formerly of Frankfurt, now at the British Museum.

have gone awry with the construction. There are marks on the equinoctial circle surely intended for the hours. These are at solar altitudes:

$$16° \; 22° \; 32° \; 40° \; 45° \; 47° \; 48°,$$

measured on the outer scale. These seven markings are unhappy, since there should be only six, but we can be sure that the underlying latitude was about 42°.

Astronomical commentary I (King)

On the basic design of the instrument and especially the rete

The distinctive design of the throne is not attested on any Islamic or other early European astrolabe.[15]

One of the most distinctive features of the rete is the rectangular frame inside the upper horizon with a quatrefoil at the middle of the upper side. Such a frame is not at all common on Islamic astrolabes. In fact, it is attested only on one astrolabe with inscriptions in Judaeo-Arabic (Hebrew written in Arabic script) (#3915). This even has the curved protrusion at the lower corners of the rectangular frame and part of a quatrefoil at the middle of the upper side. The fact that the design is Islamic and not European is attested to by the astrolabes of Abū Bakr ibn Yūsuf of Marrakesh, *ca.* 1200.[16] Some five retes by him survive and they all have a distinctive "rectangular" frame embellished by curved sides. On the quatrefoil see further below.

The *miḥrāb*s or prayer-niches standing on the outer frame and below the ecliptic ring are copied directly from Islamic models. We find them on several 11th-century Andalusian astrolabes, such as #117, #118,

---

[15] We are far from controlling the basic designs of thrones. Another example of a most unusual throne is that of the Picard astrolabe with monastic ciphers (#202) - see King, "Picard Astrolabe", p. 56. When one considers astrolabes from the same workshop clear patterns may emerge: see the examples illustrated side by side in *idem*, "Astronomical Instruments", p. 160, and *idem*, "Making Instruments Talk", p. 10.

[16] See Gunther, *Astrolabes of the World*, I, p. 264, for an illustration of one example (#124).

#121, #1099 and #2527, and on some early European astrolabes, for example, #191 (possibly Spanish - Gunther has English -, with two *miḥrābs* asymmetrically placed), #428 and #2041 (both French, with three *miḥrābs*).

The counter-changes on the horizontal (equinoctial) axis:

are not found on any surviving Andalusian astrolabes known to us. Simpler patterns, with no counter-changes outside the ecliptic are found on 11th-century and early-13th-century Andalusian astrolabes (none are known from the 12th), and yet simpler patterns, with one or two counter-changes only, are characteristic of the earliest (pre-11th century) Andalusian astrolabes (such as #110) and the Carolingian astrolabe (#3042).[17] Such a complicated pattern is rare on early European astrolabe retes. On the undated French piece #2041 and also on an early Italian astrolabe #493, for example, we have the arrangement:

No other instrument known to us, Islamic or European, has similar pointers. The star-pointers mostly have two lobes at the base, recalling some of those on the "Carolingian" astrolabe (#3042) (the lobes at the bases of pointers on Andalusian instruments were often fitted with silver studs), but here some of them have wavy and curved forms reminiscent of those on the Catalan astrolabe #416.

On the quatrefoil on the rete

One of the features that makes this a "Gothic"-type astrolabe is the

[17] In King, "Earliest European Astrolabe", various simpler patterns are presented in order to classify that on the earliest European astrolabe (#3042).

charming quatrefoil on the rete. The quatrefoil and the trefoil were standard motifs of Gothic art and architecture.[18] But the quatrefoil on this astrolabe is not an exclusively "Christian" motif; it was copied from a similar motif on an Islamic astrolabe. Already on the magnificent astrolabe (#111) of the astronomer Ḥāmid ibn Khiḍr al-Khujandī, the leading instrument-maker of the early Islamic period, there is an identical quatrefoil in the same position, that is, in the middle of the upper part of the ecliptic. This instrument was made in Baghdad in the year 984/85. The inspiration is doubtless Byzantine, for the quatrefoil occurs rarely in Islamic art generally (in fact, only on astrolabe retes). Whilst al-Khujandī's distinctive rete design, which also included zoomorphic representations for the star-pointers, was influential in the Islamic East, it is not known to have been influential in al-Andalus (except for the quatrefoil on the Society's astrolabe). But there were Western Islamic astrolabes of one sort or another decorated with quatrefoils: witness one made in 1240/41 (#154), which reflects considerable advances over al-Khujandī's astrolabe with regard to the exploitation of the quatrefoil. There is a degenerate quatrefoil on the rete of the astrolabe with inscriptions in Judaeo-Arabic (#3915), which is a clear indication that there was a quatrefoil on the Islamic astrolabe from which it was copied. There is a lot more that can be written on astrolabe rete-design in general and the use of the quatrefoil in particular.[19]

---

[18] See, for example, Stafford & Ware, *Dictionary of Ornament*, p. 176 *sub* quatrefoil: "A characteristic device in Byzantine decoration [?] and Gothic tracery and carving ... . Sometimes stated to be based on the four-leafed clover; reliable authorities [!] believe the quatrefoil to be a strictly Christian motif [!] - a form of the Greek cross with rounded ends, or of the nimbus with the four arcs representing the four Evangelists. ... "

Clearly medieval astrolabes, both Islamic and European, constitute a corpus of historical material that has never been investigated by historians of art.

[19] In passing we may mention an isolated rete which was auctioned at Christie's of London in 1995 (#4201) that is very similar to the one on the universal astrolabe of Ibn al-Sarrāj of Aleppo dated 1328/29 and now in the Benaki Museum, Athens (#140). This new rete, rather remarkable in the precision of its execution, can be associated with the Lahore school of the mid 17th century; it was surely copied from a genuine rete of Ibn al-Sarrāj. What is important for the present study is that it bears a quatrefoil on the solstitial axis. Thus Ibn al-Sarrāj, the leading instrument-maker of the Islamic Late Middle Ages, was also influenced by al-Khujandī, at least for the quatrefoil on the rete from which this one was copied; there is no quatrefoil on the rete of his astrolabe that is now in the Benaki Museum.

Since we have established above that the basic design of the Society's astrolabe - the rectangular frame and the *miḥrāb*s - is Islamic in origin, and since it is clear that the quatrefoil is also a motive on Islamic retes of a certain tradition, we may be confident that the design of the rete of the Society's astrolabe, including the quatrefoil, was copied *in toto* from an Islamic astrolabe.

The small frames connecting the ecliptic to the circumferential frame are worthy of comment: they remind one of the more developed half-quatrefoil cutouts on the supports of the lower equatorial frame on the rete of another very early European instrument (#300), possibly Spanish and dating from the 13th or even 12th century.[20] Yet more complicated and ornate designs are used, for example, on the supports of the equatorial bar on the rete of an English astrolabe dated 1342 (#292).

On the cock's head

The star-pointer for Vega (*caent*) is in the form of the head of a cock, whose comb is distinctly visible. The representation of this star with the head of a bird, common on late Eastern Islamic astrolabes, is less common on early Islamic pieces, and is not found on any other known medieval European astrolabe. It has been thought that the idea was Islamic since the star is called *lyra* in Greek, without immediately obvious associations with any bird, but in Arabic it is called *al-nasr al-wāqi*ᶜ, 'the falling eagle', whence our name Vega. But it seems that the idea is indeed Byzantine, and that it dates from the time before the advent of Islam (and of the emergence of Arabic as the new language of science). The evidence is provided first by a Byzantine astrolabe dated 1062 (#2), which in spite of its relatively late date (three centuries after the first exposure of the Muslims to the astrolabe), shows not a trace of Islamic influence, and by the earliest known Islamic astrolabe (#3702), on which the star-positions

---

[20] Gunther has an English provenance, probably because the latitudes of the plates include 52°, 55° and 60°. But the complete set - 30°, 35°, 40°, 45°, 48°, 52°, 55° and 60° - represents an attempt to render the instrument universal.

are for *ca.* 850. Both of these have the head of a bird for Vega.[21] Yet not even the astrolabe of al-Khujandī (#111), which has various zoomorphic pointers, shows a bird's head for Vega. (It is, however, possible that other instruments of his did have the bird's head.) No Western Islamic astrolabes at all display this feature. Thereafter the earliest Islamic astrolabe with a bird (actually an eagle) dates from Isfahan in 1223/24 (#5), the rete on this instrument being otherwise inspired by that of al-Khujandī. The bird for Vega was not common thereafter until much later. The conclusion is that the bird is of Byzantine origin, that it was copied on a minority of the earliest Islamic astrolabes, and that it was later slowly revived.

The fact that a bird's head appears on the Society's astrolabe, and on no other known European instrument, is a clear indication that it was copied from one of these rare Islamic pieces, of which none survive from the Islamic West, doubtless an Andalusian astrolabe inspired by the earliest Eastern Islamic ones.

On the names of the stars

The star-names include several transcribed Arabic names (nos. 1-7, 11-12, 14, 18-19), one mixture of Latin and transcribed Arabic (no. 20), four incorrectly-written Latin forms (nos. 10, 13, 16), and one correctly written Latin name (no. 8). Unusual transcriptions of Arabic, indeed forms not yet attested elsewhere, include *aldaberan* (from Arabic *al-dabarān*, *aldebaran* is more usual), *mencer* (which one might think was from *minkhar*, *menkar* being more usual, though see below), *algomic* (from *al-ghumayṣā'*, *algomeisa* is more usual), and *bennaz* (contracted from *benenaz*, a common rendering of Arabic *banāt naᶜsh*). The Arabic letter *ᶜayn* has been suppressed in *alabor* (from *al-ᶜabūr*, often rendered *alhabor*), but not in *ahzel* (from *al-aᶜzal*). *Elayye* (with the same form of *y* as used in *caytos*) is a most unusual rendering of Arabic *(ra's) al-ḥawwā'*, not attested elsewhere. Of particular interest are the forms *caent* and *volant*, and *corb*[9] - see the historico-linguistic commentary below.

---

[21] Stautz, "Früheste Formgebung", is the first attempt to establish the forms firstly in which the astrolabe was known to the Greeks and secondly in which it was transmitted to the Muslims. On the bird see already King, "Astronomical Instruments", p. 170.

On the selection of stars

Not only the names of the stars, but also the set of stars selected can provide useful historical information. We are concerned to know which of several available star-catalogues was used by the maker. In this case, the inclusion of the "ghost-star" *corbus* is the first indication that the star-table labelled VIII by Paul Kunitzsch lies behind the selection. The Arabic star *janāḥ al-ghurāb*, the raven's wing, was unhappily separated by medieval Europeans into two stars, *algorab* (and the like) and *corvus*. Table VIII, with names in transliterated Arabic and some Latin equivalents, is a combination of stars from a table by the Andalusī astronomer Maslama al-Majrīṭī (*ca.* 1000) and John of London (Paris, early 13th century);[22] it was very popular amongst medieval European astrolabe-makers, who concocted a series of new names for each of the stars, most of which are not attested in the manuscript tradition so carefully documented by Kunitzsch. Indeed, 18 stars in the list correspond more or less to 18 of the 49 stars of Table VIII. The exceptions are *algol* and *man⁹ ursi⁹*, which are not in VIII. Also it should be noted that *ahzel* (no. 11 in the list) is referred to by its other name *alchimec* in Type VIII (both from the Arabic *al-simāk al-aᶜzal*).

Type VIII was in part based on Kunitzsch's Type III, a list of 27 stars derived from a non-extant Arabic table to be associated with Maslama and his school.[23] Type III (with only transliterated Arabic names for the stars) does include *algol*, but no star equivalent to *man⁹ ursi⁹*, which is clearly a translation of the Arabic *yad al-dubb*, found on various Andalusian astrolabes.[24]

In fact, *manus ursi* (*sic* for *ursae*) is found only in Kunitzsch's Type IV, the star-catalogue of Johannes Hispalensis of Toledo (mid 12th

---

[22] Kunitzsch, *Typen von Sternverzeichnissen*, pp. 47-58. See also *idem*, "John of London".

[23] Kunitzsch, *Typen von Sternverzeichnissen*, pp. 23-30.

[24] Kunitzsch, *Arabische Sternnamen in Europa*, p. 75, where this star-name on #162 is mentioned.

century), which in its original form gave only Latin names.[25] All 20 stars on the astrolabe are found in Type IV, albeit with different names and with the reservation that the "ghost-star" *corbus* does not correspond to the "real" star *ala corui* in Type IV.

To investigate the selection of stars more closely we need to consider the positions of the star-pointers and to see whether they correspond to any of these medieval star catalogues.

On the star-positions on the rete and their accuracy

The positions of the star-pointers are, with a few exceptions, remarkably accurate for an epoch of 1200-1300. They are, however, apparently not taken directly from any of the star-catalogues documented by Paul Kunitzsch.[26] A few pointers exhibit problems of placement: nos. 1 (problem with Δ/β, α/μ correct for *ca.* 1200), 3 (all coordinates slightly in error), 14 (problem with α/μ, Δ reasonably correct), 17 (problem with α/μ, β correct), 19 (problem with α/μ, Δ correct). The position of the ghost-star no. 9 has been discussed above. For an explanation why no. 4 *mencer* has been placed close to the position of *alguese* see below.

On the calendrical scale

It would be unwise at this stage of our investigations of medieval astrolabes to try to use the fact that the equinox on the Society's astrolabe is at March 12 towards dating the piece.[27] The equinox on a contemporary (?) Catalan astrolabe (#416) is at March 11.5. And on the

---

[25] Kunitzsch, *Typen von Sternverzeichnissen*, pp. 31-33.

[26] See Stautz, *Untersuchungen*, p. 92 and fig. 4.5g, where the positions of the star-pointers are compared with those of Kunitzsch's Type VIII. We are grateful to Burkhard Stautz of Frankfurt for the information in this paragraph. Here ecliptic coordinates are denoted by λ and β, equatorial by α and Δ, and mediation by μ.

[27] On this subject see Poulle, "Dater les astrolabes?", and, more recently, Dekker, "Astrolabe Quadrant", pp. 37-38. Dekker notes: "Considering the uncertainty involved in this parameter (*sc.* the date of the equinox) and the complete lack of knowledge of the way in which this parameter was handled in medieval times it would be sheer speculation to use this information for the purposes of dating."

astrolabe made in Barcelona in 1375 by Petrus Raimundus (#3053) we find March 12.[28]

Before investigating other aspects of the astronomical markings we turn to the clues to the provenance provided by the inscriptions.

Historico-linguistic commentary (Maier)

In medieval Latin there are numerous examples illustrating that phonetic laws and grammatical rules were not taken very seriously. Thus even a *savant* of the calibre of Ramon Llull (d. 1315 or 1316), a native of Mallorca, used various spellings for, say, one single zodiacal sign, and, besides, has combinations like "De Saturnus" and "De Acàrius" in chapter-headings.[29] Also there are no constellations or star-names mentioned in the available etymological dictionaries; the forms of the zodiacal signs, however, are usually documented (see below).

Star-names

The first star-name which is of particular interest for us is that for Vega, namely, *caent*. Its Latin etymon is (*vultur*) *cadens*, the "falling eagle". The spelling of this form is identical with that of the modern Catalan active participle of *caer* and *caure* ("to fall"), derived from the Latin *cadent(em)*.[30] The only form to be found in etymological dic-

---

[28] Dekker, *op. cit.*, p. 37, note 65, observes "all known dated instruments with 13 March for the vernal equinox were made in the fourteenth century or later". Here is a dated 14th-century instrument with the equinox at March 12.

[29] See his *Tractat*, pp. 223 and 222. The range of different spellings of individual terms in this work shows that in those times there was no orthography as we understand it. Besides, there is the problem that in many modern editions of medieval Romance texts, editors modify the spelling (including caps and small initials) to render it closer to the modern forms, which does not help us in our search for the original spellings of forms that we find in inscriptions or on instruments.

[30] Coromines, *Diccionari*, ii, p. 642b, *sub caure*, mentions *caer* for the 2nd half of the 13th century. For the whole variety of medieval and modern forms see Alcover, *Diccionari*, iii, p. 67f., *sub caure*, where *caent* appears for Valencia, West Catalan and the Balearic Isles.

tionaries which comes near to our form is *recaent* (Eiximenis, Valencia, d. 1409), in the sense of "falling back into an evil or a disease".[31] So the Society's astrolabe provides the first documented attestation of this form in a medieval source.

Additonal evidence for a Catalan provenance of our instrument is the form for Altair, namely, *uolant*. Its Latin etymon is (*vultur*) *volans*, the "rising eagle". The spelling of this form is identical with that of the modern Catalan active participle of *volar* ("to fly, to rise"), derived from the Latin *volant(em)*. We find the form *volant*, "que vola", *i.a.* in Bernart Metge (d. 1410).[32]

The third form of interest is the name for the constellation of the raven, namely, *corb(us)*. The spelling of this medieval form is identical with the modern Catalan form *corb*. This form is recorded neither in the *Thesaurus* (*sub corvus*) nor in Du Cange nor in Forcellini's *Lexicon*.[33] Both the Mallorquine Ramon Llull and Joanot Martorell, the author of *Tirant lo Blanc* (born in Gandía near Valencia, d. 1468), have *corp*. Alcover (*sub corb*) says: "antigament escrit també corp i corv". Alcover gives an attestation for *corp* in Llull and for the plural form *corbs* in Metge; besides he mentions *i.a. ala de corp* ("raven's wing") for 1491 and *ales de corb* without a date.[34] For 1612 we have an attestation in Franciosini's *Vocabolario*, where the Spanish forms "corbo o cueruo" appear side by side with the Italian *coruo*. Simonet has *corbo* for Spain

---

[31] Coromines, *Diccionari*, ii, p. 643b, *sub escaure*; *ibid.*, p. 644a.

[32] Alcover, *Diccionari*, x, p. 861b, *sub volant*; Coromines, *Diccionari*, ix, p. 353b, *sub volar*, with the date 1398 for Metge.

[33] In medieval Italy we also find *corbo* in the region of Tuscany: in Bono Giamboni's (d. 1292) translation of Brunetto Latini's (d. 1294 or 1295) *Li livres dou tresor* - as a variant for *corvo*; see also *GDLI*, i, p. 766bc; *cf. Crusca*, iii, p. 737f.; DLI, i,2, p. 1725ab; *DEI*, ii, p. 1104a.

[34] Alcover, *Diccionari*, iii, p. 515b, *sub corb*, together with an attestation for *corbs marins* (plural of *corb marí*, "cormoraunt"), which, however, might be from a modern source; Coromines, *Diccionari*, ii, p. 922a, *sub corb*, where the first attestation for *corp* is given in Llull.

besides the Italian forms *corbo* and *corvo*.[35] For 13th-century Spain, Moreno Fernández mentions *corbo* and *coruo*.[36] So here again, we have an attested Catalan form to which the grammologue -⁹ (here for the ending -us) has been added. In any case, this is the first attestation of this form, not known from medieval astronomical manuscripts, on a medieval instrument.

The star-name *mencer*, in the position of *algeuse*, is apparently a regional orthographical variant for *menceb*, the Arabic name for the star in question being *mankib al-jawzā'*. A parallel to this phenomenon is the month-name Octorer on this instrument (see below).

Names of zodiacal signs

With the exception of Sagitarius for Sagittarius and Piscis for Pisces, the forms of the zodiacal signs on the Society's astrolabe are the Classical Latin standard forms (with Scorpius which is attested besides Scorpio). None of these forms, therefore, can be used for localizing the instrument to any part of medieval Europe. Some of the forms on our instrument, however, appear in sources from medieval Catalonia.

Llull has Àries (besides Àrias), and the former also appears in Mestre Aleix's Catalan translation of the Latin version of Aristotle's *De regimine principum* by Egidi Romà published in Barcelona in 1498: *Regiment de Prínceps*.[37] Llull has Taurus (with the variant Thaurus) which Alcover calls an "old variant", but like most of Llull's zodiacal signs it is rather a "gelehrte Form" than a vernacular form. Coromines has an attestation of *taur* for the end of the 13th century (not, however, designating the zodiacal sign in question). He also mentions this form for the Roussillon for "the archaic epoch" and for the work of Ausiàs March

---

[35] See Gili Gaya, *Tesoro*, fasc. iii, p. 637, *sub corbo*. Simonet, *Glosario*, p. 131, *sub corbách*, has *corbo* for Spain besides the Italian forms *corbo* and *corvo*.

[36] Moreno Fernández, "Palabra", p. 37 (no details are given): both forms are derived from *-rb-*.

[37] Alcover, *Diccionari*, i, p. 861a, *sub àries*: *cf. ibid.*, *sub arietes*; *cf.* Coromines, *Diccionari*, i, p. 387a, *sub àriet*, where we find *àries* besides *arietes* for Llull. Llull, *Tractat*, p. 218f. has Àries besides Àrias. The reader will find numerous further examples of variant spellings of the names of individual signs in these texts.

(d. 1459).[38] We find Gèminis in Egidi Romà's translation and, as an "old variant", in Llull.[39] Llull also mentions Càncer, Leo and Virgo.[40] The earliest Catalan attestation for Libra is in Llull, that for Sagitarius in Canals. Llull also has the earliest attestation for Capricorn.[41] We find Aquarius in Llull's *Tractat* and Aquari in a rather late source (1660) mentioned by Coromines.[42]

The earliest Catalan attestation for Piscis is in Llull.[43] It is attested in other medieval Catalan sources, including the *Tractat* of Raymond Llull,[44] as well as on at least half a dozen medieval Italian astrolabes (for example, #4523).

---

[38] Alcover, *Diccionari*, x, p. 180a, *sub taure*; Coromines, *Diccionari*, viii, p. 615a *sub toro*, and *ibid.*, p. 614a; Llull, *Tractat*, pp. 218-220 has Taurus.

[39] Alcover, *Diccionari*, vi, p. 252a, *sub gèminis*, who calls Romà's *gèmini* an old variant; cf. Coromines, *Diccionari*, iv, p. 456ab, *sub geminat*; *cf.* Llull, *Tractat*, p. 218, where we find Gèmini on p. 220.

[40] Càncer: Alcover, *Diccionari*, ii, p. 907b, *sub cancer*; *cf.* Coromines, *Diccionari*, ii, pp. 1029b-1030a, *sub cranc*: He calls *càncer* a Latinized doublette. Leo: Alcover, vi, p. 835b, *sub leo*. Virgo: *ibid.*, x, p. 832a, *sub virgo*, all giving more sources; cf. Llull, *Tractat*, p. 221.

[41] Libra: Alcover, *Diccionari*, vi, p. 841b, *sub libra*; *cf.* Llull, *Tractat*, p. 218, 222. Sagitarius: Alcover, *Diccionari*, ix, p. 668a, *sub sagitari*; *cf.* Alcover, Diccionari, vi, p. 841b, *sub libra*; *cf.* Coromines, *Diccionari*, vii, p. 589ab, *sub sageta*. Capricorn: Alcover, *Diccionari*, ii, p. 983b, *sub capricorn*, *capricorni*; he also mentions this form in a 14th-century translation of the history of Troy, the *Histories Troyanes*, and he adds the variant *capricornus* for Llull; *cf.* Llull, *Tractat*, pp. 218 and 222 for this form and further variants.

[42] Llull, *Tractat*, pp. 222 and 300: Aquàrius besides Aquàri (on the latter form see Alcover, *Diccionari*, i, p. 801a; Coromines, *Diccionari*, i, p. 95a, *sub aigua*.

[43] Alcover, *Diccionari*, viii, p. 607a, *sub piscis*, where *pissis* is also given for Llull. Another variant is Pisses which we find in Llull's *Tractat* (p. 223, besides Pices). Alcover mentions *piscis* for Llull, but the *Tractat* (*l.c.*) has Pisses besides Picis (p. 218). For *piscis*, Alcover mentions Andreu Febrer's Catalan translation of Dante's *Divina Commedia* (15th century).

[44] Alcover, *Diccionari*, viii, p. 607a, *sub piscis* . Llull, *Tractat*, p. 223, has Pisses; *cf.* King, "Picard Astolabe".

Month-names

The forms of the month-names as we find them on our instrument are the Classical Latin standard forms with the exception of Marcius, Madius and Octorer. The first and second forms cannot be used for localizing an instrument or an inscription, because they are Late Latin spellings not typical of a certain epoch or of a certain region of medieval Europe; they are in fact widely used on medieval astrolabes and in manuscripts.[45] As to the form Octorer, the *r* for *b* (and also *p* - see below) is not an engraver's error but rather a remarkable regional (perhaps even sub-regional?) scribal curiosity. Another attestation of the same phenomenon is the star-name *mencer* for *menceb* on the Society's astrolabe (see above). Yet another example is found on the earliest known astrolabe from Catalonia (#3042), which is probably from the 10th century: in the later markings on the rete of that instrument, dating perhaps from the 12th or 13th century, we find Rises for Pises (from Pisces).[46] This remarkable phenomenon, which defies explanation in linguistic terms, is not attested elsewhere in the sources known to us.

Astronomical commentary II (King)

On the latitudes used for the plates

The maker of this piece could have saved us a great deal of trouble had he informed us of the localities he had in mind for the plates. Read thought that the plates for 42° could serve Rome but "no considerable town in Spain", that for 41° Salamanca and Segovia, that for 38;30° "no city in Spain", that for 39;40° Valencia, and that for 32;30° Madeira "which would agree with a Spanish origin"; Gunther wrote that the three plates for 42°, 39;40° and 32;30° served Rome, Valencia and Morocco. But 32;30° is not attested as a latitude in Morocco in any

---

[45] We find the forms *marcius* and *madius* together on several astrolabes, for example, #161 and #3096 (both of uncertain provenance) and #2041 (probably French), and also #247 and #3037 (both Italian).

[46] See King, "Earliest European Astrolabe". More material on regional variants of the Latin month names on Islamic astronomical instruments can be found in Maier, "Zeugen".

medieval Islamic textual source,[47] although it is given as the latitude of Fez on an unsigned Maghribi astrolabe (#3551).[48] Valencia is fairly well-served by 39;40°, although 39;30° is not only a better value but has the added advantage that it is attested on three 11th-century Andalusian astrolabes (see below), if not in any Islamic geographical tables.[49] Rome is featured on several Andalusian astrolabes and also on the earliest European astrolabe (#3042), and it may well have been in the mind of the maker to have a plate to serve Rome, but only secondarily.

It is far more difficult to search for relevant values in the medieval European literature, and less fruitful. As we shall see, it is more rewarding to investigate latitudes and localities represented on instruments. The values found in the following sources[50] are shown in the table:

TT	The *Toledan Tables*;
V I	MS Vienna 2452 (13th-14th century);
V II	MS Vienna 2332 (after 1415);
M	MS Munich lat. 7021 (14th-15th century) (the "minutes" given here need closer investigation);
RT	The *Rudolphine Tables*.

Locality	*TT*	*V I*	*V II*	*M*	*RT*
Jerusalem	32; 0	32; 0	32; 0	32; 0/31;50	32;10
Cordova	38;30	38;30	27;30*	38;50	38;10
Menorca	-	39;10	-	-	39;24
Valencia	-	39;30	43;30**	-	39;30
Rome	41;50	41; 0	41;50	41;83 (!)/41;70 (!)	42; 2

* an obvious misreading of Arabic 37;30°!
** same as Montpellier!

47 Kennedy & Kennedy, *Islamic Geographical Coordinates*, pp. 679-680. The value is attested in the Islamic sources for Mahdiyya and Tripoli (Libya), and it should be borne in mind that some medieval European astrolabes included plates for "Africa", Sicily and Ceuta, as well as Jerusalem - see, for example, #416, discussed in note 60 below.

48 More common values for Fez were 33° or 33;40°.

49 Kennedy & Kennedy, *Islamic Geographical Coordinates*, p. 372.

50 For these five sources see Toomer, "Toledan Tables", pp. 134-139; and Sezgin *et al.*, *European Geographical Tables*, pp. 57-58, 68-69, 85-89, and 108-111.

The latitude 32;30° on the Society's astrolabe in fact serves Jerusalem,[51] even though it is not particularly accurate (31;50° would be better). It is attested for Palestine only on one rather late Maghribi astrolabe (#3643). Jerusalem was very much in the mind of the medieval European. Yet whilst no single medieval European astrolabe gives the impression that it was made specifically for taking on a crusade, this one could have been used by a crusader had he reached the ultimate goal of the Crusades. Here we should bear in mind that it was more likely that a crusader would bring *back* an Islamic astrolabe from the Crusades; one that seems to have come to Europe in this way is a Syrian astrolabe made in the 12th century (#137) and located in Nuremberg at least since the time of Regiomontanus (mid-15th century). On the other hand there is another Syrian astrolabe, made in 1222/23 (#4050) and now preserved in Istanbul, which has a distinctive feature on the rete (a short bar perpendicular to the vertical axis in the upper part of the ecliptic), unique of its kind on Islamic instruments, that seems to result from European (*i.e.* Crusader) influence. (If this is not the case, then the entire class of 13th- and 14th-century Franco-Italian astrolabes,[52] which displays this feature, is subject to Islamic influence.)

The value 38;30° is the standard medieval Islamic value for Cordova,[53] rather than for any specific locality in the Islamic East, and we can be sure that Cordova was intended for this plate. Our confidence is underlined by the fact that this popular value is in fact quite inaccurate: 38° with no minutes would have been better.

The distinctive value 39;40° is not attested on any Western Islamic astrolabe or on any other early European instrument currently known to us. It is given as the latitude of Menorca in a single late-13th-

---

[51] Two 11th-century Andalusian astrolabes (#3622 and #117) have 32° for Jerusalem, one (#1139) has 33°. Kennedy & Kennedy, *Islamic Geographical Coordinates*, pp. 679-680, shows no entry for Jerusalem under latitude 32;30°. Islamic values vary between 31° and 33°, although already Ptolemy had 31;40° (*ibid.*, pp. 160-161).

[52] See #168, one of several dozen known examples. They are still labelled "Franco-Italian" because it is not yet clear whether they are French or Italian.

[53] Attested on at least eleven 11th-century Andalusian astrolabes, including #110 and #3622, and in several sources listed in Kennedy & Kennedy, *Islamic Geographical Coordinates*, p. 95.

century Maghribi source,[54] but this may be too late to be relevant. Only rarely is Valencia specifically mentioned on two 11th-century Andalusian astrolabes: #110 has latitude 40°, #3622 has 38;30°, and #1099 and two others have 39;30°. The value 39;40° is trying to tell us something. We suggest that it has two messages for us: firstly, that it is so special that it can be persuaded to reveal the place of manufacture, and secondly, that it results from a deliberate attempt to determine the latitude of that city. Valencia is the only candidate. Now no observations for such a purpose are known from medieval Valencia, and although this does not mean that none took place, it does mean that we should perhaps investigate other possibilities. One presents itself, namely, that the latitude was *calculated* from the length of longest daylight. Elsewhere the first author has underlined the importance and influence, explicit and implicit, of the geographical climates of Antiquity in medieval instrumentation.[55] These are defined in terms of the maximum length of daylight, and, since the obliquity of the ecliptic is slowly changing, the latitudes of the climates change also with the centuries. This disturbed medieval instrument makers so much that they virtually unanimously adopted Ptolemy's values for these latitudes. (They also mainly used his value of the obliquity, 23;51°, even though more up-to-date values had been available since the 9th century.[56]) Now Valencia is close to the middle of the fifth climate, in fact it is between the boundary with the fourth and the middle of the fifth. This could be easily established by measuring the length of daylight at the summer solstice (or of night at the winter solstice) with a water-clock. In fact, if one takes the arithmetic mean of these two latitudes as given by Ptolemy, one arrives at a value of *ca.* 39;40°, and it was thus, we propose, that this value was derived.

The latitudes 41° and 42° were probably intended to serve such

---

[54] Kennedy & Kennedy, *Islamic Geographical Coordinates*, p. 229, citing the *Kitāb Basṭ al-arḍ* of Ibn Saʿīd al-Maghribī (d. 1286). The value is used for Erzerum on the gazetteer of #70, a late Eastern Islamic astrolabe, but this is not significant for our present purpose.

[55] See King, "Astronomical Instruments", pp. 152, 168-169, and *idem*, "Earliest European Astrolabe", for the first discussions of this.

[56] See the articles "Mayl" (declination) and "Minṭaḳa" (ecliptic) in *EI*².

localities as Barcelona and Gerona, respectively.[57] The former is very close to the middle of the 5th climate. Either of them would also have served Rome, the latitude of which was taken as 41;30° or 42° by medieval Andalusian astrolabists.[58]

Finally, the inclusion of curves for twilight below the horizon is unusual on an early European astrolabe and is indicative of Islamic influence, for these were standard on Western Islamic astrolabes if not on those from the Islamic East. The parameter used here is a solar depression of about 18°, which seems to be standard on the few European astrolabes on which twilight curves are featured.[59]

On the missing plates

We may presume that the two missing plates served latitudes that would have made the instrument even more "universal". Even the oldest surviving European astrolabe (#3042 - see above) has a plate serving the latitude of the seventh climate, which also happens to serve Paris. One might expect then, that the latitudes of climates VI and VII were represented (*ca.* 45° and 48°). It is not clear what else could have been represented: perhaps two of 16°, 23° and 30°, for example, for climates I, II and III. Or perhaps one of those and a universal plate of some kind? Speculative though these remarks may appear, they do result from the newly-gained insight that medieval astrolabists made deliberate attempts to produce instruments that were "universal".[60]

---

[57] No early Andalusian astrolabes mention these cities, although several have 42°, or less frequently 41;30°, for Saragossa.

[58] Two 11th-century Andalusian astrolabes have 41;30° (#121) and 42° (#2527) for Rome; two later Maghribi astrolabes have 41° (#1069) and 41;30° (#1160). The "Carolingian astrolabe" (#3042) has 41;30° for "ROMA ET FRANCIA" - see King, "Earliest European Astrolabe".

[59] On twilight in the Islamic sources see the article "Shafak" (twilight) in *EI²*.

[60] We may compare the latitudes used on a contemporary Catalan astrolabe (#416), which are as follows (dialectal forms of place-names are printed bold and letters underlined are written as a ligature):

On the horary quadrant

The horary quadrant - both the universal variety serving all latitudes (with a movable or fixed cursor for feeding in the solar longitude or date) and the specific variety for a single latitude - was invented in Baghdad in the 9th century. Both varieties were known in al-Andalus. It was there that the Europeans learned of the universal quadrant with movable cursor - the so-called *quadrans vetus* - which then, in the 13th century, inspired the so-called *quadrans novus* of Profatius of Montpellier. Also one medieval Andalusian treatise identifies several different varieties of horary quadrants for specific latitudes.[61]

Of the horary quadrant on the Society's astrolabe only the equinoctial quarter-circle appears on the lithograph published by Read and Gunther. The other markings are indeed engraved only lightly, and they appear to be original even though they are both incomplete and incompetently engraved. For example, the hour-curves are unfinished and, as noted above, there are clearly seven marks for the hours along the equinoctial quadrant (instead of six). The ensemble appears to serve a latitude of *ca.* 42°, which could have been intended for, say, Barcelona, or alternatively to serve roughly the middle of the fifth climate (in which

---

Locality	*Identification*	*Climate*	*Latitude*
DAMIATA	(= Damietta)	4	31°
GERUZALEM	(= Jerusalem)	3	32
AFRICA	(= Maghrib)	4	33
TRIPOLI	(= Tripoli, Libya)	4	34
CAPTA	(= Ceuta)	4	35
CECILIE	(= Sicily)	5	38
VALENCIA	(= Valencia)	**CLIMA 5**	39
SEGOVIA	(= Segovia)	5	40
BARCELO~A	(= Barcelona)	6	41
PANPLONA	(= Pamplona)	6	42
MACEDONIA	(= Macedonia) (!)	3 (!)	43
GENOVA	(= Genoa)	3 (!)	44
MILA	(= Milan)	2 (!)	45
unnamed	latitude actually [41°]	7 (!)	46

[61] *Cairo ENL Survey*, p. 135 (no. F6), and article "Rub^{c}" (quadrant) in *EI^2*. On the *quadrans novus* see now Dekker, "Astrolabe Quadrant".

case a latitude closer to 41° would have been more appropriate.

The date of the instrument

It is extremely difficult to date medieval European instruments, most of which (in fact all but seven from before 1450) are undated (and mostly unsigned).[62] In general it appears that estimates of such dates by colleagues in the past tended to be too late. Read and Gunther dated this piece to the 14th century (Read: "the golden age of astrolabes in the West"), D. Price *et al.* to *ca.* 1360; we tend to think, but cannot prove, that it is from the late 13th century. In other words, that it was made in Catalonia a few decades after Abū Bakr and al-Khamā'irī were active in Marrakesh and Seville, respectively. Or again, to relate it to other astrolabes by Europeans, that it was made close to 300 years after the "Carolingian astrolabe" (#3042), both pieces stemming from the same region. Or again, that it is roughly contemporary with another Catalan astrolabe (#416). Or yet again, that it precedes by about a century the astrolabe of Petrus Raimundus of Aragon, made in Barcelona in 1375 (#3053).

Conclusion

We can now identify five surviving astrolabes made or modified in Catalonia in the Middle Ages:[63]

1 The oldest European astrolabe (#3042), made in the late 10th century. Here the inscriptions are in Latin, albeit demonstrating various

---

[62] These are: an English astrolabe (#291) dated 1326; the astrolabe of Blakeney (#292), dated 1342, also English; the astrolabe of Petrus Raimundus of Aragon (#3053), made in Barcelona in 1375; and a Northern Italian astrolabe dated 1420 (#4523). To this list may be added two quadrants (#5521 and #5522) from the same English workshop dated 1398 and 1399, respectively - the authenticity of these quadrants has been incorrectly called to question (see Price, "Instruments in the BM", p. 131) but the latter has later additional markings and inscriptions, now identifiable as being due to the Elizabethan Charles Whitwell (see Turner, "Quadrant") - and a Viennese ivory horary quadrant dated 1438 (#5510).

[63] Five years ago none of these could have been listed under this rubric. The "Carolingian astrolabe" was benighted by suspicions about its authenticity, and the others had never been studied in any detail and their Catalan connection had never been noted.

unusual forms, with a numeral system derived from the Western Arabic alphanumerical system.

2 An unsigned Andalusian astrolabe made in Cordova in 446 H [= 1054/55] (#3622), with additional inscriptions, including month-names and some place-names in Catalan, dialect of Valencia.

3 An astrolabe from *ca.* 1300 with month-names in Catalan, dialect of Valencia (#416). This piece is important not least because it proves that not only the Y- and V-shaped retes of medieval England[64] were inspired from this milieu but also the flame-shaped star-pointers attested on 14th-century French astrolabes and particularly on those from the workshop of Jean Fusoris (Paris, *ca.* 1400).[65]

4 An astrolabe signed by Petrus Raimundus of Aragon, made in Barcelona in 1375 (#3053), with inscriptions in Latin. This piece is important not least because it has a frame on the rete for the celestial equator, not attested on any known earlier Islamic astrolabe or any other earlier or later medieval European astrolabe. (Only a small-subset of late Indo-Persian and Indian astrolabes show this feature.)

And fifth but not least the astrolabe of the Society of Antiquaries, now recognized for the first time for what it is. Since these five astrolabes hail from the region where Europeans first learned of the instrument they are all extremely important historical documents (and this even though they are not English). We can, on the other hand, feel ourselves fortunate in the knowledge that this precious piece is in a collection with the appropriate facilities and attitude to display it, to photograph it and to make it available for research; not many repositories of medieval instruments can claim as much.

---

[64] See Gunther, *Astrolabes of the World*, II, pp. 473-477, for some examples.

[65] Not one of these is properly published, save #192, described by Gunther, and #476, described by García Franco. See Poulle, *Fusoris*, pp. 19-25, for some brief remarks on a few of the 30 known examples of Fusoris astrolabes and related pieces.

Appendix 1

Text of Read, "Astrolabe"

[From the *Proceedings of the Society of Antiquaries*, May 18, 1893.]

C. H. READ, Esq., Secretary, also read the following notes on a planispheric astrolabe, given to the Society by the Rev. I. G. Lloyd, F.S.A.:

"The planispheric astrolabe presented to the Society by the Rev. I. G. Lloyd, F.S.A., is a good and complete example of what may be called the golden age of astrolabes in the West, that is to say, the fourteenth century. It possesses all its parts, and they seem to be all of the same date, though the rule and the "horse" (the wedge to fasten the whole together), being quite plain, may possibly be latter additions.

Professor Skeat's edition of "Chaucer's Treatise on the Astrolabe" is so accessible and so exhaustive that it is unnecessary to give any general description of the instrument. In addition to Chaucer, the learned work of Morley, which the liberality of our Director has added to our library, a charmingly written paper by Mr. Robert Taylor, and the detailed account of an interesting astrolabe of English make in *Archaeologia* , vol. lii., by our Fellow Chancellor Ferguson, may be said to cover the ground very thoroughly.

The instrument before us cannot claim to be of such interest as either of the two described by Mr. Taylor and Chancellor Ferguson. In the first place it is certainly not of English work, and in the second the calendar of festivals which figures upon both of these astrolabes is wanting here. For all this it is a well-made little instrument, and has its interesting features.

It consists of the following members, their names being those made familiar by Chaucer, viz.: the mother with its ring; within the mother two plates and the rete; and on the back the rule with its sights, all held in place by the central pin, which is itself secured by the horse. There is no indication that it ever possessed a label or *novella*. The mother is 3.9 inches in external diameter, the bordure being 0.18 of an inch wide. The ring is only 0.84 of an inch in external diameter, so that Chaucer's direction to put it on the thumb of one's right hand can scarcely be carried out. The mother itself is formed of two pieces, a thin circular plate to form the back, and the stout rim, or "brinkes," which has been made separately and rivetted to the plate. The bordure is divided into 360 degrees as in Chaucer, but numbered in tens consecutively, not in quadrants of 90 degrees, and they begin at the meridian point under the ring. The twenty-four hours of the day also are not upon the bordure, but upon the edge; from 1 to 10 are marked by Arabic numerals, 11 to 24 inclusive being left blank and indicated only by lines. The back has an outer ring of five concentric circles (*see illustration*). The outermost of these is divided into 360 degrees, in quadrants of 90 degrees; within this the circles of the zodiacal signs with their degrees. Then comes a circle of ornament, and next the months, divided into days, amounting, however, only to 365 in all, so that the instrument would not record observations with any great accuracy. Above the name of each month is a number, thus Januarius 3, Februarius 6, Marcius 5, Aprilis 1, Madius 3, Junius 6, Julius 1, Augustus 4, September 7, October 2, November 5, December 7; these numbers are the solar regulars. Within the circles is the square of the shadows, presenting no peculiar features. Innermost of all is a table, composed of three concentric circles, the outermost divided and numbered up to 28, and representing the years of the solar cycle, the next giving the concurrent for each year of the cycle and the innermost containing only the letter B seven times repeated. I take this letter B to stand for ' Bissextilis ' to indicate the leap years, which would of course occur seven times in the cycle of 28 years. It cannot I think be the Dominical letter B, as this would not recur at intervals of four years. The table would, however, serve, as our Fellow Mr. Lewis Evans has pointed out, for finding the Dominical letters, as these correspond with the concurrents.

This table is used, among other purposes, for ascertaining the day of the week for any given day of the month, in the manner explained in Sir Harris Nicolas' *Chronology of History*. There is, however, a discrepancy between the table as given on the astrolabe and that of Sir Harris Nicolas, but I am not sufficiently versed in the mystery of solar cycles to explain why it should exist. Upon the astrolabe the solar regulars for January and February are 3 and 6 respectively, while Nicolas gives 2 and 5, and the numbers for the other months correspond.

A similar but more elaborate table is given on the Blakeney astrolabe in the British Museum. This has an extra line in which the Dominical letters are given, but the concurrents are wanting.

I have said that there are two plates in the mother of the astrolabe. Each of these is engraved upon its two faces with projections of the sphere, and the base of the mother is also engraved in the same manner. All of these five projections are inscribed in Arabic numerals with the degrees of latitude for which they were constructed. When it is ascertained within what degrees of latitude an astrolabe is intended to be used, it may fairly be assumed that the place of manufacture lies somewhere between the two extremes, north and south. In the present case the most southern parallel is 32°30', and the most northern 42°0', and the three intervening lines are 38°30', 39°40', and 41°0'. It is more probable that the astrolabe was made for a latitude midway between the two extremes, so as to be of use in travelling either north or south, and we must therefore look for some city upon or near one of the intermediate parallels. The only considerable city near 39°40' is Valencia, the line passing eastward through Majorca and Sardinia; 38°30' touches no city in Spain, passes north of Sicily, through Lipari, the toe of Italy, and in Greece through Negropont; 41°0' passes near Salamanca and Segovia in Spain, somewhat north of Naples, and through Constantinople in the east. These are the intermediate parallels, and I think prove conclusively that the instrument was made for use in the Mediterranean, and I am inclined to select Valencia as the place of its manufacture, though I admit that the evidence is not of the strongest. The two extremes of latitude are of less interest, but they show the range of the instrument, and the southern one, 32°30', passing from Madeira, through the Atlas mountains, and touching or nearly touching Tripoli, would agree very well with a Spanish origin. At the same time, the most northern, that of 42°0', passes near Rome, but touches no considerable town in Spain.

There now remains only the rete, the pierced plate with pointers indicating the position of certain well-known stars. This is engraved with the names of the stars indicated, Algol, Alredan, Alfaras, etc., but I am not aware that there is any peculiarity in the stars named, and it is scarcely worth while to trouble you with the list. I should like, however, to call your attention to the design and execution of the rete. It is quite unusual to find such architectural features as are seen upon it.

It is possible, when an astrolabe is made with great accuracy, and upon a fairly large scale, to fix its date, by ascertaining the precise moment of the sun's entry into Aries. This Mr.Taylor endeavoured to do for the astrolabe that he described, working from Chaucer's account of how he performed this operation at midday on the 12th March, 1391. The instrument before us is, however, both of too small a size, and too much out of gear for so nice an adjustment."

Appendix 2

List of instruments cited

#2 Brescia, Museo dell'Età Cristiana, inv. no. 36 - Byzantine astrolabe dated 1062 - see Gunther, *Astrolabes of the World*, I, pp. 104-108 (no. 2), after a detailed study by O. Dalton; and Stautz, "Früheste Formgebung".

#5 Oxford, Museum of the History of Science, inv. no. IC 5 - Geared astrolabe by Muḥammad ibn Abī Bakr al-Iṣfahānī dated 1223/24 - see Gunther, *Astrolabes of the World*, I, pp. 118-120 (no. 5).

#70 Washington, D.C., National Museum of American History, inv. no. 316755 - an astrolabe made by ᶜAlī ibn ᶜIwaḍ in N. Iraq in 1020 H [= 1611/12] - see *Washington NMAH Catalogue*, pp. 123-124 (no. 70).

#110 London, British Museum, inv. no. OA+371 - early Andalusian astrolabe, unsigned and undated - see Gunther, *Astrolabes of the World*, II, pp. 244 (no. 110) and 280 (no. 135).

#111 Private collection - astrolabe by Ḥāmid ibn Khiḍr al-Khujandī dated 374 H [= 984/85] - see King, "Kuwait Instruments", pp. 80, 82, and 83-89.

#117 Madrid, Museo Arqueológico Nacional, inv. no. 50762 - astrolabe made by Ibrāhīm ibn Saᶜīd al-Mawāzīnī al-Sahlī in Toledo in 459 H [= 1066/67] - see Gunther, *Astrolabes of the World*, I, pp. 252-253, no. 117 (author misidentified), and pl. LX (front and back); García Franco, *Astrolabios en España*, pp. 229-235 (no. 12); *Santa Cruz 1985 Exhibition Catalogue*, pp. 80-81 (superb colour illustrations of the front and back); and *Madrid MAN 1992 Exhibition Catalogue*, pp. 57 and 224.

#118 Oxford, Museum of the History of Science, inv. no. IC118 - astrolabe made by Ibrāhīm ibn Saᶜīd al-Sahlī in Toledo in 460 H [= 1067] - see Gunther, *Astrolabes of the World*, I, pp. 253-256 (no. 18) and fig. 121 (two plates); *Santa Cruz 1985 Exhibition Catalogue*, pp. 82-83 (with colour plates of the front and back); *Madrid MAN 1992 Exhibition Catalogue*, p. 223; and on the month-names Maier, "Zeugen".

#121 Kassel, Staatliche Kunstsammlungen, inv. no. A38 - astrolabe by Ibrāhīm ibn al-Sahlī, made in Valencia in 478 H [= 1086] - see Gunther, *Astrolabes of the World*, I, p. 263 (no. 121).
#124 Strasbourg, Observatoire Ancien, inv. no. D53a - astrolabe by Abū Bakr ibn Yūsuf dated 605 H [= 1208/09] - see Gunther, *Astrolabes of the World*, pp.263-267 (no. 124), after a detailed study by F. Sarrus.
#135 see #110.
#137 Nuremberg, Germanisches Nationalmuseum, inv. no. WI 20 - undated (12th-century) astrolabe by al-Sahl al-Nīsābūrī - see *Nuremberg GNM 1992-93 Exhibition Catalogue*, II, p. 570-574 (no. 1.71); *Paris IMA 1993-94 Exhibition Catalogue*, pp. 432-434 (no. 329).
#140 Athens, Benaki Museum, inv. no. A. α. 28 - universal astrolabe of Ibn al-Sarrāj, dated 729 H [= 1328/29] - see King, *Islamic Astronomical Instruments*, IX; *Paris IMA 1993-94 Exhibition Catalogue*, pp. 434-435 (no. 330); and *Madrid MAN 1992 Exhibition Catalogue*, p. 236.
#154 Chicago, Adler Planetarium, inv. no. M-36 - an astrolabe by Muḥammad ibn Yūsuf ibn Ḥātim dated 638 H [= 1240/41] - see Gunther, *Astrolabes of the World*, I, p. 300 and pl. LXVII (no. 154), misdated to 1747, and *Chicago AP Catalogue*, s.v.
#161 London, British Museum, inv. no. 1961.12-1.1 - an early European astrolabe of uncertain provenance - see Gunther, *Astrolabes of the World*, II, p. 306 (no. 161), *London BM Catalogue*, pp. 110 and 112 (no. 323), and pl. XLIX (front and back).
#162 London, Society of Antiquaries, inv. no. Cat. 559 - unsigned, undated Catalan astrolabe - see Read, "Astrolabe"; Gunther, *Astrolabes of the World*, II, pp. 306-309 (no. 162); Price *et al.*, *Checklist*, p. 13, and the present study.
#168 Oxford, Museum of the History of Science, inv. no. IC 168 - "Franco-Italian" astrolabe - see Gunther, *Astrolabes of the World*, II, p. 317-319 (no. 168).
#191 Oxford, Museum of the History of Science, inv. no. IC 191 - medieval Spanish or French astrolabe (with plates for Saragossa and Paris) - see Gunther, *Astrolabes of the World*, II, pp. 340-341 (no. 191).
#192 Oxford, Museum of the History of Science, inv. no. IC 192 - standard astrolabe of Jean Fusoris - see Gunther, *Astrolabes of the World*, II, pp. 341-342 (no. 192).
#198 London, Science Museum, inv. no. 1880.32 - see Gunther,

Astrolabes of the World, II, p. 347 (no. 198), and Maier, "Romanische Monatsnamen", no. B.

#202 Private collection - 14th-century Picard astrolabe bearing monastic ciphers - see Gunther, *Astrolabes of the World*, II, p. 349 (no. 202), King, "Picard Astrolabe", and Maier, "Romanische Monatsnamen", no. A.

#247 London, Science Museum, inv. no. 1880-49 - late medieval Italian astrolabe - see Gunther, *Astrolabes of the World*, II, p. 423 (no. 247), and pl. CIV (front and back).

#291 London, British Museum, inv. no. 1909 6-17 1 - unsigned English astrolabe dated 1326 - see Gunther, *Astrolabes of the World*, pp. 465-467 (no. 291), and *London BM Catalogue*, pp. 112-113 (no. 325).

#292 London, British Museum, inv. no. 53 11-4 1 - English astrolabe by Blakeney, dated 1342 - see Gunther, *Astrolabes of the World*, pp. 468-469 (no. 292), and *London BM Catalogue*, p. 113 (no. 326).

#300 Oxford, Museum of the History of Science, inv. no. IC 300 - early European astrolabe, unsigned, undated - see Gunther, *Astrolabes of the World*, pp. 477-478 (no. 300) and also note 20 above.

#416 Greenwich, National Maritime Museum, inv. no. A21/NA36-21c - unsigned, undated (13th century?) Catalan astrolabe - Unpublished; illustrated in King, "Earliest European Astrolabe"; on the month-names see Maier, "Romanische Monatsnamen", no. C.

#428 Greenwich, National Maritime Museum, inv. no. A39/NA38-1661C - medieval French astrolabe (= #625) - unpublished; see *Greenwich Astrolabe Booklet*, p. 48, fig. 58 (front).

#476 Barcelona, Real Academia de Ciencias y Artes (formerly in the Observatorio Fabra) - French astrolabe related to the Fusoris tradition - see García Franco, *Astrolabios en España*, pp. 295-300 (no. 22).

#493 Florence, Museo di Storia della Scienza, inv. no. 1107 - an undated Italian astrolabe with a distinctive rectangular frame on the rete - Unpublished; the rete is illustrated in King & Turner, "Bessarion's Astrolabe", fig. 10.

#625 See #428.

#1069 Madrid, Instituto de Valencia de Don Juan - astrolabe by Abū Bakr ibn Yūsuf dated reworked by a European with maker's name and date obliterated - see García Franco, *Astrolabios en España*, pp. 277-282 (no. 19), and pls. 37 and 38 on pp. 277-278 (front and back); and *Madrid MAN 1992 Exhibition Catalogue*, p. 56 (front and back).

#1099 Nuremberg, Germanisches Nationalmuseum, inv. no. WI 353 - astrolabe by Aḥmad ibn Muḥammad al-Naqqāsh, made in Saragossa in 472 H [= 1079/80] - see *Nuremberg GNM 1992-93 Exhibition Catalogue*, II, p. 568-570 (no. 1.70).
#1139 Munich, Bayerisches Nationalmuseum, inv. no. 33/243 - astrolabe by Muḥammad ibn Saʿīd al-Ṣabbān in the year 466 H [= 1073/74] - see Mayer, *Islamic Astrolabists*, p. 77, and Stautz, *Munich Astrolabes* (forthcoming), no. 1.
#1148 Cairo, Museum of Islamic Art, inv. no. 15371 (formerly Harari 151) - astrolabe by Muḥammad ibn Fattūḥ al-Khamā'irī dated 628 H [= 1230/31], with later inscriptions in a dialect of N. Spain - unpublished; on the month-names see Maier, "Romanische Monatsnamen", no. E.
#1160 Present location unknown - "Marcel's astrolabe", often incorrectly associated with Abū Bakr ibn Yūsuf - see Mayer, *Islamic Astrolabists*, p. 32 (no. III) for details.
#2041 Oxford, Museum of the History of Science, inv. no. 57-84/173 - French astrolabe with a zoomorphic rete - see *Oxford MHS Billmeir Supplement Catalogue*, pp. 30-31 (no. 173).
#2527 Oxford, Museum of the History of Science, inv. no. Billmeir 157 - Andalusian astrolabe by Muḥammad ibn Saʿīd al-Ṣabbān, dated 474 H [= 1081/82] - see *Oxford MHS Billmeir Supplement Catalogue*, pp. 18-19 (no. 157).
#3037 Milan, collection of Dr. Tullio Tomba: 14th(?)-century Italian astrolabe - see Tomba, "Astrolabio italiano", with illustrations.
#3042 Paris, Institut du Monde Arabe, inv. no. AI86-31 - the so-called "Carolingian astrolabe", from 10th-century Catalonia - see Destombes, "Astrolabe carolingien"; *Madrid MAN 1992 Exhibition Catalogue*, pp. 192-193; King, "Earliest European Astrolabe", for an analysis of this piece in the light of several dozen other early astrolabes; and other articles in the same volume of *Physis* (notably on the epigraphy, by Anscari Mundó) and in this *Festschrift* (on the star-positions, by Paul Kunitzsch and Elly Dekker).
#3053 Boston, Museum of Fine Arts, inv. no. 88.654 - astrolabe signed by Petrus Raimundus in Barcelona in 1375 - unpublished; illustrated in King, "Earliest European Astrolabe".
#3096 Salem MA, Peabody Museum, inv. no. M10977 - an astrolabe of uncertain provenance with a remarkable rete - see Culver, "Mathematical Instruments" p. 36 (illustration of the front); and *Salem PM Catalogue*,

p. 5 (no. 5) and pls. II.5 and III.5 (front and back).
#3551 Naples, Museo di Capodimonte, inv. no. 4994 - unsigned, undated Maghribi astrolabe - see Marra, "Due astrolabi", pp. 280-295 (= pp. 10-25 of the separatum) (detailed description), and pls. I-X (complete set).
#3622 Cracow, Jagellonian Museum, inv. no. 4037-35/V - unsigned Andalusian astrolabe dated 446 H [= 1054/55], with additional inscriptions, including month-names and some place-names in Catalan - see Burczyk-Marona, "Cracow Astrolabe" (cites two earlier studies in Polish), and Maier, "Romanische Monatsnamen", no. D.
#3643 Washington, D.C., National Museum of American History, inv. no. 316753 - unsigned, undated Maghribi astrolabe with a replacement throne and Eastern Islamic rete - see *Washington NMAH Catalogue*, pp. 177-179 (no. 3643).
#3702 Baghdad, Archaeological Museum, inv. no. 9723 - astrolabe signed by Aḥmad ibn Kamāl, copied from or modified from a 9th-century Islamic astrolabe - see Fransis & Naqshbandi, "Astrolabes", and Stautz, "Früheste Formgebung".
#3915 Present location unknown - unsigned, undated astrolabe with Judaeo-Arabic inscriptions - see *Christie's (Amsterdam) 15.12.1988 ('Important Judaica') Catalogue*, lot 247.
#4050 Istanbul, Deniz Müzesi - astrolabe by ʿAbd al-Raḥmān ibn Sinān al-Baʿlabakkī dated 619 H [= 1222/23] - see *Paris IMA 1993-94 Exhibition Catalogue*, p. 480 (no. 386), and a more detailed study in King, "Syrian Astrolabe".
#4201 Present location unknown, in early 1995 with Christie's of London - unsigned, undated rete for a universal astrolabe, copied in Lahore in the mid 17th century from one by Ibn al-Sarrāj - see *Christie's (London) 4.10.1995* Catalogue, pp. 20-21 (lot 61).
#4523 Germany, private collection - Northern Italian astrolabe dated 1420 - see Stautz, "Norditalienisches Astrolab" for a detailed description and analysis.
#5510 Vienna, Kunsthistorisches Museum, inv. no. 166 - ivory horary quadrant for Vienna dated 1438 - see King, "Astronomical Instruments", pp. 161, 182, 187 and 191.
#5521 Dorchester, Dorset County Museum - English horary quadrant dated 1398 - see *London BM Catalogue*, pp. 55-56 (*ad* no. 146); Price, "Instruments in the BM", p. 131 (misguided); and Turner, "Quadrant".
#5522 London, British Museum, inv. no. 60 5-19 1 - English horary

quadrant dated 1399 - see *London BM Catalogue*, pp. 55-56 (no. 146); Price, "Instruments in the BM", p. 131 (misguided); and Turner, "Quadrant".

Bibliography and bibliographical abbreviations

Note: We have divided the bibliography in accordance with the two main aspects of this paper.

(a) Works dealing with historical instruments and history of science

Burczyk-Marona, "Cracow Astrolabe": Danuta Burczyk-Marona: "Problem datowania dawnych instrumentów naukowych", *Zeszyty Naukowe Uniwersytetu Jagiellonskiego*, 1022 (1991), pp. 95-116 (*Opuscula Musealia*, 5).

Cairo ENL Survey: David A. King, *A Survey of the Scientific Manuscripts in the Egyptian National Library*, (American Research Center in Egypt, Catalogs, vol. 5), Winona Lake IN: Eisenbrauns, 1986.

Chabàs & Bosch, *L'astrolabi pla*: Josep Chabàs and Daniel Bosch, *L'astrolabi pla - Guia per a la construcció i utilització*, Barcelona: Institut de Ciències de l'Educació de la Universitat Politècnica de Catalunya, 1987.

Chicago AP Catalogue: A catalogue of the instruments in the Adler Planetarium, Chicago IL, edited by R. and M. Webster, forthcoming.

Culver, "Mathematical Instruments": H. B. Culver, "Early Mathematical Instruments of Europe", *International Studio: Associated with the Connoisseur*, December, 1928, pp. 34-36. [Contains illustrations of some instruments in the Collection of Mr. Whitney Warren.]

Dekker, "Astrolabe Quadrant": Elly Dekker, "An Unrecorded Medieval Astrolabe Quadrant from *ca.* 1300", *Annals of Science* 52 (1995), pp. 1-47.

Destombes, "Astrolabe carolingien": Marcel Destombes, "Un astrolabe carolingien et l'origine de nos chiffres arabes", *Archives internationales d'histoire des sciences* 58-59 (1962), pp. 1-45, reprinted in G. Schilder, P. van der Krogt, S. de Clercq, eds., *M. Destombes (1905-1983): Selected Contributions to the History of*

- A Good Year for Medieval Instruments", *ibid.*, 36 (March, 1993), pp. 17-18.

King, *Islamic Astronomical Instruments*: D. A. King, *Islamic Astronomical Instruments*, London: Variorum, 1987, reprinted Aldershot: Variorum, 1995.

King, "Kuwait Instruments": D. A. King, "Early Islamic Astronomical Instruments in Kuwaiti Collections", in Arlene Fullerton and Géza Fehérvári, eds., *Kuwait: Arts and Architecture - A Collection of Essays*, Kuwait (no publisher mentioned), 1995, pp. 76-96.

King, "Making Instruments Talk": D. A. King, "Making Instruments Talk - Some Medieval Astronomical Instruments and Their Secrets", *Bulletin of the Scientific Instrument Society*, no. 44 (March, 1995), pp. 7-12.

King, "Picard Astrolabe": D. A. King, "Rewriting History through Instruments: The Secrets of a Medieval Astrolabe from Picardy", in *Turner Festschrift*, pp. 42-62.

King, "Syrian Astrolabe": D. A. King, "The Monumental Syrian Astrolabe in the Maritime Museum, Istanbul", to appear in a special memorial issue of *Erdem* (Ankara: Atatürk Cultural Centre) for Aydın Sayılı.

King & Turner, "Bessarion's Astrolabe": D. A. King and Gerard L'E. Turner, "The Astrolabe Presented by Regiomontanus to Cardinal Bessarion in 1462", *Nuncius: Annali di Storia della scienza* (Florence) 9:1 (1994), pp. 165-206.

Kunitzsch, *Arabische Sternnamen in Europa*: Paul Kunitzsch, *Arabische Sternnamen in Europa*, Wiesbaden: Otto Harrassowitz, 1959.

Kunitzsch, "John of London": P. Kunitzsch, "John of London and his Unknown Arabic Source", *Journal for the History of Astronomy* 17 (1986), pp. 51-57, reprinted in *idem*, *Studies*, XXI.

Kunitzsch, *Typen von Sternverzeichnissen*: P. Kunitzsch, *Typen von Sternverzeichnissen in astronomischen Handschriften des zehnten bis vierzehnten Jahrhunderts*, Wiesbaden: Otto Harrassowitz, 1966.

Kunitzsch, *Studies*: P. Kunitzsch, *The Arabs and the Stars*, Northampton: Variorum, 1989.

London BM Catalogue: F. A. B. Ward, *A Catalogue of European Scientific Instruments in the Department of Medieval and Later Antiquities of the British Museum*, London: British Museum

Cartography and Scientific Instruments, Utrecht: Hes & Paris: A. G. Nizet, 1987, pp. 153-200.

EI[2]: *Encyclopaedia of Islam*, new ed., 7 vols. to date, Leiden, E. J. Brill, 1960 to present.

Frankfurt IGN Festband: Anton von Gotstedter, ed., *Ad radices - Fest band zum fünfzigjährigen Bestehen des Instituts für Geschichte der Naturwissenschaften Frankfurt am Main*, Stuttgart: Franz Steiner, 1994.

Fransis & Naqshbandi, "Astrolabes": B. Fransīs and N. al-Naqshbandī, "*Al-asṭurlābāt fī Dār al-Āthār al-ᶜArabiyya fī Baghdād* (in Arabic)", *Sumer* (Baghdad) 13 (1957), pp. 9-33 and 3 pls.

García Franco, *Astrolabios en España*: S. García Franco, *Catálogo crítico de astrolabios existentes en España*, Madrid: Instituto Histórico de la Marina, 1945.

Greenwich Astrolabe Booklet: *The Planispheric Astrolabe*, Greenwich: National Maritime Museum, 1976.

Gunther, *Astrolabes of the World*: Robert T. Gunther, *The Astrolabes of the World* , 2 vols., Oxford: Oxford University Press, 1932, reprinted in 1 vol., London: The Holland Press, 1976.

Hill, *Numerals*: G. F. Hill, *The Development of Arabic Numerals in Europe Exhibited in Sixty-Four Tables*, Oxford: Clarendon Press, 1915.

Kennedy & Kennedy, *Islamic Geographical Coordinates*: Edward S. Kennedy and Mary Helen Kennedy, *Geographical Coordinates of Localities from Islamic Sources*, Frankfurt am Main: Institut für Geschichte der Arabisch-Islamischen Wissenschaften, 1987.

King, "Astronomical Instruments": David A. King, "Astronomical Instruments between East and West", in Harry Kühnel, ed., *Kommunikation zwischen Orient und Okzident*, Vienna: Österreichische Akademie der Wissenschaften, 1994, pp. 143-198.

King, "Earliest European Astrolabe": D. A. King, "The Earliest European Astrolabe in the Light of Other Early Astrolabes", in Wesley Stevens with Anthony J. Turner and Guy Beaujouan, eds., *The Oldest Latin Astrolabe*, to appear in a special issue of *Physis - Rivista di storia della scienzia* (Rome) in 1996 (dated 1995).

King, "Instrument Catalogue": D. A. King, "Medieval Astronomical Instruments: A Catalogue in Preparation", *Bulletin of the Scientific Instrument Society* 31 (Dec., 1991), pp. 3-7, and "1992

Publications, Ltd., 1981. [See also Price, "Instruments in the BM".]

Madrid MAN 1992 Exhibition Catalogue: Juan Vernet, Julio Samsó, *et al.*, *El legado científico andalusí*, Madrid: Ministerio de Cultura, 1992. [Catalogue of an exhibition held during April-June, 1992 at the Museo Arqueológico Nacional.]

Maier, "Romanische Monatsnamen": Kurt Maier, "Bemerkungen zu romanischen Monatsnamen auf mittelalterlichen Astrolabien", in *Frankfurt IGN Festband*, pp. 237-254.

Maier, "Zeugen": K. Maier, "Zeugen der Mehrsprachigkeit: mittelalterliche romanische Monatsnamen auf islamischen astronomischen Instrumenten", to appear in Jens Lüdtke, ed., *Romania Arabica, Festschrift für Reinhold Kontzi*, in press.

Marra, "Due astrolabi": Ornella Marra, "Di due astrolabi ispano-moreschi conservati nel Museo di Capodimonte", *Annali del Istituto Universitario Orientale di Napoli* 44 (1984), pp. 271-306, also published as a separatum.

Mayer, *Islamic Astrolabists*: Leo A. Mayer, *Islamic Astrolabists and Their Works*, Geneva: Albert Kundig, 1956.

Nuremberg GNM 1992-93 Exhibition Catalogue: Gerhard Bott, ed., *Focus Behaim-Globus*, 2 vols., Nuremberg: Germanisches Nationalmuseum, 1992.

Oxford MHS Billmeir Supplement Catalogue: Francis R. Maddison, *A Supplement to a Catalogue of Scientific Instruments in the Collection of J. A. Billmeir, Esq., C.B.E.*, Oxford: Museum of the History of Science & London: Frank Partridge & Sons, 1957.

Paris IMA 1993-94 Exhibition Catalogue: Sophie Cluzan, Eric Delpont and Jeanne Mouliérac, eds., *Syrie, Mémoire et Civilisation*, Paris: Flammarion (Institut du Monde Arabe), 1993.

Poulle, "Dater les astrolabes?": Emmanuel Poulle, "Peut-on dater les astrolabes médiévaux?", *Revue d'histoire des sciences* 9 (1956), pp. 301-322.

Poulle, *Fusoris*: Emmanuel Poulle, *Un constructeur d'instruments astronomiques au 15e siècle - Jean Fusoris*, Paris: Honoré Champion, 1963.

Price, "Instruments in the BM": Derek J. de Solla Price, "Instruments in the British Museum", *Journal for the History of Astronomy* 13 (1982), pp. 129-132. [A review of *London BM Catalogue*.]

Price *et al.*, *Checklist*: Sharon Gibbs, Janice A. Henderson and Derek J. de Solla Price, *A Computerized Checklist of Astrolabes*, New Haven CT: Yale University, Department of History of Science and Medicine, 1973. [See also Price, "An International Checklist of Astrolabes", *Archives internationales d'histoire des sciences* 8 (1955), pp. 243-263, and pp. 363-381.]

Read, "Astrolabe": C. H. Read, "Notes on a Planispheric Astrolabe ... ", *Proceedings of the Society of Antiquaries* 14 (1893), pp. 361-364. [See Appendix 1.]

Salem PM Catalogue: M. V. Brewington, *The Peabody Museum Collection of Navigating Instruments with Notes on Their Makers*, Salem MA: Peabody Museum, 1963.

Samsó: see *Madrid MAN 1992 Exhibition Catalogue* and *Santa Cruz 1985 Exhibition Catalogue*.

Santa Cruz 1985 Exhibition Catalogue: Juan Vernet, Julio Samsó *et al.*, *Astronomical Instruments in Medieval Spain - Their Influence in Europe*, Santa Cruz de la Palma: Ministerio de Cultura, 1985. [Catalogue of an exhibition held during June-July, 1985.]

Sezgin *et al.*, eds., *European Geographical Tables*: Fuat Sezgin *et al.*, eds., *Islamic Geography*, vol. 23: *European Geographical Tables Based on the Arabic Tradition ...* , Frankfurt am Main: Institut für Geschichte der Arabisch-Islamischen Wissenschaften, 1992.

Stafford & Ware, *Dictionary of Ornament*: Maureen Stafford and Dora Ware, *An Illustrated Dictionary of Ornament*, London: George Allen & Unwin, 1974.

Stautz, "Früheste Formgebung": Burkhard Stautz, "Die früheste bekannte Formgebung der Astrolabien", in *Frankfurt IGN Festband*, pp. 315-328.

Stautz, *Munich Astrolabes*: B. Stautz, *Die Astrolabien des Deutschen Museums und des Bayerischen Nationalmuseums in München*, forthcoming.

Stautz, "Norditalienisches Astrolab": B. Stautz, "Ein norditalienisches Astrolab aus dem Jahr 1420", to appear in *Internationale Zeitschrift für Geschichte und Ethik der Naturwissenschaften, Technik und Medizin* in 1997.

Stautz, *Untersuchungen*: B. Stautz, *Untersuchungen von mathematisch-astronomischen Darstellungen auf mittelalterlichen Astrolabien islamischer und europäischer Herkunft*, doctoral dissertation,

Frankfurt (Institut für Geschichte der Naturwissenschaften, Johann Wolfgang Goethe-Universität), 1995.

Tomba, "Astrolabio italiano": Tullio Tomba, "Un astrolabio del XIV secolo di probabile origine italiana", *Physis* 12 (1970), pp. 82-87.

Toomer, "Toledan Tables": Gerald J. Toomer, "A Survey of the Toledan Tables", *Osiris* 15 (1968), pp. 5-174.

Turner, "Quadrant": Gerard L'E. Turner, "Charles Whitwell's Addition, *c.* 1595, to a Fourteenth-Century Quadrant", *The Antiquaries Journal*, 65 (1985), pp. 454-455 and 476, and pls. XCIXb and C-CI.

Turner Festschrift: Robert G. Anderson, James A. Bennett, and Will F. Ryan, eds., *Making Instruments Count - Essays on Historical Scientific Instruments presented to Gerard L'Estrange Turner*, Aldershot (U.K.): Variorum, 1993.

Vernet: see *Madrid MAN 1992 Exhibition Catalogue* and *Santa Cruz 1985 Exhibition Catalogue*.

Washington NMAH Catalogue: Sharon Gibbs with George Saliba, *Plani spheric Astrolabes from the National Museum of American History*, Washington DC: Smithsonian Institution, 1984.

(b) Works dealing with Romance linguistics, with particular reference to medieval Catalan

Alcover, *Diccionari*: Antoni Maria Alcover (with Francesc de B. Moll), *Diccionari català-valencià-balear*, 10 vols., 2nd edn., Palma de Mallorca: Editorial Moll, 1978-79.

Coromines, *Diccionari*: Joan Coromines, *Diccionari etimològic i complementari de la llengua catalana*, 9 vols.(various edns.), Barcelona: Curial, 1980-91.

Crusca: *Vocabolario degli Academici della Crusca*, 5th edn., vols. 1 (1863) - 11,1 (1914), Florence: Tipografia Galileiana di M. Cellini e. c.

DEI: Carlo Battisti and Giovanni Alessio, *Dizionario etimologico italiano*, 5 vols., Florence: Università degli Studi, G. Barbera, 1950-57.

DLI: Niccolò Tommaseo and Bernardo Bellini, *Dizionario della lingua italiana*, 4 pts. in 7 vols., Torino: L'Unione Tipografico-Editrice, 1865-79.

Du Cange, *Glossarium*: Charles Dufresne Du Cange, *Glossarium mediae et infimae latinitatis*, 10 vols., reprint of the 1678 edn. in 5 vols., Graz: Akademische Druck- und Verlagsanstalt, 1954.

Forcellini, *Lexicon*: Egidio Forcellini, *Totius latinitatis lexicon* ..., 4th enlarged edn. (Giuseppe Furlanetto, ed.), 6 vols., Prato: Aldiniani, 1858-75.

Franciosini, *Vocabolario*:Lorenzo Franciosini, *Vocabolario español e italiano*, Rome 1620

GDLI: Carlo Battaglia, *Grande dizionario della lingua italiana*, Torino: Unione Tipografico-Editrice Torinese, 1961 - (vol. 12: 1994).

Gili Gaya, *Tesoro*: Samuel Gili Gaya, *Tesoro lexicográfico 1492-1726*, fasc. 1-4 (A-E; all published), Madrid: Consejo Superior de Investigaciones Científicas, 1947-57.

Llull, *Tractat*: Ramon Llull, *Tractat d'astronomía* (segons el ms. Add. 16.434 del British Museum), (Jordi Gayà and Lola Badia, eds.), in Vernet, *Textos*, pp. 205-323.

Maier, "Romanische Monatsnamen": Kurt Maier, "Bemerkungen zu romanischen Monatsnamen auf mittelalterlichen Astrolabien", in *Frankfurt IGN Festband*, pp. 237-254.

Maier, "Zeugen": K. Maier, "Zeugen der Mehrsprachigkeit: mittelalterliche romanische Monatsnamen auf islamischen astronomischen Instrumenten", to appear in Jens Lüdtke, ed., *Romania Arabica, Festschrift für Reinhold Kontzi*, in press.

Moreno Fernández, "Palabra": Francisco Moreno Fernández, "B y v en interior de palabra ... durante los siglos XIII, XIV y XV", *Revista de Filología Española* 67 (1987), pp. 35-48.

Simonet, *Glosario*: Francisco Javier Simonet, *Glosario de voces ibéricas y latinas usadas entre los Mozárabes* ... , Madrid, 1888.

Thesaurus: *Thesaurus linguae latinae*, vols. 1 - , Leipzig, *etc.*: Teubner *et al.*, 1900 - (vol. X, fasc. 7: praesuscipio - pragmaticus, 1993).

Vernet, *Textos*: Juan Vernet, ed., *Textos y estudios sobre astronomía española en el siglo XIII*, Barcelona: Universidad Autónoma de Barcelona, Facultad de Filosofía y Letras, 1981.

Figure 1
The elegant rete of the Catalan astrolabe. The 'Gothic' quatrefoil was copied from a quatrefoil on an Islamic astrolabe in the tradition of al-Khujandī (Iraq and Iran, late 10th century): the reaction of art historians to this phenomenon is awaited with anticipation! (Photos for Figs. 1 and 3 courtesy of the Society of Antiquaries, London.)

Figure 2a
The front of the trone. (Photos for Figs. 2a-b and 5a-d generously provided by Professor Gerard L'E. Turner, Oxford.)

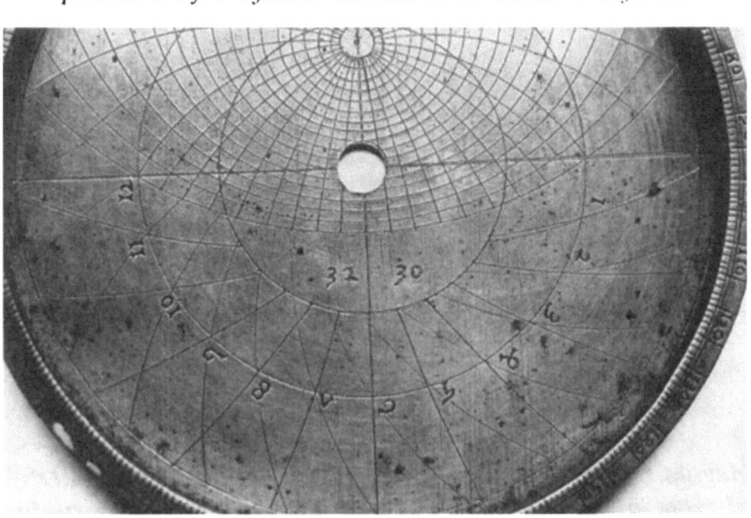

Figure 2b
A detail of the mater showing the markings for latitude 32;30° [Jerusalem].

Figure 3
The back of the Catalan astrolabe.

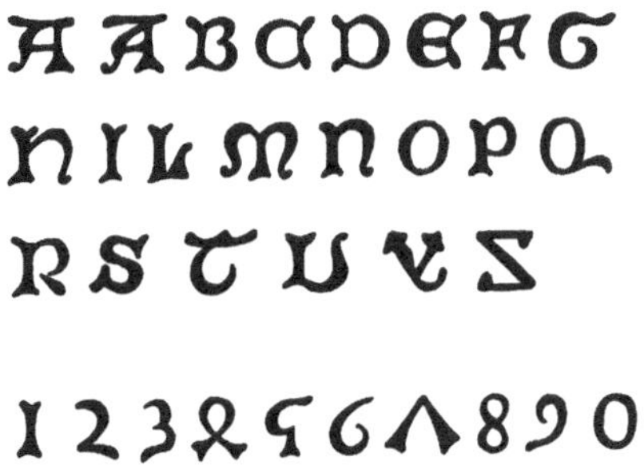

Figure 4
The inscriptions on the front and back of Catalan astrolabe. Note in particular the form of the 'Z'. (Courtesy of Reinhard Glasemann, Historisches Museum, Frankfurt, prepared from photos.) Some different forms on the rim are shown in Fig. 5a.

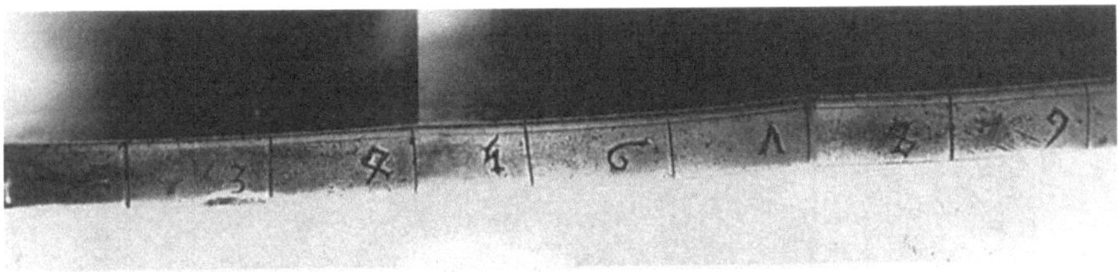

Figure 5a
Numeral forms on the rim different from those on the body of the astrolabe.

Figure 5b
The alidade and paraphernalia.

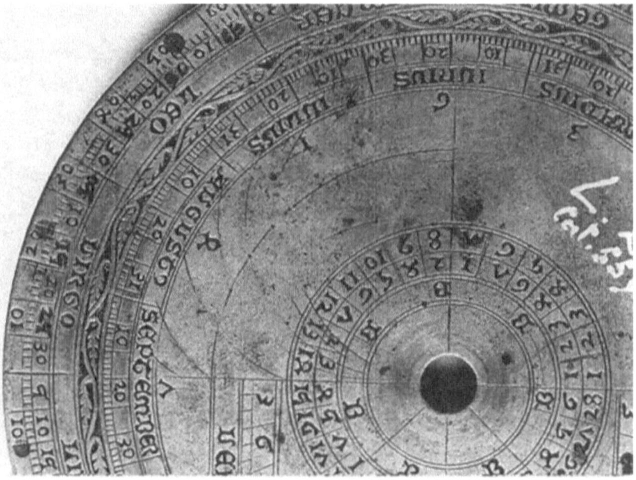

Figure 5c
The horary quadrant on the back of the Catalan astrolabe.

Figure 5d
Detail of the back of the rete. Technological aspects of medieval astrolabes, a subject beyond the competence of either author of this study, have yet to be addressed by any author.

Figure 6

An astrolabe with inscriptions in Judaeo-Arabic (#3915). Note the very similar rete design to the Catalan astrolabe, and the degenerate quatrefoil. It is extemely difficult to date this instrument since it is unique of its genre and its provenance is unknown. But the maker was not completely in control of his subject - the latitudes mentioned in the inscriptions mentioning the latitudes on the plates do not correspond to the latitudes (plates and inscriptions have been shuffled). In any case, it is clear that this instrument, like the Society of Antiquaries astrolabe, was copied - not necessarily directly - from an Islamic astrolabe, of a type of which no Islamic examples survive. (Photo from the archives of the Institut für Geschichte der Naturwissenschaften, Frankfurt.)

V

A REMARKABLE ITALIAN ASTROLABE FROM *CA.* 1300 – WITNESS TO AN INGENIOUS TRADITION OF NON-STANDARD ASTROLABES

A. Introductory remarks

1. *The problem*

A charming tiny medieval astrolabe with highly unusual features, preserved for the last century in the Museum of the History of Science at Oxford (Figs. 1-2), presents us with a series of challenges with regard to its geographical provenance, historical context, technical function, and authenticity.[1] The instrument has some standard features,[2] but it is those which are non-standard that are of particular historical interest. Lewis Evans

[1] This paper came to being during the course of the Frankfurt-based project to catalogue all medieval Islamic and European instruments to *ca.* 1500. On some of the results of this see David A. King, *Astronomical Instruments between East and West*, in Harry Kühnel (ed.), *Kommunikation zwischen Orient und Okziden. Alltag und Sachkultur*, Vienna, Österreichische Akademie der Wissenschaften, 1994, pp. 143-198, and also Id., *The Ciphers of the Monks. A Forgotten Number Notation of the Middle Ages*, (*Boethius: Texte und Abhandlungen zur Geschichte der Mathematik und der Naturwissenschaften*, Band 44), Stuttgart, Franz Steiner, 2001, pp. 364-405. A list of such instruments ordered by provenance and date is available on the website *Catalogue of Medieval Instruments* (for the table of contents see www.uni-frankfurt.de/fb13/ign/instrument-catalogue.html). Funding over several years has been generously provided by the German Research Foundation (Deutsche Forschungsgemeinschaft), Bonn. Each instrument has been assigned an International Instrument Checklist (IIC) number, preceded by the symbol #, following the tradition established in Sharon Gibbs, Janice A. Henderson and Derek de Solla Price, *A Computerized Checklist of Astrolabes*, New Haven, Conn., Yale University, Department of History of Science and Medicine, 1973.

[2] For general introductions to the astrolabe see *The Planispheric Astrolabe*, Greenwich, National Maritime Museum, 1976; John D. North, *The Astrolabe*, «Scientific American», 230 (1974), pp. 96-106, repr. in Id., John D. North, *Stars, Minds and Fate: Essays in Ancient and Medieval Cosmology*, London & Ronceverte, W.V., Hambledon, 1989, no. 14; and Paolo Trento, *L'astrolabio: storia, funzioni, costruzione*, Roma, Biblioteca del Vascello, 1989.

acquired the piece for Oxford around 1900 from the bookseller Olschki in Florence. It has not been seriously studied since the early 1930s, when Robert T. Gunther included a description in his *Astrolabes of the World*.[3] Gunther somehow derived 38° for the terrestrial latitude underlying the markings on the mater and hence designated it «Sicilian». The latitude is actually 24°, serving Aswan, *etc.*, but the instrument is not to be classified as «Syenian». Nevertheless, there may well be a *Sicilian connection.*

The preparation of regional surveys of instrumentation in medieval Europe is a worthwhile and feasible task for the future, but for Italy the task is somewhat more daunting than for other regions.[4] This is not because of any lack of sources. For over two dozen Italian astrolabes pre-dating 1500 are known but not all published, and also various medieval treatises of Italian provenance on the construction and use of the astrolabe await study. The earliest dated Italian astrolabe (#4523) is signed by Antonius de Pacent in Lanzano in the year 1420.[5] Some undated pieces, alas

---

[3] ROBERT T. GUNTHER, *The Astrolabes of the World*, Oxford, Oxford University Press, 1932, repr., London, The Holland Press, 1976, II, pp. 319-320 (no. 169); SHARON GIBBS, JANICE A. HENDERSON and DEREK DE SOLLA PRICE, *A Computerized Checklist...*, cit., #169. My first reflections on this instrument are recorded in DAVID A. KING, *Some Medieval Astronomical Instruments and Their Secrets*, in RENATO MAZZOLINI (ed.), *Non-Verbal Sources in Science before 1900*, Florence, Leo S. Olschki, 1993, pp. 29-52, esp. pp. 50-52, and ID., *Astronomical Instruments...*, cit., pp. 168-169. An anonymous description in the website *Oxford MHS Database* (www.mhs.ox.ac.uk/database/index.htm), accessible under *Sicily*, gives the date as «*ca.* 1460?» and leads also to another instrument which is not Sicilian at all. The more recent website *EPACT - Scientific Instruments of Medieval and Renaissance Europe* (featuring instruments preserved in the Museo di Storia della Scienza, Florence; the British Museum, London; and the Museum of the History of Science, Oxford, located at www.mhs.ox.ac.uk/epact/catalogue) has a fine colour illustration of the front, as well as a description by Jim Bennett partly based on Gunther. The instrument, which now bears the inventory number 40829 in the Oxford collection, here becomes «later 15th century» and «Italian?», but with the latitude still 38°.

[4] The problems of instrumentation in medieval Italy are well reflected by the fact that Gunther in 1932 could provide no introduction to his section on Italian astrolabes, and included in that section some astrolabes that are actually German and French. Major steps to improve the documentation of Italian astrolabes were taken by Tullio Tomba in a series of articles on individual instruments (TULLIO TOMBA, *Due astrolabi latini del XIV secolo conservati a Milano*, «Physis», 8 (1966), pp. 295-306 (A), *Nuove osservazioni sui due astrolabi latini del XIV secolo conservati a Milano*, «Physis», 10 (1968), pp. 119-122 (B); *Un astrolabio del XIV secolo di probabile origine italiana*, «Physis», 12 (1970), pp. 82-87 (C); and Tre astrolabi latini, del XIV secolo, conservati in Italia, «Rassegna di Studi e di Notizie», 20 (Anno XXII [= 1996]) (the offprint is unpaginated) (D)), and catalogues of collections in Milan. No list of medieval Italian astrolabes is available, and for the time being the reader may consult the table of contents of *Catalogue of Medieval Instruments*, cit., §§6.1-5 and 6.7.

[5] This piece, preserved in a private collection in Germany, is analysed in detail in BURKHARD STAUTZ, *Ein Astrolab aus dem Jahr 1420*, «NTM - Internationale Zeitschrift für Geschichte und Ethik der Naturwissenschaften, Technik und Medizin», N.S., 5 (1997), 3, pp. 142-159; see also

also unsigned so that none of them can be conclusively associated with Italy, may well be earlier than this. I mention only one in Oxford (#166),[6] three in Nuremberg (#558, #547 and #548),[7] one in Munich (#621),[8] and two others in private collections (#4509 and #4556).[9]

Clearly we are dealing with one of the earliest surviving Italian astrolabes, and also with the smallest surviving European astrolabe. However, as we shall see, this instrument is important for other reasons, and is surely one of the most interesting astrolabes from the European Middle Ages. Before we can even begin to understand it and to consider its authenticity,[10]

ID., *Untersuchungen von mathematisch-astronomischen Darstellungen auf mittelalterlichen Instrumenten islamischer und europäischer Herkunft*, Bassum (D), Verlag für Geschichte der Naturwissenschaften und der Technik, 1997, pp. 88-92 and 249-252. It is not without its problems – the star-positions on the rete are a disaster because the ecliptic and equatorial coordinate systems have been confused, and plates are labelled for different latitudes but all are in fact engraved for latitude 45° – but these do not detract from its historical interest.

[6] On #166 (Museum of the History of Science, inv. no. IC 166) see GUNTHER, *The Astrolabes of the World*, cit., II, pp. 316-317 (no. 166), and STAUTZ, *Untersuchungen*, cit., pp. 85 and 244. On this piece, purchased in 1899 in Venice, the latitudes serve the climates and there are no clearly Italian features, but I am inclined to think that it is Italian not least because the rete-design bears no relation to the Western Islamic / Iberian tradition.

[7] On #558 (Germanisches Nationalmuseum, inv. no. WI 282), an early medieval European astrolabe of uncertain provenance but possibly Italian, with plates for latitudes between 16° and 52°, see DAVID A. KING, *Die Astrolabiensammlung des Germanischen Nationalmuseums*, in GERHARD BOTT (ed.), *Focus Behaim-Globus*, Nuremberg, Germanisches Nationalmuseum, 1992, I, pp. 101-114, and II, pp. 568-602, 640-643, esp. II, pp. 574-576 (no. 1.72).

On #547 (inv. no. WI 21), a 14th-century Italian quatrefoil astrolabe, with a single plate for 42° and 45°, see *ibid.*, II, pp. 576-578 (no. 1.73).

On #548 (inv. no. WI 6), a 14th-century Italian (?) astrolabe with plates for latitudes between 30° and 55° and additional 15th-century markings, probably added in Paris, see *ibid.*, II, pp. 578-581 (no. 1.74).

[8] On #621 (Deutsches Museum, inv. no. 5178) see BURKHARD STAUTZ, *Die Astrolabiensammlungen des Deutschen Museums und des Bayerischen Nationalmuseums*, Munich, Oldenbourg, 1999, pp. 161-176 (no. 2), also illustrated in DAVID A. KING, *The Earliest European Astrolabe in the Light of Other Early Astrolabes*, in STEVENS et al., *The Oldest Latin Astrolabe*, pp. 359-404, fig. 13. This is a composite piece with a very early limbus, rete and plates, and a back from another astrolabe with inscriptions in Hebrew.

[9] On #4509, which has a replacement plate from a Byzantine astrolabe and additional markings in Hebrew, see *Sotheby's 18.6.1986 Catalogue*, pp. 24-25 (lot no. 125); and ANTHONY J. TURNER (ed.), *Time*, [Catalogue of an exhibition at the Nieuwe Kirk, Amsterdam, in 1990], Amsterdam, Tijd voor Tijd Foundation, 1990, p. 101 (no. 186, dated *ca.* 1300, described as French) with a colour plate of the front on p. 106; also KING, *The Ciphers of the Monks*, cit., pp. 290-291.

On #4556 see the detailed description in *Christie's South Kensington, Fine Scientific Instruments, ... 29 September, 1994 ...*, London, 1994, pp. 34-39 (lot 136, for the provenance «both Italy and England are contenders» (!)); the front is illustrated in KING, *The Earliest European Astrolabe*, cit., fig. 21.

[10] There is an unfortunate trend in our field that is motivated by the belief that if a histor-

we need to survey the possibilities relating to its historical context, first introducing some general concepts. We also need to be aware that astronomical instrumentation in medieval Europe was largely indebted to Islamic instrumentation.[11] Indeed, for most technical innovations in European instrumentation up to *ca.* 1600 we can identify Islamic precedents, although this does not, of course, always imply direct transmission or exclude independent initiative.[12]

2. *Universal and latitude-dependent instruments in medieval astronomy*

In the history of spherical astronomy we can distinguish between solutions that are universal and those that are restricted to a single latitude. The underlying formulae are of course by their nature universal, and the universal solutions to which I refer can be in the form of astronomical tables or astronomical instruments. Already in Greek astronomy, certain solutions were proposed for each of the seven geographical *climates* of Antiquity (see Fig. 3).[13] These included, for example, tables of oblique ascensions for each of the climates, and astrolabes fitted with plates for each climate. This explains why medieval European astrolabes were sometimes fitted with a set of plates for an extensive range of latitudes, say, 16°-52°. Likewise, an armillary sphere that can be adjusted for any latitude is universal. On the other hand, an astrolabic quadrant derived from a single astrolabic

---

ical instrument cannot be understood in the light of our present knowledge, it *must* be a fake. See n. 27 and §B7 below.

[11] The best overview of medieval European instruments remains EMMANUEL POULLE, *Les instruments astronomiques du Moyen Âge*, Oxford, Museum of the History of Science, 1969, [reprinted from «Le ruban rouge», 32 (1967)], published over 30 years ago. For an overview of Islamic instrumentation see DAVID A. KING, *Strumentazione astronomica nel mondo medievale islamico*, in GERARD L'E. TURNER (ed.), *Gli strumenti*, Torino, Einaudi, 1991, pp. 154-189 and 581-585. An updated version of the original English text is to appear as DAVID A. KING, *Studies in Astronomical Timekeeping in Medieval Islam*, Leiden, E.J. Brill, in press, VIII.

[12] The same holds for other scientific disciplines. To the list in F. JAMIL RAGEP, *Freeing Astronomy from Philosophy: An Aspect of Islamic Influence on Science*, «Osiris», 16 (2001), pp. 49-71, esp. p. 51, add astronomical timekeeping and astronomical instrumentation.

[13] On the climates in instrumentation see DAVID A. KING, *Bringing Astronomical Instruments Back to Earth: The Geographical Data on Medieval Astrolabes (to ca. 1100)*, in ARJO VANDERJAGT and LODI NAUTA (eds.), *Between Demonstration and Imagination: Essays in the History of Science and Philosophy Presented to John D. North*, Leiden, E.J. Brill, 1999, pp. 3-53, esp. pp. 6-9; ID., *World-Maps for Finding the Direction and Distance to Mecca: Innovation and Tradition in Islamic Science*, Leiden, E.J. Brill, and London, Al-Furqan Islamic Heritage Foundation, 1999, pp. 24, 27-28 and 230-234; and ID., *The Ciphers of the Monks...*, cit., pp. 356-357, 360-361 and 411-415.

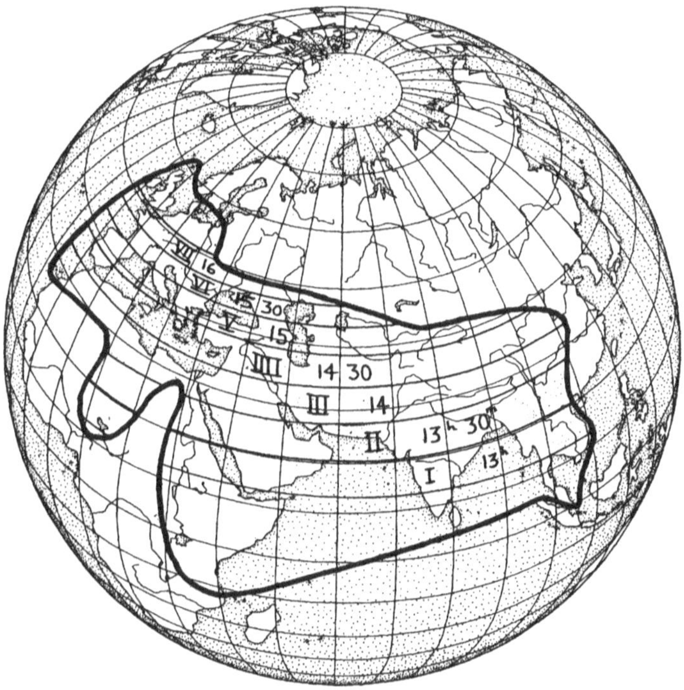

Fig. 3. The seven climates of Antiquity, defined in terms of the length of longest daylight. By a quirk of history, the latitudes of the central bands of the climates – rounded to 16°, 24°, 30°, 36°, 41°, 45° and 48° – correspond roughly to several localities of significance in the history of astronomy. [From KING, *Mecca-Centred World-Maps*, p. 24].

plate serves only a specific latitude. Universal solutions were very popular in Islamic astronomy.[14]

3. *Mixed north-south astrolabe projections*

The standard astrolabe is based on what is traditionally called a northern stereographic projection, that is, a projection of the *northern* sky in the plane of the celestial equator with the *south* celestial pole as the pole of the projection. The resulting projection of the ecliptic is unhappy, because the

---

[14] See DAVID A. KING, *Astronomy in the Service of Islam*, Aldershot, Variorum, 1993, VI and VII, repr. as KING, *Studies in Astronomical Timekeeping*..., cit., VIa-b.

northern signs of the ecliptic end up on a small arc and the southern signs on an arc about twice the size (see Fig. 4a). Muslim astronomers in the 9th and 10th centuries addressed this problem and devised a series of solutions involving combinations of northern and southern projections. With a mixed rete, one must have appropriate combinations of north-south markings on the plate. It helps to keep in mind that in the case of problems involving solar longitude and solar altitude, the same problems can be solved using the opposite solar longitude on the ecliptic and the same solar altitude

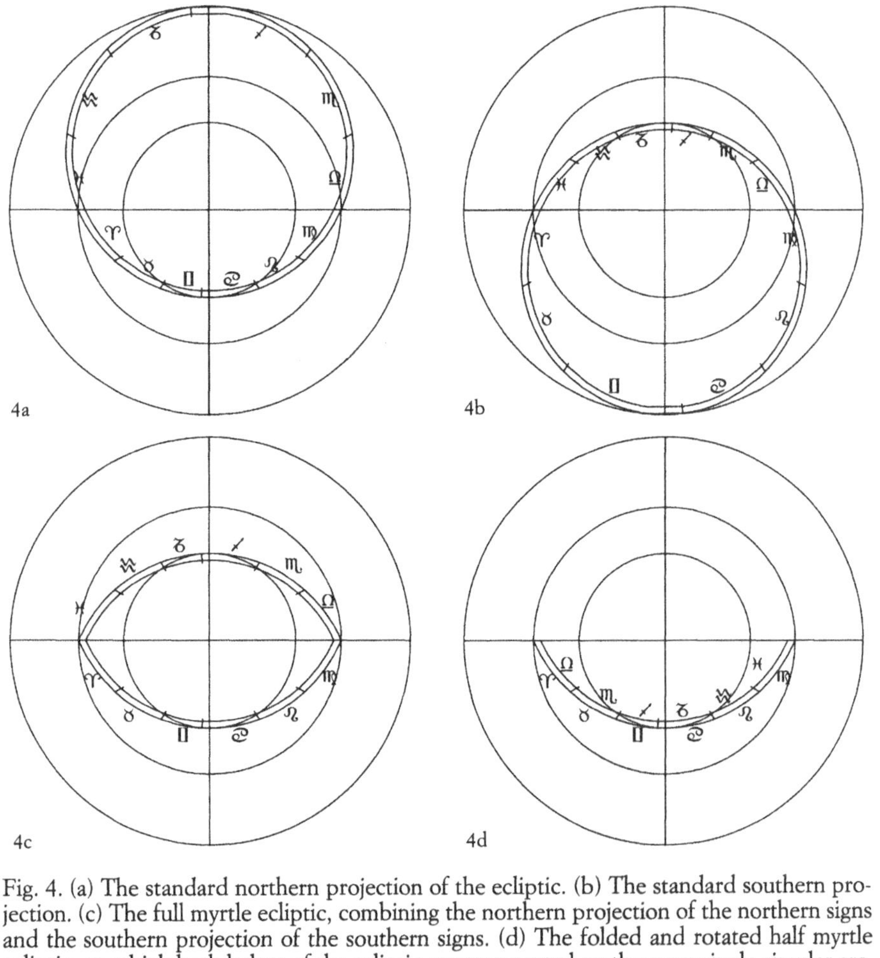

Fig. 4. (a) The standard northern projection of the ecliptic. (b) The standard southern projection. (c) The full myrtle ecliptic, combining the northern projection of the northern signs and the southern projection of the southern signs. (d) The folded and rotated half myrtle ecliptic, on which both halves of the ecliptic are represented on the same single circular arc. [Graphics by François Charette.]

now *below* the horizon. To put it another way, if one has a mixed ecliptic, one needs a mixed plate with altitude markings below the horizon equivalent to markings above the horizon in the opposite projection (see Fig. 5a-c).

The most useful, and at the same time the most aesthetically pleasing of these solutions was the so-called *myrtle-leaf* ecliptic, comprising a northern projection of the northern signs and a southern projection of the southern

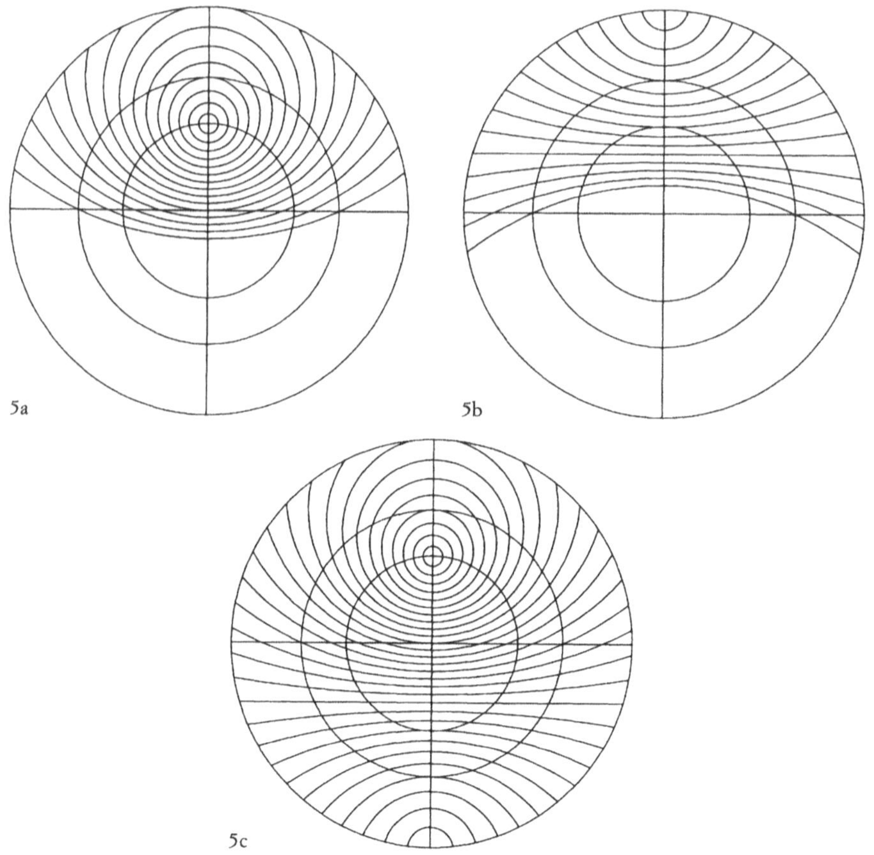

Fig. 5. (a) A set of standard astrolabic markings, that is, based on a northern projection, for latitude 24°. (b) A set of markings for the same latitude based on a southern projection. (c) The combination of such markings, which could be equally be considered as representing altitude circles above and below the horizon for either projection. On the myrtle astrolabe, the markings bounded by the equatorial circle would suffice. See further Fig. 9. [Graphics by François Charette.]

signs and bounded by the equinoctial circle (see Fig. 4c).[15] With such an ecliptic, one needs only a complete set of altitude circles above and below the horizon to solve all of the standard problems of spherical astronomy. (There is no longer empty space for the markings for the seasonal hours below the horizon, but these can be included nevertheless). The *drum-shaped* ecliptic was the made up of the two longer arcs of the ecliptic.

With such a combination, one has, however, lost the function of the astrolabe as an 'analogue computer' in the sense of a model of the universe, and produced more of an abstract instrument, or less of a representational one. The fact that the myrtle ecliptic is symmetrical about the horizontal (equinoctial) axis was not lost on medieval astronomers: one-half would actually suffice to represent the entire ecliptic – see Fig. 4d – with a consequent further loss of *direct* representation of the heavens.

The astrolabe with a myrtle rete was the invention of some unidentified 9th- or early-10th-century Muslim astronomer in Baghdad, for both it and the astrolabe with the drum-shaped ecliptic are mentioned by the mathematician Ibrâhîm ibn Sinân (907-946):[16]

> Some people make the astrolabe with respect to the northern celestial pole, and this is the most common case. But a minority make it with respect to the southern pole ... Other people are less conventional: they make the astrolabe in two halves. One of them is (projected) from the northern pole, and the other from the southern pole [...]. [The author then presents illustrations of the myrtle and drum ecliptics].

The myrtle rete was known to the major Muslim authors of instrument texts thereafter: al-Sijzî in the late 10th century, al-Bîrûnî in the early 11th (see Fig. 6), al-Marrâkushî in the late 13th, and Najm al-Dîn al-Misrî in the early 14th.[17]

---

[15] I maintain the English equivalent of the Arabic *âs*, «myrtle», as in *al-asturlâb al-âsî*, literally «the myrtle astrolabe», meaning «the astrolabe with an ecliptic shaped like a myrtle-leaf on its rete». The reader should consult the works cited in the next note to learn about «buffalo», «crab», «tortoise» and «fish» astrolabes, to name just a few.

[16] FRANÇOIS CHARETTE, *Mathematical Instrumentation in Fourteenth-Century Egypt and Syria – The Illustrated Treatise of Najm al-Dîn al-Misrî*, Leiden, E.J. Brill, 2003, §2.3.3.

[17] See JEAN-JACQUES SÉDILLOT (père), *Traité des instruments astronomiques des Arabes composé au treizième siècle par Aboul Hhassan Ali de Maroc ...*, Paris, Imprimerie Royale, 1834-1835, repr. Frankfurt am Main, Institut für Geschichte der Arabisch-Islamischen Wissenschaften, 1985; and LOUIS-AMÉLIE SÉDILLOT (fils), *Mémoire sur les instruments astronomiques des Arabes*, «Mémoires de l'Académie Royale des Inscriptions et Belles-Lettres de l'Institut de France», 1 (1844), pp. 1-229, repr. Frankfurt am Main, Institut für Geschichte der Arabisch-Islamischen

Alas, very few Islamic instruments featuring these non-standard features survive. One, preserved in Oxford (#140), was made in the year 1327/28 by 'Alî ibn Ibrâhîm al-Jazzâr, a muezzin in Taza (now in Morocco): it has a full myrtle ecliptic and a universal plate of the type similar to, but not identical with, the one designed by the Andalusî Ibn Bâso in the early 13th century (see Fig. 7).[18] Evidence of other instruments in this tradition is not lacking, but they were clearly not widespread. On the other hand, they were well enough known to have been part of the corpus of Islamic astronomical notions transmitted to medieval Europe, if only to languish there, first being truly appreciated in the Renaissance.[19]

In this connection we should mention a 14th-century Northern French astrolabe, preserved in the Science Museum in London, which is unique amongst European astrolabes not only because it has a gear mechanism for representing the relative motions of the sun and moon, but also because it is based on a southern stereographic projection.[20] This piece is clearly of

---

Wissenschaften, 1989 (on al-Marrâkushî, late-13th-century Cairo), pp. 181-182; and CHARETTE, *Mathematical Instrumentation*..., cit., §2.3.3 (general survey and details of the instruments by Najm al-Dîn al-Misrî, late-14th-century Cairo); also JOSEPH FRANK, *Zur Geschichte des Astrolabs*, Erlangen, Universitäts-Buchdruckerei, 1920, repr. in FUAT SEZGIN et al. (eds.), *Arabische Instrumente in orientalistischen Studien*, Frankfurt am Main, Institut für Geschichte der Arabisch-Islamischen Wissenschaften, 1990-91, repr. as «Islamic Astronomy and Mathematics», 85-90 (1998), with 6 further volumes *ibid.*, 91-96 (1998) [reprints of studies from the 19th and early 20th century: a monumental total of over 5,000 pages], IV, pp. 31-62, esp. pp. 9-18; RICHARD P. LORCH, *Mischastrolabien im arabisch-islamischen Kulturbereich*, in ANTON VON GOTSTEDTER (ed.), *Ad radices – Festband zum fünfzigjährigen Bestehen des Instituts für Geschichte der Naturwissenschaften Frankfurt am Main*, Stuttgart, Franz Steiner, 1994, pp. 231-236; and DAVID A. KING, *World-Maps*, cit., pp. 345-349. Richard Lorch has prepared a study of the writings of al-Sijzî, which is as yet unpublished. For an overview see also HENRI MICHEL, *Traité de l'astrolabe*, Paris, Gauthier-Villars, 1947, repr. Paris, Librarie Alain Brieux, 1989, pp. 69-71, although Michel erred when he dismissed these mixed retes as «une coquetterie géométrique».

[18] On the maker see LEO A. MAYER, *Islamic Astrolabists and Their Works*, Geneva, Albert Kundig, 1956, with a supplement in Richard Ettinghausen (ed.), *Aus der Welt der islamischen Kunst*, Berlin, Gebrüder Mann, 1959, pp. 293-296, supp., p. 294. On this instrument see already KING, *Strumentazione*..., cit., p. 167. On the universal plate of Ibn Bâso and its afterlife in Europe see EMILIA CALVO LABARTA, *Abû 'Alî al-Husayn ibn Bâso (m. 716/1316): Risâlat al-safîha al-ŷâmi'a li-ŷamî' al-'urûd (Tratado sobre la lámina general para todas las latitudas)*, Madrid, Consejo Superior de Investigaciones Científicas and Instituto de Cooperación con el Mundo Árabe, 1993; CALVO, *Ibn Bâso's Universal Plate and its Influence on European Astronomy*, «Historia scientiarum», (Brussels) 18.1 (1992), pp. 61-69 (B); and *A Study of the Use of Ibn Bâso's Universal Astrolabe Plate*, «Archives Internationales d'Histoire des Sciences», 50 (2000), pp. 264-295 (C), and also KING, *Astronomical Instruments*..., cit., p. 156.

[19] See n. 12.

[20] On #198 (Science Museum, inv. no. 1880.32) see GUNTHER, *The Astrolabes*..., cit., II, p. 347 (no. 198) and pl. LXXXI; JOHN D. NORTH, '*Opus quarundam rotarum mirabilium*', «Physis», 8 (1966), pp. 337-372, repr. in ID., *Stars, Minds*..., cit., A, no. 11, pp. 369-371 (pp. 167-169 of the

great historical interest not least because the first monumental astrolabic clocks in Europe featured the same.[21] But we perhaps need to be reminded that some astrolabists in 9th- or 10th-century Iraq also made astrolabes based on a southern projection.[22]

4. *Mixed north-south ecliptic scales on universal astrolabes*

The 11th-century Andalusî astronomer Ibn al-Zarqâlluh devised a single plate – based on a stereographic projection from the equinox onto the solstitial colure – which could be used universally since it was independent of latitude. From the same milieu came the universal astrolabe, associated with 'Alî ibn Khalaf, in which at least a semicircle of such universal markings can rotate over another complete set: this serves the transformation of coordinates from one system – equatorial, ecliptic or horizon-based – to another.[23] The ecliptic would be represented on a diameter and the stars based on a universal projection. However, in order to add to the universal astrolabe some of the functionality of the standard astrolabe, later astronomers devised other representations of the stars and ecliptic on the rete of the universal astrolabe that would fit on the other one-half of the rete. The *folded half myrtle ecliptic*, that is, a half myrtle ecliptic with a double scale, so the northern and southern signs are represented on the same circular arc, was the optimal solution, and now the stars could be in either a northern or southern projection. Their positions in some cases will no longer correspond to the appropriate signs of the ecliptic scale.

Only two Islamic universal astrolabes have survived. One, already

---

reprint); and KING, *The Ciphers...*, cit., pp. 398-399, 402-403, 406, 410, 416-417, also p. 138, n. 6. This piece is not without its problems: for example, the ecliptic on the rete is improperly constructed (see STAUTZ, *Untersuchungen...*, cit., pp. 103-104 and 262), as are the scales of the shadow square on the back (see KING, *The Ciphers...*, cit., pp. 416-417).

21 JOHN D. NORTH, *Monasticism and the First Mechanical Clocks*, in J.T. FRASER and N. LAWRENCE (eds.), *The Study of Time*, New York, Springer, 1975, pp. 381-393, repr. in ID., *Stars, Minds...*, cit., no. 12, pp. 391-392 (pp. 181-182 of the reprint).

22 See the text to n. 16 above.

23 CHARETTE, *Mathematical Instrumentation...*, cit., §2.6.4, provides new sources and new insights to both supplement and correct the discussions in KING, *On the Early History of the Universal Astrolabe ...* (1979), repr. in ID., *Islamic Astronomical Instruments*, London, Variorum, 1987, repr. Aldershot, Variorum, 1995, VII; and the article *Shakkâziyya* in *The Encyclopaedia of Islam*, new edn., 11 vols. to date, Leiden, E.J. Brill, 1960 to present; which cites the voluminous publications of the Barcelona school.

known to Gunther, was made in the year 1328/29 by the astronomer Ibn al-Sarrâj of Aleppo (#140): it is universal in five different ways and easily qualifies, from a mathematical point of view, as the most sophisticated astrolabe ever made.[24] The other (#4201) was discovered only in the late 1990s: it is from 17th-century Lahore, but is of the kind known from the treatise by 'Alî ibn Khalaf in its Castilian manifestation, though fitted with a folded half myrtle ecliptic.[25] This is not surprising, since it is a copy of a simpler instrument also made by Ibn al-Sarrâj.

Suffice it to say at this stage that not a single *medieval* European universal astrolabe has survived. Folded half-myrtle representations of the ecliptic were known in Europe, namely, on universal plates described by Jean de Linières (*ca.* 1325) and Regiomontanus (*ca.* 1535).[26] We cannot doubt that such instruments, even universal astrolabes, were made in medieval times, but the earliest examples are from 16th-century England and Flanders.[27]

---

[24] On #140 see GUNTHER, *The Astrolabes...*, cit., I, pp. 284-285 (no. 140), and KING, *Islamic Astronomical Instruments*, cit., IX. A detailed study has been prepared: see FRANÇOIS CHARETTE and DAVID A. KING, *The Universal Astrolabe of Ibn al-Sarrâj* ... (forthcoming).

[25] On #4201 see *Christie's South Kensington, Fine Scientific Instruments, ... 4 October 1995...*, London, 1995, pp. 20-21 (lot 61), for the rete, and *Christie's Catalogue. Exceptional Scientific and Engineering Works of Art, Instruments and Models, ... 5 April 2001...*, London, 2001, pp. 43-45 (lot 32), for the mater. The two parts are now in the hands of the same dealer in London.

[26] See EMMANUEL POULLE, *Un instrument astronomique dans l'occident latin – la 'saphea'*, «Studi Medievali», 10 (1969), pp. 491-510, esp. pp. 508-509. On the *saphea* see also ERNST ZINNER, *Deutsche und niederländische astronomische Instrumente des 11.-18. Jahrhunderts*, 2nd edn., Munich, C.H. Beck, 1967, repr. 1972, pp. 145-149.

[27] Only around 1450 do instruments appear in Italy and Germany (Vienna) that could be labelled 'Renaissance' as opposed to 'Medieval'.

Also the earliest Renaissance Italian astrolabes present some problems of their own. The earliest one, dated Urbino, 1462, has been stolen: see n. 38. The authenticity of the most spectacular Renaissance Italian astrolabe (#171), associated with Piervincenzo Danti de' Rinaldi *ca.* 1490 (GUNTHER, *The Astrolabes...*, cit., no. 171), which is preserved in Hamburg, has been questioned. On this piece, which has the double burden of being partially gilt and having passed through the highly-dubious Frédéric Spitzer collection in the late 19th century, see the diplomatic description by Francis Maddison in JAY A. LEVENSON (ed.), *Circa 1492, Art in the Age of Exploration*, Washington, D.C., National Gallery of Art, and New Haven & London, Yale University Press, 1991, pp. 224-225 (no. 123), and also G.R. LEVI-DONATI, *Uno strumento ritrovato: l'Astrolabio perugino dell'anno 1498*, «Bollettino della Deputazione di Storia Patria per L'Umbria», 90 (1993), pp. 79-107. The situation is not facilitated by the fact that there is a «real» late-19th-century copy of that very astrolabe in Chicago: see RODERICK and MARJORIE WEBSTER, *Western Astrolabes*, in BRUCE CHANDLER (gen. ed.) and SARA SCHECHNER GENUTH (ed.), *Historical Scientific Instruments of the Adler Planetarium & Astronomy Museum*, Chicago, Ill., Adler Planetarium & Astronomy Museum, 1998, vol. I, pp. 114-116 (no. 28). Also, the earliest Renaissance astrolabe from the Vienna school, dedicated by Regiomontanus to his patron Bessarion in 1462

More or less out of the blue, we find John Blagrave of Reading in 1585 proposing a universal astrolabe that in its essence closely resembles a simple version proposed by Ibn al-Sarrâj, with a full myrtle ecliptic, the upper part superposed on the semicircle of universal markings, as well as a calendrical scale around the front of the limbus.[28] The universal astrolabe of Charles Whitwell (*ca.* 1597) preserved in Florence is in this style,[29] as is the unsigned piece preserved in Chicago.[30] A full myrtle ecliptic is found on a *nautical hemisphere* made by Whitwell *ca.* 1597, and on a quadrant made by William Senior in 1600.[31] A folded half myrtle ecliptic rotating over a universal projection features on an astrolabe made by Adrian Zeelst in Louvain *ca.* 1600.[32] Recently we have discovered a more complicated instrument by an unidentified Spanish instrument-maker working in the Louvain tradition possibly in El Escorial *ca.* 1560; the source of the technical inspiration behind this piece has not yet been established.[33] Suffice it to say, the fate and fortunes of the universal astrolabe in Europe have yet to be properly documented.

With this background in the climates and notions of universality in instrumentation, mixed north-south projections and unusual types of astrolabes, as well as myrtle retes on universal astrolabes, we now turn to a very remarkable artefact.

(#640), was labelled suspect after it had come up for auction at Christie's in 1989. It was returned to Christie's and eventually passed to a private collection, after research had established that it is one of several fine instruments that had not been previously identified, all of which stem from the same workshop or group of workshops: see DAVID A. KING and GERARD L'E. TURNER, *The Astrolabe Presented by Regiomontanus to Cardinal Bessarion in 1462*, «Nuncius: Annali di Storia della Scienza», 9 (1994), 1, pp. 165-206.

28 GUNTHER, *The Astrolabes...*, cit., II, pp. 492-500 (no. 308).

29 GERARD L'E. TURNER, *Elizabethan Instrument Makers: The Origins of the London Trade in Precision Instrument Making*, Oxford, Oxford University Press, 2000, pp. 187-190 (no. 43).

30 GUNTHER, *The Astrolabes...*, cit., II, p. 501 (no. 309) and pl. CXLI, and TURNER, *op. cit.*, pp. 86-89.

31 *Ibid.*, pp. 196-198 (no. 47) and 249-250 (no. 77).

32 KOENRAAD VAN CLEEMPOEL, *A 'Catalogue Raisonné' of Scientific Instruments from the Louvain School, 1530 to 1600*, Turnhout (B), Brepols, 2002 (*De diversis artibus*: Collections of Studies from the International Academy of the History of Science, vol. 65 (N.S. 28), pp. 218-221 (no. 74), also colour plates on pp. 264-265).

33 ROBERTO MORENO, DAVID A. KING and KOENRAAD VAN CLEEMPOEL, *A Sixteenth-Century Spanish Astrolabe*, «Annals of Science», 59 (2002), pp. 331-333, esp. 362. Both Julio Samsó and François Charette have registered other interpretations of the underlying Toledan materials.

1

2

Figs. 1-2. The front and back of the Oxford astrolabe (#169). [Courtesy of the Museum of the History of Science, Oxford].

Fig. 6. Illustrations of the myrtle and drum astrolabe retes and the full plate of altitude circles that serves them both, taken from a fine copy of the treatise on astrolabe construction by al-Bîrûnî (*fl.* Central Asia, *ca.* 1025). al-Bîrûnî took most of his information on non-standard astrolabes from his early contemporary al-Sijzî, who in his treatise on astrolabe construction, rediscovered in Istanbul in the 1970s, had mentioned not only the different varieties but also the names of those who invented them and a few examples that had been presented to various dignitaries. One particular myrtle astrolabe made by al-Sijzî himself is mentioned on this page. [From MS Leiden Or. 591, fol. 97v, courtesy of the Universiteitsbibliotheek, Leiden].

Fig. 7. The front of the medieval Italian astrolabe in Florence (#493). [Photo by Franca Principe, courtesy of the Museo di Storia della Scienza].

8a

8b

8c

Figs. 8a-c. Various detailed shots of the ecliptic arc and the altitude circles and markings for the seasonal hours underneath. [Photos by the author, courtesy of the Museum of the History of Science, Oxford].

B. The Oxford astrolabe

1. *Basic description*

Our astrolabe[34] is a mere 5.9 cm in diameter, and 0.8 cm thick, thus qualifying as the smallest surviving European astrolabe.[35] It barely needs saying that any astrolabe of this size has no serious practical use.[36] The engraving is carefully executed in a bold hand, which appears to be Italian and looks as though it can be dated *ca.* 1300 – see further §B2 below. Or maybe the instrument is a 19th-century Italian copy – see §B7.

The throne is of a simple tri-lobe design – three-quarters of a quatrefoil – such as is found on various very early European instruments.[37] The outer scales on the front and back are divided and labelled for each 10° and subdivided for each 2°. The rete bears a complete circumferential ring and two complete horizontal and vertical diameters, the latter extended outwards at the top to serve as a marker on the outer scale. On the ecliptic, whose scale is improperly divided, and the four unlabelled star-pointers see §B3-4. The mater bears unusual astrolabic markings for a specific latitude – see §B5. The markings and alidade on the back are standard, but merit comment – see §B6. The rete and the alidade are riveted in place by an unusual bolt with an eight-pointed star on the front; the rete can rotate freely but could not be removed without breaking the bolt. Some details are shown in Fig. 8.

2. *The engraving*

The engraving, which is bold and heavy, but not inelegant, is of a kind not attested on any other astrolabe known to me. It bears some resemblance to that on another, possibly contemporaneous Italian astrolabe (#493) preserved in the Museo di Storia della Scienza in Florence: see Fig. 7. I have recently studied this other piece in detail, as part of an investigation of a series of astrolabes connected with Urbino.[38]

---

[34] On previous publications featuring this piece see n. 3.

[35] A decorative astrolabe from 17th-century Isfahan has diameter 4.1 cm!

[36] See Gibbs, Henderson and Price, *A Computerized Checklist...*, cit., pp. 69-84, for the sizes of many surviving astrolabes.

[37] King, *The Earliest European Astrolabe...*, cit., p. 365, where this piece is mentioned.

[38] The Florence astrolabe, in gilded brass, is of a type that provided the original inspiration for a Renaissance astrolabe (#4506), dated Urbino, 1462, which was stolen from a museum in Moulins (Allier, France) in 1977. This Renaissance astrolabe was of a type that in turn inspired

The numerals on the Oxford piece are a curious mix (and different from those on the Florence piece). For '1' we find a capital 'I'. The forms for '4', '5', and '7' are old-style *Gothic*, whereas the forms for '2', '3', '6', '8' and '9' are more developed and well rounded.[39]

My dating of both the Oxford and the Florence astrolabes to *ca.* 1300 is based on a feeling that they are not 15th century, nor are they 14th century. Only a comparison of the engraving with that on *dated* Italian metalwork from these periods, which I have not pursued, or the (unlikely) rediscovery of a related instrument that is dated, could confirm or disprove my tentative dating.[40]

3. *The ecliptic on the rete*

The ecliptic is of the folded half myrtle variety (see Fig. 4d) with double graduation, so that the northern projection of the northern ecliptic and the southern projection of the southern ecliptic are represented as a single

---

the astrolabe depicted in the intarsia of the *studiolo* of Archduke Federico of Urbino, completed *ca.* 1476. The Florence astrolabe shows signs of having been influenced by the tradition of the prolific Abû Bakr ibn Yûsuf of Marrakesh *ca.* 1200. For detailed descriptions of all of these see DAVID A. KING, *The Astrolabe Depicted in the Intarsia of the Studiolo of Archduke Federico in Urbino*, in FLAVIO VETRANO (ed.), *La scienza del Ducato di Urbino – The Science of the Dukedom of Urbino*, Urbino, Accademia Raffaello, 2001, pp. 101-139. On the star-positions see STAUTZ, *Untersuchungen...*, cit., pp. 115-116 and 279-280.

39 The examples of Hindu-Arabic numerals from Italy recorded in G.F. HILL, *The Development of Arabic Numerals in Europe Exhibited in Sixty-Four Tables*, Oxford, Clarendon Press, 1915, pp. 108-121, are mainly too late to be of any use here. For a new look at the Hindu-Arabic numerals in medieval Europe see KING, *The Ciphers of the Monks...*, cit., pp. 309-317.

40 In 1956 the French expert on historical instruments Marcel Destombes claimed that an astrolabe that he had stumbled across was from 10th-century Catalonia. His detractors maintained that the piece was at the earliest of 12th, 13th, 14th, 15th century provenance, if not a 20th-century fake. Comparison with several dozen other early astrolabes reveals that the piece could *only* be from the 10th century, although the names of the signs on the rete may be two or three centuries later: see KING, *The Earliest European Astrolabe...*, cit. Also the unusual engraving is attested elsewhere *only* in 10th-century Catalonian inscriptions: see ANSCARI M. MUNDÓ, *Analyse paléographique de l'astrolabe 'carolingien'*, in WESLEY STEVENS, GUY BEAUJOUAN & ANTHONY J. TURNER (eds.), *The Oldest Latin Astrolabe*, «Physis - Rivista di storia della scienza», Nuova serie 32 (1995), 2-3 [published in September, 1996], pp. 189-450, esp. pp. 303-317.

There has been a tendency to date medieval European instruments too late. For example: the well-known 15th-century French miniature of the monk Henricus Suso with a series of astronomical instruments includes an astrolabe that I would have said was 14th-century French. However, a manuscript copied in Paris in 1276-77 contains an illustration of an astrolabe with the same rete-design. See further KING, *The Ciphers of the Monks...*, cit., pp. 42, 393 and 397.

minor arc of a circle. The corresponding rete and plate beneath are restricted to the domain within the celestial equator.

The two halves of the ecliptic are superposed in such a way that both the northern and the southern signs run counter-clockwise from the left-hand side. The arrangement of the two half sets of signs is to some extent arbitrary, and has no consequences for operations with the sun. For operations with the stars we are confronted by the fact that the stars do not correspond to the appropriate signs of the ecliptic scale. One could live and work with this – with some stretching of the imagination and an occasional *saltus stellae* – if the stars were named, but they are not.

The names of the signs are abbreviated as follows:

ARIĒ	TAV̄	GĒ	CĀ	LEO	VIRḠ
LIBRA	SCOR̄	SAḠ	CAP̄	AQR̄	PISCĒ

These forms alas tell us nothing about the provenance. (In recent years regional forms of medieval Latin and dialects of vernaculars on instruments have been exploited to establish provenance).[41]

The scale of the rete is divided for each pair of signs into three parts, each subdivided into four parts. This is a mistake: each 10° would be more happily divided into *five* sections. The placement of the ecliptic with respect to the outer frame is accurate. However, the lower part of the vertical bar is about 1° to the right of where it should be: see Fig. 1.

4. *The star-pointers on the rete*

There are four unnamed star-pointers shaped like *fleurs-de-lys* with long, limpid stems. No other medieval astrolabe comes to mind with pointers of this shape.

There are only four pointers, one in each of the four quadrants. This use of only one star in each quadrant is not without parallel: the *Toledo astrolabe*, dating from *ca.* 1300, is a piece prepared by a Jew, partly finished by a Christian, and then finished off by a Muslim Arab: the last-mentioned added the Arabic names of some stars, choosing only four, one in each quadrant.[42]

---

[41] See KING, *The Ciphers...*, cit., p. 365, n. 6, where various philological studies of inscriptions on astrolabes by Kurt Maier are mentioned.

[42] DAVID A. KING, *An Astrolabe from 14th-Century Christian Spain with Inscriptions in La-*

The omission of the star-names is not tragic on a standard astrolabe with just a few star-pointers: every potential serious user would have known which of 20- or 30-odd *astrolabe stars* were intended,[43] and one could not read star-names at night anyway. The earliest European astrolabe, from 10th-century Catalonia, bears no star-names, probably because the maker could not write the names.[44] However, the omission of star-names on an instrument of this non-standard variety – so that one has to decide for oneself first whether the pointer serves a star derived by a northern or southern projection – can only strengthen the impression that the maker was barely in control of his task.

The stars on the Oxford astrolabe have been chosen so that their positions are roughly symmetrical with respect to the vertical diameter. The stars in question are so well known as medieval astrolabe stars that any medieval astronomer worth his salt could identify them immediately, but only by realizing that their positions are for a northern projection, as on a standard astrolabe. (When medieval instrument-makers combined star-pointers based on both of the northern and southern projections they had to be *very* careful).[45] To include any stars without their names was perhaps a little unkind, but there is not much space available for the names, so presumably this was deliberate. Already Gunther recognized the stars in question,[46] namely:

lower left	Aldebaran	α Tauri
lower right	Regulus	α Leonis
upper right	Alhave	α Ophiuchi
upper left	Altair	α Aquilae

Note that Regulus is appropriately shown on the ecliptic at longitude

tin, Hebrew and Arabic – A Unique Testimonial to an Intercultural Encounter, «Suhayl - Journal for the History of the Exact and Natural Sciences in Islamic Civilisation», 3 (2003), in press, §3i and §4o. See already *Christie's South Kensington, Exceptional Scientific and Engineering Works of Art, Instruments and Models, 15 April 1999*, London, 1999, pp. 98-107 (lot 52).

43 On the most commonly used stars on astrolabes see PAUL KUNITZSCH, *Arabische Sternnamen in Europa*, Wiesbaden, Otto Harrassowitz, 1959, pp. 59-96.

44 KING, *The Earliest European Astrolabe...*, cit., p. 366.

45 Compare the retes on the spectacular 14th-century universal astrolabe of Ibn al-Sarrâj of Aleppo (#140 - refs. in n. 24) and the problematic 14th-century geared astrolabe from Northern France (#198 - refs. in n. 20), both discussed in STAUTZ, *Untersuchungen...*, cit., pp. 69 and 213, and 103-104 and 262, respectively.

46 On these four standard *astrolabe stars* see KUNITZSCH, *Arabische Sternnamen in Europa*, cit., no. 18 on p. 70; no. 30 on p. 76; no. 51 on p. 81; and no. 54 on pp. 81-82.

Leo 20°, which does correspond to *ca.* 1300, but not too much should be concluded from this alone. In fact, however, the four pointers correspond to the accurate positions of these four stars *for a northern projection* for epoch 1300 to within 1-2 mm on the instrument itself.[47] For a medieval European instrument this may be regarded as something of an achievement.

5. *The latitude-dependent markings on the mater*

The main astrolabic markings on the mater represent the altitude circles for each 6° above and below the horizon for a latitude that is not specified, but which is, by inspection, 24°. This is most easily demonstrated by the fact that the small circle for the zenith lies on the inner circle for the (summer) solstice. The altitude circles are not properly engraved for they are not uniformly spaced. Also, there are not 14 circles between the horizon and the zenith as there should be for such a 'sextile' plate.

In addition, there are curves for the seasonal hours below the horizon that are appropriately drawn as arcs of circles. Above the horizon, the arcs for the first and eleventh hours and the rectilinear segments engraved more heavily for the other hours are unlikely to be a later addition for the same hand also highlighted the horizon. For the 'upper' hours there are two sets of markings, the other set lightly engraved as arcs of circles, like the altitude circles, and dotted. Both sets are restricted to the domain between the equinoctial and solstitial circles. It seems that our maker has tried to engrave the markings for the astrological doctrine known as «casting the rays»,[48] which were an occasional feature of Islamic astrolabes from the 10th century onwards and European instrument some centuries later.

---

[47] See the techniques outlined in STAUTZ, *Untersuchungen*..., cit.

[48] See, for example, the article *Tasyîr* by OTTO SCHIRMER in *The Encyclopaedia of Islam*; and JOHN D. NORTH, *Horoscopes and History*, London, The Warburg Institute, 1986 (Warburg Institute Surveys and Texts, vol. 12), pp. 1-69. On the earliest Islamic astrolabe with markings for the houses – from Baghdad, 984/85 – see DAVID A. KING, *Early Islamic Astronomical Instruments in Kuwaiti Collections*, in ARLENE FULLERTON & GÉZA FEHÉRVÁRI (eds.), *Kuwait: Art and Architecture – Collection of Essays*, Kuwait (no publisher stated), 1995, pp. 76-96, or ID., *World-Maps*..., cit., p. 18 (*ad* #111). The oldest European astrolabe with such markings – Italy (?), 13th (?) century – is probably #558 (see n. 7 above). They were particular popular on Renaissance instruments, such as those mentioned in nn. 27 and 38: see KING and TURNER, *The Astrolabe Presented to Bessarion*, pp. 171, or KING, *The Astrolabe Depicted in Urbino*..., cit., p. 115 (*ad* #640); and *ibid.*, pp. 107 and 132 (*ad* #4506).

The resulting *ensalata mista* of curves and lines on an instrument of this size was only calculated to confuse. See the graphics in Fig. 9.

In places, the markings on the mater appear to be worn down, as if by excessive use – see further §B7 below.

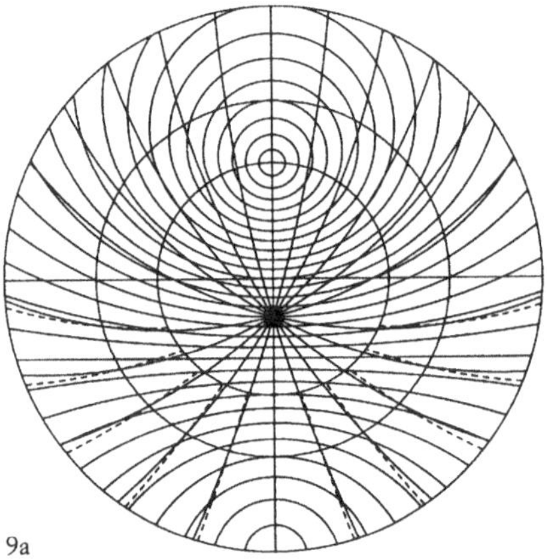

9a

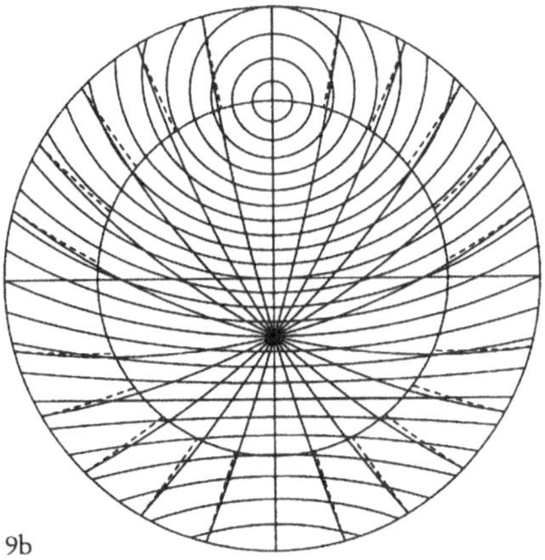

9b

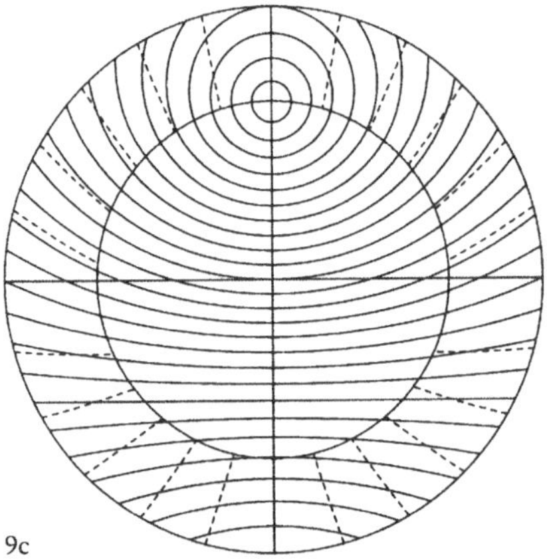

9c

Fig. 9. (a) A set of standard astrolabic markings for latitude 24° (see Fig. 5) with curves for the seasonal hours (dotted) below the horizon and additional curves for the astrological doctrine known as "casting the rays". (b) Altitude circles, together with complete sets of curves for the hours and casting the rays, now bounded by the celestial equator to serve a myrtle astrolabe. (c) A reconstruction of the markings on the Oxford astrolabe, with a complete set of altitude circles above and below the horizon, and curves for the hours (dotted) and casting the rays, now restricted to the space between the equatorial and solstitial circles. [Graphics by Dr. François Charette.]

6. *The back of the astrolabe*

On the back of the instrument, within the smallest of several circular arcs inside the outer scale, the upper half bears a double universal horary quadrant and the lower half a double shadow square to base 12, with scales divided and labelled for each 3 digits and subdivided for each single digit – see further below on these markings. The alidade is counterchanged and bears two sights with a correctly-aligned cylindrical hole in each: it shows every sign of being original. The markings on the back have, like the mater, a worn look about them: see further §B7.

Some remarks on the universal horary quadrant are in order, not least because of the misunderstanding currently associated with these markings in some circles. The device, invented in Baghdad in the 9th century, serves to find for any latitude the time in seasonal hours from the instantaneous altitude of the sun and its meridian altitude.[49] One sets the alidade to the meridian solar altitude and makes a mark with ink or wax on the alidade at the point where it intersects the semicircle representing the 6th hour; one then moves the alidade down to the instantaneous solar altitude, and the mark indicates the time on the other hour curves.[50] The result is approximate but the error in Mediterranean and Nilotic climes (such as latitude 24°!) is minimal and more substantial in Northern Europe. In the Renaissance, some astronomers added a solar scale on the alidade of their astrolabes, which was necessarily latitude dependent. In some modern writings, this kind of scale is considered indispensable;[51] in fact, a radial solar scale is not only unnecessary but also inappropriate, because it compromises the universality of the quadrant.

It is perhaps significant that there is no solar scale on the back of the Oxford astrolabe, as there are on the vast majority of medieval European astrolabes. Such scales were uncommon on early *Eastern* Islamic astrolabes, although they are mentioned by al-Bîrûnî in the early 11th century,

---

49 DAVID A. KING, *A 'Vetustissimus' Arabic Treatise on the 'Quadrans Vetus'*, «Journal for the History of Astronomy», 33 (2002), pp. 1-19, and the more detailed version in KING, *Studies in Astronomical Timekeeping...*, cit., IXa.

50 The only modern work on the astrolabe in which this procedure is correctly described is SALVADOR GARCÍA FRANCO, *Catálogo crítico de los astrolabios existentes en España*, Madrid, Instituto Histórico de la Marina, 1945, pp. 218-221, esp. p. 219.

51 NORTH, *The Astrolabe...*, cit., figure on p. 105 (fig. 16 on p. 216 of the reprint); and ID., *Astrolabes and the Hour-Line Ritual*, «Journal for the History of Arabic Science», 5 (1981), pp. 113-114, repr. in ID., *Stars, Minds...*, cit., no. 15.

and, for example, one is to be found on the back of the universal astrolabe of Ibn al-Sarrâj (#140 – see §A4). The fact that they appear on the earliest surviving *Western* Islamic astrolabe (#110 = #135, 10th century?),[52] on most Andalusî and Maghribî astrolabes from the 11th century onwards, as well as on the earliest European astrolabe (#3042), has led to some speculation concerning their origin.[53]

7. *Is the Oxford astrolabe a fake?*

In recent times, there have been some serious cases of misidentification of genuine astrolabes as fakes, and of fake astrolabes as genuine.[54] We tread on particularly dangerous ground when we dismiss as a fake a given piece that we do not fully understand, not least because it is never possible to fully restore a lost reputation.

Now Gunther records that the Oxford astrolabe was «formerly gilt», and somehow a rather high proportion of gilt instruments do tend to be fake. Also, we know it surfaced in Italy only at the end of the 19th century, so it has no earlier attested provenance. I am nevertheless inclined to dismiss outright the possibility that the Oxford astrolabe is, say, a 19th-century Italian «fake». First, it feels very much like a medieval astrolabe. Second, there were plenty of standard medieval astrolabes available in Italy for a 19th-century faker to copy. Third, one needs to know a lot about astrolabes to even begin copying an instrument of this potential complexity, even if one does make a mess of it. And fourth, if an instrument is a 19th-century copy of a genuine medieval instrument, this is usually immediately apparent. But not always ...

One thinks of the astrolabe (#4507), inevitably in gilt brass, in the Museo Bargello in Florence, which could be 15th-century or 19th-century Italian.[55] Whatever it may be, the rete on the Bargello astrolabe was originally inspired, not necessarily directly, by the kind of rete found on an 11th-cen-

---

52 Inadvertently listed twice in GUNTHER, *The Astrolabes...*, cit., I, p. 244 (no. 110: *Early Arabian astrolabe – Undated*), and fig. 117 (front), and p. 280 (no. 135: *Spanish astrolabe – XIIIth-XIVth Cent.*).

53 KING, *The Earliest European Astrolabe...*, cit., pp. 376 and 384.

54 See n. 27.

55 See MARA MINIATI (ed.), *Strumenti scientifici della Collezione Carrand*, Florence, Museo Nazionale del Bargello, 1991, pp. 48-54 (no. 22), with a full set of illustrations of the Bargello piece.

tury Andalusî astrolabe now preserved in Berlin (#116),[56] and its historical importance lies mainly in that fact. But one thinks also of the earliest European astrolabe from 10th-century Catalonia. One reason I proposed for its not being a fake was that:

> any faker who made this astrolabe knew more about the history of the astrolabe and about Latin epigraphy than all of the scholars who have ever looked at this piece ... put together.[57]

The same holds for the Oxford astrolabe.

A different situation holds for the Florence astrolabe fondly called «the astrolabe of Pope Sylvester II» (#101).[58] Here we are dealing with a genuine Islamic astrolabe from 10th-century Iraq, to which someone has added a medieval-looking solar scale on the back inscribed in Latin. The markings on the back are faked, probably 19th century Italian.

Now, metal analysis might eventually establish that the Oxford piece is indeed a 19th-century production, *deliberately* made to look as though it had been extensively used. In this case, the maker *surely* copied it from a genuine medieval instrument, *possibly* one with a complete set of plates for the climates. That genuine instrument would have been copied from an Islamic astrolabe, most *probably* one with such a complete set of plates.

It makes more sense that a 19th-century faker would provide an instrument too small to use properly with a *useless* set of markings for latitude 24° rather than a 13th- or 14th-century astrolabist. Certainly a 19th-century faker could not have copied this piece from an instrument with inscriptions in Arabic. Another impediment to the evaluation of this instrument is that, unlike the vast majority of medieval instruments, it gives the impression of having been used. Substantial parts of the altitude markings on the mater and the surface of the back are worn away, apparently by excessive 'use'. Given the fact that the front of this pretty little instrument is quite useless

---

56 Compare the two retes illustrated in KING, *The Earliest European Astrolabe...*, cit. figs. 6 and 19. On the star-positions on the Andalusî and Italian retes see STAUTZ, *Untersuchungen...*, cit., pp. 57 and 191, and 86-87 and 247.

57 KING, *The Earliest European Astrolabe*, cit., p. 362.

58 GUNTHER, *The Astrolabes...*, cit., I, pp. 230-231 (no. 101) and pl. LIIb; also GARCÍA FRANCO, *Catálogo crítico...*, cit., pp. 131-152 (no. 3); KING, *Some Medieval Instruments...*, cit., p. 35; and ID., *Astronomical Instruments...*, cit., p. 171. The instrument has been featured in several exhibitions, although already Gunther recognized the dubious character of the markings on the back.

for any sensible purpose in Italy, it seems highly unlikely that anybody other than a faker would have made it look used.

In any case, if the instrument is a 'fake', it has the unusual characteristic of now being of greater historical value than any 'real' instrument from which it was copied, for that is apparently lost for all time. Under no circumstances does it decrease in historical value if it *is* a 19th-century copy of a medieval instrument, and I suggest that we keep on regarding it as 'medieval', whether it is or not.

8. *Concluding remarks*

The Oxford astrolabe is remarkable as the sole surviving *medieval* European instrument with a folded half myrtle ecliptic with double graduation. Furthermore, it is one of only two medieval European astrolabes featuring the equivalent of altitude circles *below* the horizon, although this may not be historically significant.[59] (This, as is well known, is a feature of some of the earliest astronomical clocks in medieval Europe.) The piece also has the kind of rete usually associated with universal astrolabes, but the 'universality' in this case, as with some of the earliest European astrolabes,[60] was originally achieved by a series of plates for each climate.

Our astrolabe is degenerate in the sense that, from a practical point of view, it is quite useless in European latitudes. Whether it is a genuine piece from *ca.* 1300, or a copy of such a piece from the 19th century, the question arises: why would any European make such an instrument *ca.* 1300? The answer appears to be that he was a fairly competent craftsman who had, however, only a very vague idea about non-standard astrolabes. He was clearly copying, reproducing, imitating – call it what you will – another instrument, and the only sensible explanation would appear to be that he was copying an instrument which had plates for each of the seven climates of Antiquity: from this he copied only the plate for the second climate.

It is tempting to suppose that our maker was copying an Eastern Isla-

---

[59] See the text to n. 21. On the 14th-century French geared astrolabe (n. 20) the altitude circles are above the horizon for a southern projection, and hence correspond to those below the horizon for a standard northern projection. On an 11th-century Andalusî astrolabe we find altitude circles for latitude 16½° *south*, this being the lower limit for Ptolemy's world-maps: see KING, *Bringing Astronomical Instruments Back to Earth*, cit., pp. 19 and 53.

[60] KING, *Bringing Astronomical Instruments...*, cit., pp. 27-28; and ID., *Ciphers of the Monks*, cit., pp. 411-415.

mic instrument with inscriptions in Arabic. This would explain why the star-pointers are not named, the latitude is not specified, and there is no solar scale on the back. There is no known Islamic precedent for precisely such an instrument, but there is evidence of both components. Indeed, it is not unreasonable to hypothesize that some Muslim skilled in astrolabe design had proposed a double half myrtle rete in the 'new' Islamic tradition of the 9th and early 10th century, together with a set of plates for each of the climates in the 'old' tradition of Antiquity, also attested on the earliest Islamic astrolabe from the 8th century. Since only the very earliest Islamic astrolabes had markings for each of the climates, it is further reasonable to suspect an early, say, early-10th-century, origin for this hypothetical combination.[61] The resulting instrument would be universal in the same sense as the standard astrolabe with a set of plates for the climates is universal (see §A2): it is not a *universal astrolabe* in the traditional sense, working for any latitude. The star-pointers would surely have been carefully labelled in order to show whether they corresponded to a northern or southern projection. Furthermore, the back of such a hypothetical Eastern Islamic astrolabe would probably not have had a solar scale. On the other hand, it is unlikely that a serious early Islamic astrolabe would have been made as small as the Oxford astrolabe.

The milieu of the original transmission to the maker of the Oxford astrolabe may well have been Sicily, although some Islamic scientific materials appear to have come to Italy directly from Egypt.[62] An alternative proposal, namely, that any European had first hit upon this idea and was able to execute it properly, is, in the light of our present limited knowledge of the history of the astrolabe, particularly in medieval Italy, unlikely.

In brief, this little instrument presents us with a series of problems that we are – at this time – unable to solve. And it gently reminds us how little we know about the fates and fortunes of even the standard astrolabe in medieval Italy.

---

61 KING, *Bringing Astronomical Instruments...*, cit., pp. 7-8.

62 We are reminded of the planetarium described in various medieval Italian manuscripts in Latin that John North has associated with Egypt. See NORTH, *Opus quarundam...*, cit., esp. p. 364 (p. 162 of the reprint).

Acknowledgments

It is a pleasure to acknowledge my longstanding debts to the Museum of the History of Science at Oxford and the Museo di Storia della Scienza in Florence. To Dr. François Charette of the Dibner Institute (M.I.T., Cambridge, Mass.) my thanks for his critical comments on earlier versions of this text (and most other new texts of mine) and for most of the graphics. Recent research in Frankfurt on medieval instruments has been generously funded by the German Research Foundation (DFG) in Bonn (see n. 1).

VI

An Astrolabe from Einbeck datable ca. 1330

1. Introduction

One of this author's favourite watering-places is the charming little town of Einbeck, some 80 km south-west of Brunswick and some 120 km north-west of Erfurt. Not only the medieval town centre and the Stiftskirche attract this and other visitors, but Einbeck is also famous for its brewery, which dates back to before 1378, the date of a record of a delivery of Einbeck beer to Celle.[1] The subject of this paper is an early-14th-century astrolabe from Einbeck, and I confess from the start that I have been unable to find any connection between any astrolabe-workshop and the brewery. This fortunately does not mean that there was none.

The astrolabe in question – see Figs. 1–6 – has been preserved in Cracow apparently since the 15th century. In the modern literature, it was first mentioned by Ludwik BIRKENMAJER in his 1892 study of the 15th-century Cracow astronomer Martin BYLICA.[2] A brief notice about it was included by Ernst ZINNER in his monumental survey of historical instruments.[3] The Einbeck connection had been discussed already by BIRKENMAJER, but was re-evaluated and clarified in a valuable 1987 study by Helmar HÄRTEL.[4] The same author has shown that, in all probability, the Einbeck astrolabe was taken to Cracow by the son of a Brunswick doctor, LUDOLFUS DE BRUNSWIK. The son's name was the same as that of his father, Ludolfus, and both had the same first name as the first owner of the astrolabe.

The astrolabe has never been described in detail, and this is the purpose of the present study. In fact, we are dealing with *the earliest German astrolabe that is datable*; the next batch of instruments from the German-speaking world that are datable are from the period 1457–1482 and stem from the Vienna school of PEUERBACH (1423–1461) and REGIOMONTANUS (1436–1476) and their successors, or from derivative workshops, such as one in Schmalkalden.[5] In a recent study I have outlined the importance of astrolabes as historical sources.[6] The Einbeck astrolabe is a particularly interesting and revealing

1 See www.einbeck-online.de.

2 BIRKENMAJER, *Marcin Bylica z Olkusza*, especially pp. 101–106 (in Polish). A summary in German is contained in a lengthy, almost instantaneous, review by F. KARLIŃSKI, in which the astrolabe is mentioned on pp. 109–110. It is a pleasure to thank our Institute's Librarian, Ryszard DYGA, for tracking down these two items.

3 ZINNER, *Astronomische Instrumente*, pp. 535–536, where the name is misread and the text "lost" under Ludolf von Siete (*sic*), also *idem*, *Regiomontanus*, pl. 42, figs. 78–79 (front and back).

4 HÄRTEL, "Ludolphus Borchdorp de Brunswik". See the text to n. 21 below.

5 KING & TURNER, "The Astrolabe Presented by Regiomontanus to Cardinal Bessarion in 1462", also KING, *Astrolabes and Angels, Epigrams and Enigmas*.

6 KING, *In Synchrony with the Heavens*, XIIIa-3.

example. One of the things it reveals is how little we know about astronomy in the German-speaking world, for it was surely not made in Einbeck. I shall argue that the most likely place for a workshop in which such an astrolabe could have been produced is Erfurt.

2. Description of the astrolabe

Location: Jagiellonian University Museum.
Inventory number: 41/V 4043.
The piece was formerly in the possession of the University Observatory. It was probably originally brought to Cracow in the late 15th century by LUDOLFUS (II) BORCHDORP DE BRUNSWIK.
International Instrument Checklist number: 2072.[7] Reference in the Frankfurt catalogue of medieval instruments: 6.3.10.[8]
Material: brass, with some copper-tone brass strips inlaid on the back
Dimensions: 24.9 cm in diameter and 4 mm thick.

The engraving

The letters and numbers in the inscription are engraved, not punched. All of the engraving is in the same careful and elegant and easily legible hand. The letters are in Gothic upper-case characters, with the distinctive forms such as Ā with a bar across the top for A, VV overriding each other for W, and "h", which *is* a capital. In some star-names the letters OR or OL are in a single ligature, here underlined <u>OR</u> or <u>OL</u>. Both on the rete and on the back, the two Ts in Sagittarius are written differently, the second more like the distinctive G: this we represent by <u>T</u> (see Fig. 6b). An "R" with a tilde across the lower right leg occurs in the inscription on the back and is here denoted by "R~". In others, some standard Latin abbreviations are used.[9] Omissions of letters are usually indicated by a bar above the preceding and following letters, here indicated by *, so we shall encounter, for example, CA* for CAPVT and GE*OR* for GEMINORVM. On one star-pointer and on the back, the standard raised letter resembling a "9" (a supplement, along with "&", to the medieval alphabet A-Z) is used for the ending *-us*.[10] The numerals have the standard Gothic forms for the 4, 5 and 7.[11]

---

7 PRICE *et al.*, *Astrolabe Checklist*, p. 24. There no date is given and the maker is listed as LUDOLF DE SIETE (*sic*), following ZINNER. International Instrument Checklist numbers up to 4000 are those used by PRICE *et al.* Higher numbers are those for those instruments documented thereafter: see KING, *In Synchrony with the Heavens*, XIIIa-3.

8 For a provisional table of contents see: www.uni-frankfurt.de/fb13/ign/instrument-catalogue.html. See also the acknowledgements at the end of this paper.

9 The standard work is CAPPELLI, *Lexicon abbreviaturarum*.

10 On this see CAPPELLI, *op. cit.*, pp. xxiv–xxvi and 68–85, and on a peculiar usage on a medieval Spanish astrolabe see KING, *In Synchrony with the Heavens*, XV-2.2, 2.8. 3.2, *etc.*

11 See, for example, HILL, *Arabic Numerals in Europe*.

The throne and limbus

The throne is a hollowed semi-quatrefoil frame, and to this is attached a rather primitive shackle and thin circular ring (perhaps both of steel), neither of which looks original. The front of the throne and the limbus of the mater are of one piece and the ensemble is brazed onto the mater and back of the throne, which is of equal thickness. The scale of the front of the limbus is divided and labelled for each 5° and subdivided for each 1°.

The rete

The rete bears a large quatrefoil and four equally large semi-quatrefoils, one at each end of the principal axes. In addition, the equatorial bar is attached to the circumferential frame by two elegant supports consisting of quarter-quatrefoils. The horizontal equinoctial bar is unusual in that it tapers on both sides from the centre to the ecliptic; it is, however, unduly broad at the centre, and is counter-changed once on each side at the ecliptic and again at the centre. There is a vertical solstitial bar between the central disc and the equatorial bar, otherwise this axis is taken up by the quatrefoil frames. The scale of the ecliptic is divided and labelled for each 5° of each sign (5° to 360°), and subdivided for each 1°. The names of the signs of the zodiac are the standard Latin ones:

ARIES TAVRVS GEMINI CANCER LEO VIRGO
LIBRA **SCORPIVS SAGITTARIVS** CAPRICORNVS AQVARIVS PISCES

except for SCORPIVS, an accepted variant, and the two different Ts in SAGITTARIVS. The star-pointers are of three basic shapes: sharp spikes, talons and daggers standing on bases, some composed of three circular discs. The star CAPVT GEI*OR (8 below) is represented by a hole on the ecliptic frame at Cancer 15° and is *named on the back of the rete* (see Fig. 6a); the stars DVBhE (10) and CA* LEO*IS (11) are represented by holes on the equinoctial bar. Some 21 stars are represented, some with their Latinized Arabic names, others with the Latin names: see the commentary. The 21 stars represented are a subgroup of the astrolabe-stars in list VIII of Paul KUNITZSCH's survey of medieval European star-catalogues.[12] In the following list, the stars are arranged in the four quadrants counter-clockwise from the vernal equinox, that is, according to increasing longitude and right ascension. I give the names as engraved on the astrolabe, the modern identifications, and the numbers in KUNITZSCH's lists of medieval European astrolabe-stars,[13] medieval Latinized Arabic star-names,[14] and stars of his "Typ VIII".[15]

---

12 KUNITZSCH, *Typen von Sternverzeichnissen*, pp. 47–58.

13 *Idem*, *Arabische Sternnamen in Europa*, pp. 59–96.

14 *Ibid.*, pp. 97–224.

15 See the reference in n. 12 above. On a relevant star-table with some star-names different from those presented by KUNITZSCH – notably SPICA for our no. 14 – see also n. 31 below.

1	MIRACh	β And	A7	140	1
2	VE*T' CETI	ζ Cet	A8	(68)	3
3	CAPVT ALGOL	α Per	A14	24	7
4	ALDEBARAN	α Tau	A18	16	9
5	RIGIL	β Ori	A19	164	11
6	ALhAIOT	α Aur	A20	27	10
7	AlhABOR	α CMa	A23	26	13
8	CAPVT GEI*OR	α Gem	A24	(161)	14
9	COR LEONIS	α Leo	A30	(109)	20
10	DVBhE	α UMa	A33	89	23
11	CA* LEO*IS	β Leo	A35	(86)	24
12	B*NAS	η UMa	A40	71	27
13	SPICA	α Vir	A39	(66)	26
14	ALRAMECh	α Boo	A41	46	28
15	ALFETA	α CrB	A45	43	29
16	COR SCOPIONIS (*sic*)	α Sco	A48	(110)	32
17	CAPVT DRACONIS	γ Dra	A52	(100)	33
18	WEGA	α Lyr	A53	198	35
19	ALTAIR	α Aql	A54	(52)	36
20	MVSIDA PEGASI	ε Peg	A58	(98)	43
21	HVMER⁹ EQVI	β Peg	A62	(133,177)	47

On the back of the rete is an ecliptic scale divided for each 30°, and then subdivided for each 5° and then for each 1°. There is also one star-name – see above.

The astrolabic markings on the mater

The mater is engraved with altitude circles for each 1° up to 65°, those for each 5° somewhat deeper than the rest, and then for each 5° up to the zenith, as well as with azimuth circles for each 5°, none labelled. The latitude is given in Gothic numerals as "50" degrees. There are construction marks around the equator and along the meridian, and the "shadow" of the rete is visible on the mater as the result of oxidation of the exposed parts. The markings are apparently accurate except for some azimuth circles near the southern meridian. These should be investigated more closely.[16]

16 Namely, to see whether they correspond to the non-standard approximate procedure attested in some medieval European texts on astrolabe construction: see KING, *In Synchrony with the Heavens*, XV-2.10 and 3.17. See also n. 39 below.

The back of the instrument

The back bears four altitude circles divided for each 5° (5° to 90°), subdivided for each 1° and labelled for each 5°. The degree scale serves also the solar scale, and each 5° of each sign is labelled. The names of the signs are engraved as:

ARIES TAVRVS GEMINI CANCER LEO VIRGO
LIBRA **SCORPIVS SAGITTARI**⁹ CAPRICORN⁹ AQVARI⁹ PISCES

Note the orthography for Sagittarius, with two different forms of T. There is an excentric calendrical scale divided for each 5^d and subdivided for each 1^d: the divisions are labelled for each 5^d, even when the last entry for some months should be 6^d, but with "25|3" at the end of February. The names of the months are standard, but for the third, whose orthography is very common on medieval astrolabes of widely-differing provenance:

IANVARIVS FEBRVARIVS **MARCIVS** APRILIS MAIVS IVNIVS
IVLIVS AVGVSTVS SEPTEMBER OCTOBER NOVEMBER DECEMBER

The equinox is at March 10.5 – see the commentary.

In the upper half of the central space, within a circle concentric with the rim, there is a double universal horary quadrant with unlabelled curves.[17] In the lower half there is a double shadow-square to base 12,[18] with scales divided and labelled for each 4 units and subdivided for each 1 unit. The inscriptions VMBRA VER (horizontal!) and VMBRA REC (vertical!) are engraved in a frame inside the scale which is on strips of copper-tone brass that have been inlaid in the back. It is not clear why these inscriptions have been added in this way.

Below the shadow-squares is an inscription:

LVDOLFI DE SCICTE ThESAVR~ ECCL*E EMBICE*N
in full, LVDOLFI DE SCICTE THESAVRARII ECCLESIE EMBICENSIS

I interpret LVDOLFI here as a *genetivus possessionis*, not as in [OPVS] LVDOLFI, "[a work] *by* LUDOLFUS".[19] So I suggest the meaning:

"Belonging to Ludolfus of Scicte, the Treasurer of the Church of Einbeck."

Since the engraving is the same as on the rest of the instrument, this would mean that the maker made the piece especially for LUDOLFUS and put this statement on the astrolabe at his request.

The alidade

The alidade is counter-changed at a rather large central disc (diameter 4.8 cm) decorated with a five-petal flower. Both arms of the alidade are slightly tapered, observing the

---

17 On the development of these markings, first devised in 9th-century Baghdad, and their unsuitability for higher regions (such as Northern Germany), see KING, *In Synchrony with the Heavens*, XIIa.

18 On the development of shadow scales, also first devised in 9th-century Baghdad, see *ibid.*, XIIa–B.

19 REGIOMONTANUS engraved IOANNIS OPVS on his 1462 astrolabe for Cardinal BESSARION to mean "a work made by Ioannes". See KING & TURNER, "The Astrolabe Presented by Regiomontanus to Cardinal Bessarion in 1462", p. 172, and KING, "Astrolabes and Angels, Epigrams and Enigmas".

"fiducial" diameter. The vanes have two holes each, both off-centre and on one the upper hole has been excavated and there is now a cut-out reaching almost to the lower hole (see Fig. 6c). The radial rule is not tapered and is probably a later addition, and the pin with a small circular head and the cylindrical wedge are likewise modern replacements.

3. Commentary

The Einbeck connection

Scicte is modern Sickte, a small town 10 km from Brunswick.[20] The earliest mention, as Kicthi, is from 888. Thereafter we have Siudi (1042), Xikthi and Xicthe (1060), Tsikthe (~1200), Zickede (1264), and *Major* Sickte (1308). From 1315 there is a document mentioning *inferus* Tzithe and *superius* Tzicthe.

LUDOLFUS DE SCICTE was indeed the Treasurer of the Stiftskirche in Einbeck, founded in the 11th century by Graf DIETRICH II VON KATLENBURG-EINBECK in honour of St Alexander.

In the records of the city of Brunswick for 1324 and 1342,[21] we find:

1324, January 21, at Einbeck: Amongst the six witnesses to the three Dukes Ernst, Wilhelm and Johann awarding Reinbold, assistant priest (Kaplan) at St. Katherinen, a year of repose is our *Ludolfus de Tzicte thesaurarius ecclesie Embicensis*.

1342, March 12, at Einbeck: The Dean (Dekan) of St Blasien and others proclaim that LUDOLF in his two functions has issued a bill to the monastery of St Aegidien. He is referred to as *Ludolfus thesaurarius ecclesie sancti Alexandri Embicensis* and *judicus ad infrascripta a sede apostolica delegatus*, that is, "apostolic judge" or judge answering to the Holy See (verordneter apostolischer Richter).

So the inscription on the astrolabe can date from any time before 1324 and until just after 1342. Let us say ca. 1330, or 1330 ± ca. 15 years. In a sense, these dates offer a challenge to future investigators to show that the star-positions on the astrolabe do not contradict this textual evidence.

Where was the astrolabe made?

As far as I am aware, we have no information on any 14th-century German astrolabe workshop such as the one where this piece was made. The quality of the workmanship indicates that it was not a "one-off" production of some craftsman, but rather was the kind of astrolabe that a skilled instrument-maker could produce on commission. Since we do not really have regional histories of astronomy in any part of medieval Europe, let alone Germany, it is difficult to speculate further on the context of this one piece. On the other hand, Helmar HÄRTEL has suggested that LUDOLFUS may have acquired the astronomical skills necessary to use an astrolabe in Erfurt, only 120 km from Einbeck, where there was a

20 See www.sickte.de/ and www.sickte.de/sickte/sickte_geschichte.htm .

21 *Urkundenbuch der Stadt Braunschweig* 3 (1905), pp. 84–85, and 4 (1912), pp. 62–63, from two archives in the Niedersächsisches Hauptstaatsarchiv in Hanover: Cal. Or. 100 and Einbeck, St. Marien, Ur. 19. Cited in Härtel, "Ludolphus Borchdorp de Brunswik", p. 114, n. 3.

longstanding astronomical tradition.[22] It is indeed possible that there was an instrument-workshop there and that LUDOLFUS had the astrolabe made for himself there too. The maker would have assured him that markings for 50° would work nicely for both Erfurt and Einbeck.

How did the piece come to Cracow?

LUDOLFUS BORCHDORP DE BRUNSWIK[23] began his studies in Erfurt in 1439, gaining a bachelor's degree in 1441 and a master's in 1445, and using the title *Doctor artium*. With the help of a grant from the City of Brunswick he studied in Padua and obtained a doctorate in medicine. On his return to Brunswick he practiced medicine but also studied mathematics further. In 1465/66 he copied Euclid's *Elements*, and the *Perspectiva communis* and a treatise on music by JOHANNES DE MURIS (c. 1300–c. 1350) (preserved as MS Cracow JUL 568).[24] In 1467 he corresponded with the celebrated mathematician CHRISTIAN RODER DE HAMBURG at the University of Erfurt about problems of the arithmetic of roots and fractions.

His son also bore the name LUDOLFUS. In 1485 the son began his studies at the University of Cracow, but he interrupted these when he became involved with printing and never obtained a degree. Perhaps his father had given him his own manuscript of EUCLID, which would explain why that manuscript is now in Cracow (MS 568). And perhaps his father had given him the astrolabe made for LUDOLFUS DE SCICTE, which would explain how that astrolabe came to Cracow in the late 15th century.[25]

The quatrefoil decoration

The quatrefoil was used for the decoration of astrolabe retes in medieval Spain, Italy, France and England, and here we have one isolated example from medieval Germany.[26] Yet the distinctive and extravagant design of this astrolabe rete gives the impression of belonging to a German tradition of which we have no other examples. It would be interesting to know whether quatrefoils appear in medieval treatises on the construction or use of the astrolabe from the German-speaking world. The better-known "Sloane astrolabe"

---

22 HÄRTEL, article "Ludolphus de Tzicte sive de Scicte". I have been unable to find any references to Erfurt astronomers ca. 1330 but have not looked beyond ZINNER, *Astronomische Instrumente*. Indications of interest in astronomy in Erfurt during the 14th century are presented in KLEINEIDAM, "Geschichte der Wissenschaft im mittelalterlichen Erfurt", especially pp. 161–162. Here the activities of Themo JUDAEI in physics and cosmology, and Johannes DANKOWE in calendrics, are discussed, but but there is no mention of anything so mundane as an instrument workshop. In his later *Universitas Studii Erffordensis*, II, pp. 63–73, the same author deals with astronomy in Erfurt after the foundation of the University there in 1392.

23 This information on LUDOLFUS father and son is taken from Dr. Helmar HÄRTEL'S articles in the *Braunschweigisches Biographisches Lexikon*, which the author kindly made accessible to me before their publication. See also KINTZINGER, *Das Bildungswesen in Braunschweig*, pp. 466 and 528.

24 A detailed description is in Maria KOWALCZYK *et al.*, *Catalogus codicum ... Cracoviae*, pp. 21–26.

25 On astronomy in Cracow in the 15th century see MARKOWSKI, "Astronomie an der Krakauer Universität".

26 On the development of quatrefoil decoration on astrolabe retes see KING, *In Synchrony with the Heavens*, XVII, where this piece is illustrated as fig. 3.4

in the British Museum, a spectacular English piece datable ca. 1300, has quatrefoil decoration that is even more extravagant and more aesthetically pleasing than that on the Einbeck astrolabe: see Fig. 7.[27] Here also, though, we have not a single English astrolabe that bears any relationship to the Sloane astrolabe or which can be dated earlier than it. A German astrolabic plate dated 1468, of uncertain provenance but often falsely associated with the hand of REGIOMONTANUS, also has a half-quatrefoil on the throne.[28]

The choice of stars and their positions

The names used for the stars are basically in accordance with those that KUNITZSCH found in the manuscripts that he used for "Typ VIII", of which the majority seem to be of 14th- or 15th-century German provenance. Alas, Burkhard STAUTZ did not investigate the positions of the star-pointers on this particular astrolabe in his 1997 study of astrolabe retes; he might have compared them graphically either with the coordinates of "Typ VIII" and/or modern positions for the epoch of the instrument.[29] So I unfortunately cannot refer the reader to any published graphics. The positions on the rete should be compared with the coordinates for Typ VIII published by Paul KUNITZSCH. In the meantime, Elly DEKKER has developed new techniques of investigating star positions on retes, without, however, any regard for geographical provenance or other aspects of the instruments. Her investigations provide no graphical aids for the reader, who is instead confronted with tables of numbers, and they sometimes reach conclusions at variance with those provided by the rest of the instrument.[30] At the present time, I do not have access to any computer graphics programs that would allow me to investigate and then illustrate whether or not the positions on the Cracow astrolabe correspond to either reality or to KUNITZSCH'S coordinates for "Typ VIII". Perhaps some enterprising student will undertake such an investigation.

However, there is another lead that needs to be pursued. In May, 2006, Prof. Paul KUNITZSCH kindly sent me a preliminary edition of a catalogue of ecliptic star-positions dated 1314, preserved in MS Erfurt Ampl. 4° 369, fol. 203vb.[31] This contains ecliptic coordinates for our stars 3, 4, 5, 6, 7, 8 (?), 9, 12, 13, 14, 15, 16, 18, 19, and 21. A preliminary check with a transparent folio of a set of astrolabic markings for latitude ca. 66°30′,[32] which display ecliptic coordinates, reveals that some of the star-positions on the Einbeck astrolabe, but by no means all, correspond very nicely to the coordinates in the 1314 table.

---

27 GUNTHER, *Astrolabes of the World*, II, pp. 463–465 (no. 290) and pls. 126–127, also KING, *op. cit.*, XVII, fig. 3.7.

28 On this see GUNTHER, *Astrolabes*, II, p. 280 (front only), and a detailed description in KING, "Die Astrolabiensammlung in Nürnberg", II, pp. 589–592 (no. 1.77).

29 STAUTZ, *Untersuchungen*.

30 DEKKER, "Exploring the Retes of Astrolabes".

31 On this see already KUNITZSCH, *Typen von Sternverzeichnissen*, pp. 3 and 133.

32 I used a transparent folio based on the plate by AL-KHUJANDĪ on his astrolabe of 984/5, for which see KING, *In Synchrony with the Heavens*, fig. XVI-10.2, also XIIIb-9. The markings are for each 5° of longitude and each 3° of latitude.

The choice of latitude

The markings that are stated to be for latitude 50° accord with the fact that the instrument was made for a citizen of Einbeck only if we realize that modern latitudes are largely irrelevant to a discussion of a medieval astrolabe. The accurate latitude of that town is 51°48′. We should also keep in mind that 50° is close to the latitude of the end of the 7th climate.[33]

CONRAD VON DYFFENBACH (1426) gave the following latitudes for cities in Northern Germany:[34]

> Cologne 50;20°; end of the 7th climate 51;30° [50;30° would be better for this epoch]; Erfurt 50;50°; "Riddageshusen" [Riddagshausen, a Cistercian monastery south-east of Brunswick] 51;10°; Brunswick 51;10°; Münster ∅; Lubeck 54;30° [!]

(On his maps, or at least in the only (?) surviving copy of them, carefully fitted with an orthogonal grid for longitude and latitude, Brunswick is near Oxford, and Riddagshausen is halfway between London and Erfurt!)

A 15th-century German (?) geographical table, preserved in Vienna, gives:[35]

> Prague (at the end of the 7th climate) 50;30° [for obliquity 23;30°, accurately 50;32°]; Erfurt 51;0°

Yet another, also preserved in Vienna, gives:[36]

> Cologne 49;50°; Prague (end of 7th climate) 50;7° [for the Ptolemaic obliquity it is accurately 50;4°; PTOLEMY has 54;1°]; Erfurt 51;30°; Brunswick 52;10°; Lubeck 53;0°

Regiomontanus in his *Deutscher Kalendar*, printed ca. 1474, has the following latitudes for such locations:[37]

> Erfurt 51;04°; Leipzig 51;10°; Cologne 51;13°; Brunswick 53;0°; Lubeck 56;16 [!]

It is not inappropriate to look at the latitudes presented on 16th-century Nuremberg sundials because some of these might simply be lifted from earlier sources. There we find, for example, 52°, 54° and 55° for Lubeck, accurately 53°52′.[38]

We can conclude simply that the maker – wrongly – thought that 50° was appropriate for Einbeck.[39] Or he just used a rough value for the end of the 7th climate based on

---

33 On the geographical climates of Antiquity and their importance in the history of astronomical instrumentation see KING, *In Synchrony with the Heavens*, XVI-2–3. The reader should keep in mind that the latitudes of the climates, being defined in terms of the length of longest daylight, change over the centuries (+) because the obliquity of the ecliptic changes (–).

34 KAMAL, *Monumenta cartographica*, IV, pp. 233–236 (VI, pp. 1447–1449 of the reprint). Dyffenbach appears to be modern Tiefenbach near Alzey in the Pfalz (Palatinate).

35 KAMAL, *Monumenta cartographica*, IV, pp. 131 (VI, p. 1393 of the reprint).

36 *Ibid.*, IV, pp. 132–133 (VI, pp. 1394–1395 of the reprint).

37 ZINNER, *Regiomontanus*, pl. 30.

38 GOUK, *The Ivory Sundials of Nuremberg*, p. 102.

39 Helmar HÄRTEL (*op. cit.*, p. 114 and n. 6) maintained that the markings were for 51° rather than for the 50° engraved on the plate. On the photos available to me it looks as though the equatorial circle is tangential to the altitude circle for 40°, which would favour latitude 50° as stated. On the other hand, a comparison with the correct markings for latitudes 50° and 51° would tell us more, especially about the azimuth markings (see n. 16)

PTOLEMY'S long outdated value of the obliquity of the ecliptic.[40] We can also conclude that it would be extremely useful to have a computer-assisted overview of all available medieval European lists of longitudes and latitudes, such as exists for the more copious Islamic sources.[41]

The dating of the equinox

The date March 10.5 corresponds to an epoch of ca. 1525 using the astronomical calendar (noon to noon) and ca. 1590 in the civil reckoning (midnight to midnight). March 12.0–12.5 would correspond better to the epoch of the astrolabe. The equinox on REGIOMONTANUS' 1462 astrolabe is also at March 10.5, and eight out of the ten "Regiomontanus-type" astrolabes have March 10–11.[42] As Gerard TURNER has shown, such discrepancies simply reveal how little we know about the way in which such calendar scales were constructed in practice.[43]

The Marian rosette

Possibly the flower at the centre of the alidade is to be interpreted as a Marian rosette, such as are found on astrolabes from 15th-century Vienna and 16th-century Nuremberg.[44]

Concluding remarks

I hope that this little study of one astrolabe and three men called Ludolfus and several places from Einbeck to Erfurt and from Brunswick to Cracow will give pleasure to my distinguished colleague, Menso FOLKERTS. Ever since I came to Frankfurt in 1985 it was always easy and worthwhile to pick up the phone and talk to Menso about many things, personal, professional and purely academic. We could always commiserate about the rather bleak prospects for our discipline. But it was much more fun talking together over a beer, if not about astrolabes. It would be nice to do this in Einbeck and raise a glass to the memory of LUDOLFUS DE SCICTE and then another and a few more to Menso FOLKERTS for his retirement.

Acknowledgements

It is pleasure to thank Dr. Danuta BURCZYK-MARONA and Dr. Ewa WYKA of the Jagiellonian University Museum for the privilege of inspecting this and other instruments in

40 KING, *In Synchrony with the Heavens*, XVI, Table 1 on p. 927.

41 KENNEDY & KENNEDY, *Islamic Geographical Coordinates*.

42 KING & TURNER, "The Astrolabe Presented by Regiomontanus to Cardinal Bessarion in 1462", also KING, *Astrolabes and Angels, Epigrams and Enigmas*.

43 See Gerard L'E. TURNER, "A Critique of the Use of the First Point of Aries in Dating Astrolabes", on what one expects on astrolabes and what one finds.

44 KING, "Astronomical Instruments between East and West", pl. 12 and pp. 175–176.

1991 and 1997, and to Prof. Stanisław WALTOŚ and his staff for preparing the photos used in this study. The basic description of the astrolabe was taken from the Frankfurt catalogue of medieval instruments, first prepared with the help of a generous grant (Fo 60/19–1/2) from the German Research Organization (DFG) during 1997–2000. Dr. Helmar HÄRTEL of the Herzog-August-Bibliothek in Wolfenbüttel generously provided me with copies of the Brunswick archives and the as yet unpublished text of his three articles on the three Ludolfi. As usual, Prof. Paul KUNITZSCH provided me with more information about the star-names and their sources, especially the 1314 star-table. Dr. Richard LORCH in his editorial capacity kindly expunged a few solecisms.

Figures

Fig. 1: The front of the Einbeck astrolabe. [Figs. 1–6 courtesy of the Jagiellonian University Museum, Cracow.]

Fig. 2: The rete with its quatrefoil decoration.

Fig. 3: The mater with the “shadow” of the rete on the astrolabic markings for latitude 50°.

Fig. 4: The back with the calendrical scales, universal horary markings and shadow squares.

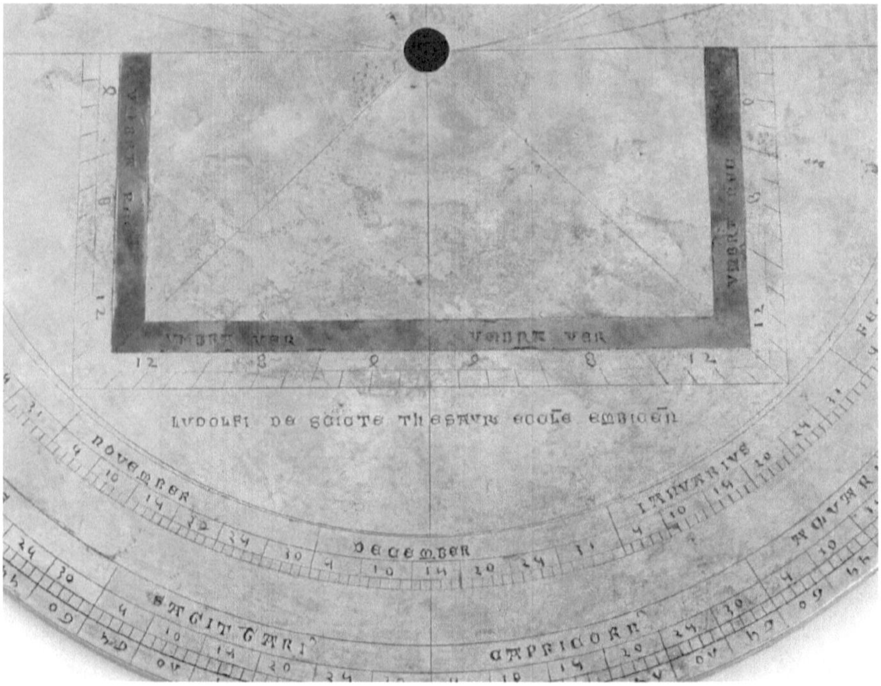

Fig. 5: The lower back with the inscription.

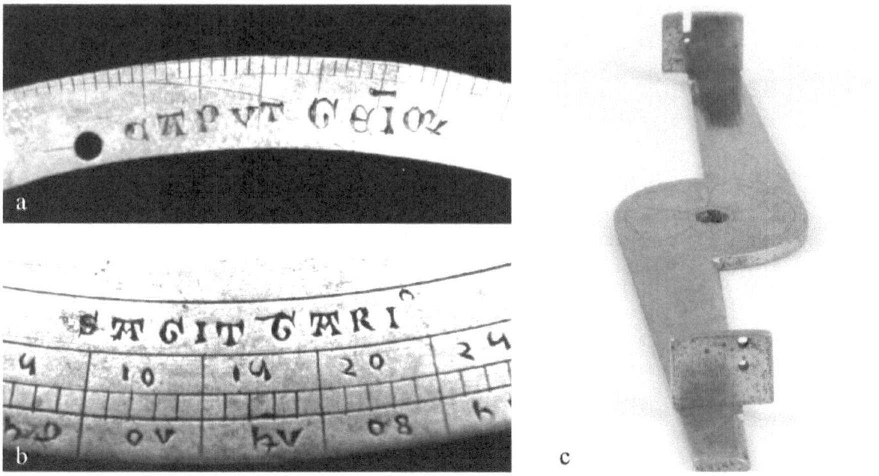

Figs. 6a–c: The marker for a single star on the back of the rete (a); the two different Ts in the name Sagittarius (b); and a side view of the alidade (c)

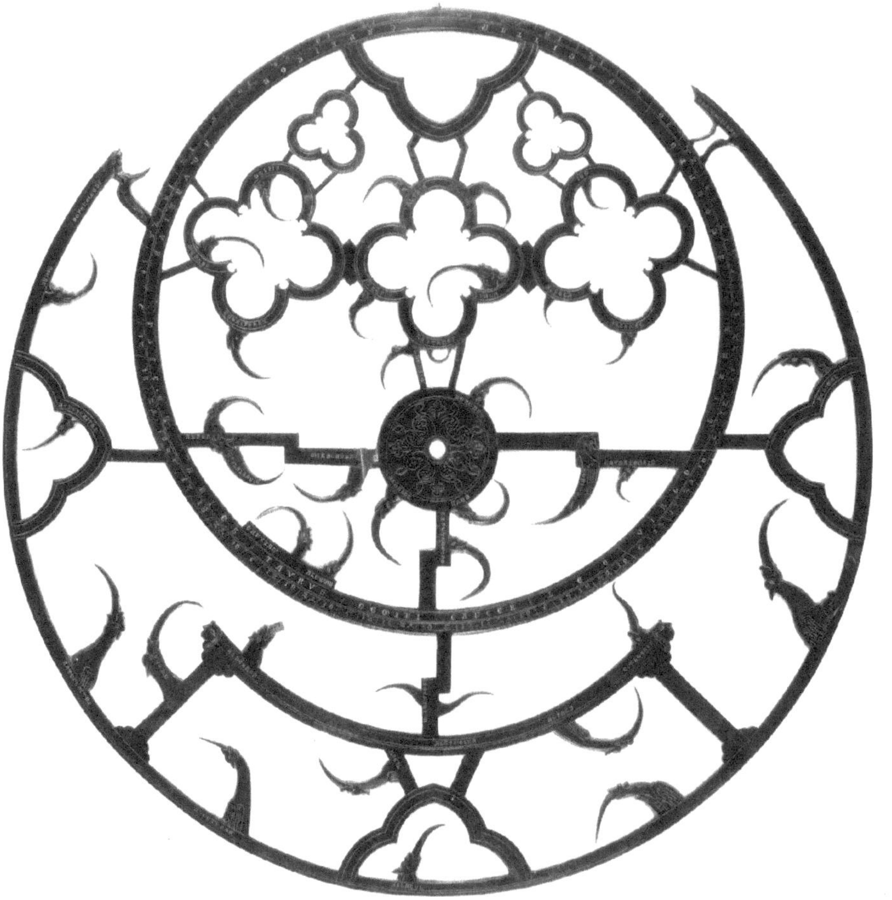

Fig. 7: The spectacular quatrefoil decoration on an English astrolabe that seems to predate the Einbeck astrolabe by a few decades. If the dating of the London astrolabe to *ca.* 1300 is correct, it would be at the same time the earliest known English astrolabe and the most sophisticated from before the Renaissance. [Courtesy of the British Museum, London.]

Bibliography

BIRKENMAJER, Ludwik Antoni: "Borchtorp, Ludolf", in *Polski Slownik Biograficzny* 2 (1936), pp. 315–316.

—, *Marcin Bylica z Olkusza [...]*, Cracow: Akademii Umiejetnosci, 1892. [See also the summary and review in German listed under KARLIŃSKI.]

Braunschweigisches Biographisches Lexikon 8. bis 18. Jahrhundert, ed. Horst-Rüdiger JARCK with Dieter LENT und Gudrun FIEDLER, Martin FIMPEL, Ulrich SCHWARZ, Silke WAGENER-FIMPEL, Brunswick: Appelhans, 2006.

Braunschweigische Urkunden: see *Urkundenbuch der Stadt Braunschweig.*

CAPPELLI, Adriano: *Lexicon abbreviaturarum – Dizionario di abbreviature latine ed italiane*, Milan: Ulrico Hoepli, 1990 (reprint of the 6th edn., 1961).

DEKKER, Elly: "Exploring the Retes of Astrolabes", in van Cleempoel *et al.*, *Astrolabes at Greenwich*, pp. 47–71.

FOLKERTS: see KUNITZSCH *Festschrift.*

GOUK, Penelope: *The Ivory Sundials of Nuremberg 1500–1700*, Cambridge: Whipple Museum of the History of Science, 1988.

GUNTHER, Robert T.: *The Astrolabes of the World [...]*, 2 vols., Oxford: The Clarendon Press, 1932, repr, in 1 vol., London: The Holland Press, 1976. [Long out of date, but still very useful.]

HÄRTEL, Helmar: "Ludolfus Borchdorp de Brunswik", *Braunschweigisches Jahrbuch* 68 (1987), pp. 113–120.

– : articles "Ludolphus de Tzicte sive de Scicte", "Ludolphus Borchdorp de Brunsvik sen.", "Ludolphus Borchdorp de Brunsvik jun.", to appear in *Braunschweigisches Biographisches Lexikon.*

HILL, George Francis: *The Development of Arabic Numerals in Europe Exhibited in Sixty-Four Tables*, Oxford: Clarendon Press, 1915.

KAMAL, Youssouf: *Monumenta cartographica Africae et Aegypti*, 16 fascs. in 5 vols., privately printed and distributed, 1926–1952, reprinted by Fuat SEZGIN, ed., in 6 vols., Frankfurt am Main: Institut für Geschichte der Arabisch-Islamischen Wissenschaften, 1987. [The pagination given in this study is to the original version, with that of the reprint given in parentheses.]

KARLIŃSKI, F.: Summary and review in German of Birkenmajer, *Marcin Bylika z Olkusza* (in Polish), in *Anzeiger der Akademie der Wissenschaften in Krakau*, 1892, (Krakau: Universitätsdruckerei, 1893), pp. 98–110. [The author is named on p. 69.]

KENNEDY, Edward S. and KENNEDY, Mary Helen: *Geographical Coordinates of Localities from Islamic Sources*, Frankfurt am Main: Institut für Geschichte der Arabisch-Islamischen Wissenschaften, 1987.

KING, David A.: (inspired by two remarkable discoveries by Berthold HOLZSCHUH), *Astrolabes and Angels, Epigrams and Enigmas: From Regiomontanus' acrostic for Cardinal Bessarion to Piero della Francesca's* Flagellation of the Christ, Stuttgart: Franz Steiner, 2007. *Preprint Series* of the Institute for the History of Science, Johann Wolfgang Goethe University, Frankfurt am Main, 4th series, 1 (2006). For a summary see http://web.uni-frankfurt.de/fb13/ign/Code.htm

– : "Die Astrolabiensammlung des Germanischen Nationalmuseums", in *Focus Behaim Globus*, Gerhard BOTT, ed., 2 vols., Nuremberg: Germanisches Nationalmuseum, 1992, I, pp. 101–114, and II, pp. 568–602.

– : "Astronomical Instruments between East and West", in Harry KÜHNEL, ed., *Kommunikation zwischen Orient und Okzident – Alltag und Sachkultur*, (*Sitzungsberichte der Phil.-Hist. Klasse der ÖAW*, Band 619), Vienna: Österreichische Akademie der Wissenschaften, 1994, pp. 143–198.

– : *In Synchrony with the Heavens – Studies in Astronomical Timekeeping and Instrumentation in Medieval Islamic Civilization*, 2 vols., I: *The Call of the Muezzin (Studies I–IX)*, and II: *Instruments of Mass Calculation (Studies X–XVIII)*, Leiden: Brill, 2004–2005.

KING, David A.; TURNER, Gerard L'E.: "The Astrolabe Presented by Regiomontanus to Cardinal Bessarion in 1462", *Nuncius – Annali di Storia della Scienza* (Florence) 9 (1994), pp. 165–206,

KINTZINGER, Martin: *Das Bildungswesen in der Stadt Braunschweig im hohen und späten Mittelalter*, Cologne & Vienna: Böhlau, 1990.

KLEINEIDAM, Erich: "Geschichte der Wissenschaft im mittelalterlichen Erfurt", in *Geschichte Thüringens*, Hans PATZE and Walter SCHLESINGER, eds., vol. 2, Cologne & Graz, 1969, pp. 150–162.

– : *Universitas Studii Erffordensis – Überblick über die Geschichte der Universität Erfurt*, II: *Spätscholastik, Humanismus und Reformation 1461–1521*, 2 vols., 2nd edn., Leipzig: St. Benno, 1992.

KOWALCZYK, Maria, *et al.*: *Catalogus codicum manuscriptorum medii aevi Latinorum qui in Bibliotheca Jagellonica Cracoviae asservantur*, vol. 4, Wratislaviae, *etc.*: Academiae Scientiarum Poloniae, 1988.

KUNITZSCH, Paul: *Arabische Sternnamen in Europa*, Wiesbaden: Otto Harrassowitz, 1959.

– : "The Stars on the Astrolabe", in VAN CLEEMPOEL *et al.*, *Astrolabes at Greenwich*, pp. 41–46.

– : *Typen von Sternverzeichnissen in astronomischen Handschriften des zehnten bis vierzehnten Jahrhunderts*, Wiesbaden: Otto Harrassowitz, 1966.

Kunitzsch Festschrift: *Sic itur ad astra – Studien zur Geschichte der Mathematik und Naturwissenschaften – Festschrift für den Arabisten Paul Kunitzsch zum 70. Geburtstag*, Menso FOLKERTS and Richard P. LORCH, eds., Wiesbaden: Otto Harrassowitz, 2000.

LORCH: see *Kunitzsch Festschrift*.

MARKOWSKI, Mieczysław: "Astronomie als Leitwissenschaft an der Krakauer Universität in der vorkopernikanischen Zeit", in *Astronomy as a Model for the Sciences in Early Modern Times*, Menso FOLKERTS and Andreas KÜHNE, eds., (*Algorismus – Studien zur Geschichte der Mathematik und der Naturwissenschaften*, Menso Folkerts, ed., Heft 59), Augsburg: Dr. Erwin Rauner, 2006, pp. 103–113.

DE SOLLA PRICE, Derek J. with Sharon L. GIBBS and Janice HENDERSON: *A Computerized Checklist of Astrolabes*, New Haven, Ct.: Yale University, Department of History of Science and Medicine, 1973. [Privately distributed. Still invaluable.]

STAUTZ, Burkhard: *Untersuchungen von mathematisch-astronomischen Darstellungen auf mittelalterlichen Instrumenten islamischer und europäischer Herkunft*, Bassum (D): Verlag für Geschichte der Naturwissenschaften und der Technik, 1997.

TURNER, Gerard L'E.: "A Critique of the Use of the First Point of Aries in Dating Astrolabes", in *Kunitzsch Festschrift*, pp. 548–554.

Urkundenbuch der Stadt Braunschweig [...], vols. III (1321–1340), Ludwig HAENSELMANN and Heinrich MACK, eds., Berlin: C. A. Schwetschke und Sohn, 1905, and IV (1341–1350 and supp. for 1067–1340), Heinrich MACK, ed., Brunswick: E. Appelhans, 1912.

VAN CLEEMPOEL et al., Koenraad: *Astrolabes at Greenwich – A Catalogue of the Astrolabes in the National Maritime Museum, Greenwich*, Oxford: Oxford University Press, & Greenwich: National Maritime Museum, 2005.

ZINNER, Ernst: *Deutsche und niederländische astronomische Instrumente des 11.–18. Jahrhunderts*, 2nd edn., Munich: C. H. Beck, 1967, repr. 1972.

– : *Leben und Wirken des Joh. Müller von Königsberg genannt Regiomontanus*, 2nd edn., Osnabrück: Otto Zeller, 1968, translated by Ezra BROWN as *Regiomontanus: His Life and Work*, Amsterdam: North-Holland, 1990. [References are given to the German version.]

VII

The Star-Names on Three 14th-Century Astrolabes from Spain, France and Italy

To Paul Kunitzsch, in gratitude
for many years' friendship and collaboration

Introduction

The surviving astronomical instruments of the Middle Ages constitute a corpus of historical sources whose potential is not yet generally appreciated.[1] The study of these instruments – notably astrolabes, the most common variety – progresses with each new piece catalogued,[2] and our understanding of their context progresses with each new instrument text edited and translated.[3] Various aspects of these instruments are now under much better control than they were just a few years ago: the design of retes and thrones, and other decorative features;[4] the positions of star-pointers;[5] vernacular and dialectal influence in inscriptions;[6] and the geographical data explicit or implicit in the latitude-dependent markings.[7] One challenge yet to be confronted is the *pot pourri* of star-names found on medieval European astrolabes.[8]

---

1 See King, "Astronomical Instruments between East and West", for an overview of Islamic and European instruments. On the astrolabe in particular see Hartner, "Astrolabe", A–B, and North, "Astrolabe" (both in English); Poulle, "L'astrolabe" (French); Trento, *L'astrolabio* (Italian); Chabàs & Bosch, *L'astrolabi* (Catalan); and *Nuremberg GNM 1992–93 Exhibition Catalogue*, I, pp. 101–114 (German).

2 See King, "Instrument Catalogue", A–B, on a long-term project supported by the Deutsche Forschungsgemeinschaft during 1992–96 and 1999–2002. Following the tradition of Price *et al.*, *Checklist*, each instrument is assigned a number, here preceded by the symbol #.

3 See the most recent study of this kind, listed as Lorch, "Rudolf of Bruges on the Astrolabe", with references to numerous earlier studies.

4 On rete-design see the various studies listed in nn. 11, 12, 53, 77 and 80 below, and also Tomba, "Astrolabi", A–D. On throne-design see, for example, King, "Astronomical Instruments between East and West", pp. 175–176 and plate XII.

5 See Stautz, *Mathematisch-astronomische Darstellungen*, for a preliminary survey of the star-positions on numerous medieval Islamic and European astrolabes. For a given instrument the actual names should be checked against the equivalents proposed by Stautz.

6 See the various studies of Kurt Maier listed in the bibliography.

7 See King, "Geography of Early Astrolabes".

8 The star-names on Islamic astrolabes have yet to be surveyed, but first impressions from the catalogue now in preparation (see n. 2 above) are that they are much more orderly and correct than those on European astrolabes. Indeed, if one finds a star-pointer incorrectly labelled it is more than likely that the instrument is a modern fake.

Anyone who looks at the star-names on a medieval European astrolabe will immediately be confronted by a series of problems: (1) they are usually in Gothic script and inevitably difficult to read; (2) one can only read them properly and begin to understand them if one knows Arabic; (3) having read them, they have to placed in the context of the generally sad fate of serious Arabic star-names at the hands of our medieval forebears, as documented in Paul Kunitzsch's numerous works on star-names in the manuscript tradition;[9] and (4) many of the star-names will not be found in those works anyway, since medieval instrument-makers were a law to themselves and did not prepare themselves for their task by consulting manuscripts.[10] The best thing to do is to make an attempt at interpreting them and send the preliminary version to Paul Kunitzsch for his corrections and insights. I have done this many times. The purpose of this contribution to Paul Kunitzsch is to record the star-names on three 14th-century astrolabes, each of singular historical importance for a variety of widely-different reasons, and to offer preliminary interpretations of some of these names. I have deliberately chosen three instruments from different parts of medieval Europe, if only to draw attention to the fact that for medieval Europe we are very far from being able to produce any regional studies of the history of astronomy, let alone of astronomical instrumentation, such as has been possible for the history of astronomy in the medieval Islamic world.

1. A 14th-century astrolabe from Christian Spain

A medieval Spanish astrolabe (#4560) which came to light in 1998 – see Fig. 1 – is remarkable for a variety of reasons.[11]

- First, it bears a set of original Latin inscriptions completed by a set of Arabic ones (mainly on the back and on the boss of the shackle), and it has an additional layer of markings scratched in Hebrew characters (identifying the latitudes served by the plates). The last-mentioned appear to predate the others, having been added during the construction and hence prior to the main engraving. In any case, this piece is not the work of a single craftsman, but of at least two and possibly three or even more.
- Second, it is the sole surviving medieval astrolabe from Christian Spain excluding Catalonia (whence some four medieval pieces are known[12]). Toledo seems the most likely locality for its provenance, not least because it was there more than any-

9 See especially Kunitzsch, *Typen von Sternverzeichnissen*, and *idem*, *Arabische Sternnamen in Europa*.

10 It is thus not unusual that the *selection* of stars represented on medieval astrolabes does not correspond to any of the various textual traditions that have been discussed by Paul Kunitzsch. It is likewise not unusual that the *names* actually used on astrolabe retes do not correspond to the names known from textual sources. However, it is usually possible to detect the stages in which a given Arabic name has been corrupted to produce an apparently-unrelated "Latin" name.

11 The astrolabe now belongs to an unidentified private collector, having been acquired by auction on April 15, 1999, at Christie's of London. A detailed description by this author is in King, "14th-Century Spanish Astrolabe"; the shorter description in *Christie's London 15.4.1999 Catalogue*, pp. 98–107, lot 52, is based on that study.

12 These are:

where else where Christians, Muslims and Jews enjoyed a *convivencia* which was occasionally fruitful.

- Third, the Latin engraving, which is otherwise standard for 13th- and 14th-century Spain,[13] exhibits a most remarkable feature which, as far as I know, is not attested in any other medieval inscription or text: the symbol resembling a '9', which, like the ampersand '&', is part of the extended medieval alphabet,[14] is used as a letter of the alphabet, namely, a hard 'C'. Thus, for example, the name of the star SPICA is engraved as SPI9A.[15] This was clearly a local usage for a limited time, and if other examples come to light it may be possible to identify the provenance of the astrolabe more precisely.
- Fourth, the forms of the Hindu-Arabic numerals used by the same, presumably Christian, engraver are the standard forms for the 14th century with one exception: the '2' is engraved – from our perspective – upside down. This is to be regarded as old-fashioned, since the earliest attested form of the the '2' in Europe was also upside down.[16] In addition, in the bar fractions used for the latitudes 32 1/2° and

---

- #3042, from 10th-century Catalonia – see Destombes, "Astrolabe carolingien"; King, "Earliest European Astrolabe"; Kunitzsch & Dekker, "Stars on the Carolingian Astrolabe"; and other papers in Stevens *et al.*, *Oldest Latin Astrolabe*.
- #162, from Catalonia *ca.* 1300 – see Gunther, *Astrolabes*, II, pp. 306–309 (no. 162), and the detailed study King & Maier, "Catalan Astrolabe".
- #416, also from Catalonia *ca.* 1300 – see King & Maier, "Catalan Astrolabe", pp. 694–695, n. 60, King, "Earliest European Astrolabe", fig. 14, and *idem*, "14th-Century Spanish Astrolabe", fig. 10.
- #3053, made by Petrus Raimundus in Barcelona in 1375 – see King, "Earliest European Astrolabe", fig. 16, and *idem*, "14th-Century Spanish Astrolabe", fig. 11.

Descriptions of each of these are in King, "Astrolabes from Medieval Catalonia" (forthcoming).

13 The resemblance to the engraving on #162, datable *ca.* 1300, and #3053, made in Barcelona in 1375, is striking. On the former see the letters and numbers reproduced in King & Maier, "Catalan Astrolabe", p. 716, fig. 4.

14 The expanded medieval "Gothic" alphabet consisted of the letters 'a' to 'z' (with or without certain letters such as 'h', 'j', 'k', 'q', 'v', 'w' and 'y', depending on the language involved), and in addition the '&', the "ampersand" that we still use today, and a letter '9' which has no name, but which, as one colleague has remarked, was "so graceful and chaste in early and middle gothic". This '9' is attested already in Roman Antiquity: it is mentioned by the grammarian Valerius Probus in the 1st century. Originally conceived as one of the symbols of the Tironian Notes, the Roman speed-writing notation devised by Tiro, amanuensis of Cicero (*ca.* 63 B.C.), in the Middle Ages it served firstly as an abbreviation for the prefix *con-* or *cum-* and, secondly, albeit upper-case and in smaller versions which sometimes resemble a comma, the ending *-us*. It is well known to all those familiar with Latin and medieval palaeography and its almost perverse systems of abbreviations: see, for example, Cappelli, *Lexicon abbreviaturarum*, pp. XXV–XXVI, where the two different forms are distinguished, and Bischoff, *Palaeography*, pp. 151–168.

15 The '9' for hard 'C' is used in the star-names and also in the word RE9TA on the back, but not on the ecliptic ring, whose engraving in in a different hand. The normal form of 'C' is found in the star-name RIGORCA (no. 8), which seems to mean that it was pronounced RIGORÇA or RIGORSA.

16 The two tables of forms in Ifrah, *Histoire des chiffres*, II, pp. 348 and 362, display respectively 23 examples of inverted '2's from between 976 and the beginning of the 13th century, and 22 examples of upright '2's from the period between the 12th century and the early 16th.
The history of the Hindu-Arabic numerals in Europe has yet to be written. Studies such as Lemay, "Arabic Numerals", and Beaujouan, "Rotation des chiffres", like all other modern studies, suffer from the fact that the authors are familiar only with the Latin tradition. Before looking at the forms attested in early medieval Europe it is instructive to note the divergent forms existing

49 1/2° on two of the plates, serving [Jerusalem] and [Reims], respectively, the numerator is on the bottom and the denominator on the top.[17] Such a usage is not known to be attested elsewhere in the medieval sources, but is found on a single astrolabe from 16th-century Louvain.[18] There, as on the Spanish astrolabe, this is probably to be interpreted as a distinctive usage of the engraver rather than a local phenonomenon.

- Fifth, the Muslim who completed the engraving in *naskhī* script and added his name, Masʿūd, in a inscription identifying himself as the owner, was a *mudéjar*. He clearly had plans for leaving Christian-dominated Spain for the *Dār al-Islām*, more specifically, Algiers, for he reworked one of the original plates and provided it with markings for Mecca on one side and for Algiers on the other. Perhaps it was love for a woman which drove a sensible man to Algiers, an intellectual backwater in the 14th century: I have identified an ʿAbd al-ʿAzīz ibn Masʿūd, an Andalusī from Tlemcen, who appears to be his son. This Ibn Masʿūd was active in Fez, Tunis, Cairo and Damascus *ca.* 1370–1390 and he not only wrote on astronomical instruments but also constructed them.[19] But it is time to move on to the star-names.

The rete is of elegant and distinctive design.[20] There are three half-quatrefoils attached to the inside of the ecliptic ring, each supporting a single star-pointer. This is an indication that the design is original, the quatrefoils serving their original function. The larger star-pointers have three decorative holes in the base and the smaller ones a single hole. Another pointer at the left-hand end of the lower equatorial frame is a dummy,[21] serving not only the exigencies of symmetry but also to strengthen the rete.

There are some 21 star-pointers, and the stars seem to have been taken – appropriately enough – from the catalogue associated with the celebrated translator Johannes Hispalensis (Toledo, mid 13th century),[22] that is, Paul Kunitzsch's Typ IV.[23] At least, all of the stars but one in Pegasus are found in his catalogue.[24] In this the

---

in the Islamic world before some of them were transmitted to Europe. For new light on this topic see Kunitzsch, "Hindu-Arabic Numerals". Suffice it to say that first, there existed various forms of the nine symbols and zero-symbol in use in each of the Islamic and Byzantine worlds, as well as in each of Spain and Italy, and second, the upside-down variety of the '2' was already old-fashioned by the 14th century. It should be borne in mind that the numerals were engraved on this instrument less than two centuries after they had been introduced in Spain, and that in the vast majority of Spanish manuscripts (not necessarily scientific ones) from the 13th to 15th century Roman numerals rather that Hindu-Arabic numerals are used for dating.

17 There is no trace of such usage in Cajori, *Mathematical Notations*, I, pp. 309–314, or in Benoit *et al.*, *Histoire des fractions*.

18 See Nuremberg GNM 1992–93 Exhibition Catalogue, p. 599 and fig. 1.82.7.

19 See *Cairo ENL Survey*, p. 66 (no. C46); and the two publications of Maravillas Aguiar Aguilar listed in the bibliography, which deal with his treatise on the trigonometric quadrant. An astrolabic quadrant with a trigonometric quadrant made by him in 1372/73 is preserved in the National Museum, Damascus.

20 R. T. Gunther would have labelled it "Hispano-Mauresque" – see his *Astrolabes*, II, p. 306.

21 Namely, between nos. 1 and 2, with a support symmetrically placed to the pointer for COR LEONIS (no. 7), as on various Western Islamic astrolabes.

22 See Sarton, *IHS*, II:1, pp. 169–172.

23 Kunitzsch, *Typen von Sternverzeichnissen*, pp. 31–33.

24 I refer to ϵ Peg, and there is also seems to be a problem with α and β Pegasi.

Fig. 1: The front of the 14th-century Spanish astrolabe #4560. (Photo by Christie's of London, courtesy of the owner.)

Latin names of 29 stars are presented, and Kunitzsch has suggested that the star-names may have been translated from those on an Andalusī astrolabe.[25] The position of Regulus, which being on the ecliptic serves as an indicator of precession and hence of the date, is at Leo 22°, corresponding to *ca.* 1425, but this is insufficient to date the piece. The remaining star-positions point to a date in the range 1300–1350.[26]

The names of the stars on the rete are mainly Europeanised versions of the Arabic names, as was standard in the Middle Ages. There are also several examples of Latin names, which we can associate with the introduction of the Latin translation of the *Almagest* of Ptolemy in al-Andalus in the 12th century.[27] But there are also three examples of bastardised Latin names, as if dictated by someone who could not read them properly; this is most unusual, but not entirely without precedent. Table 1 presents the original "Latin" names of the stars, the Arabic additions (in two layers), the corresponding numbers in Kunitzsch's Typ IV, and the Arabic or Latin forms from which the names are derived, as well as the modern designation of the star.

Some comments on the more unusual names are in order. Firstly, some forms derived from the Arabic:

- PANTA (no. 1) from *baṭn*, "belly (of the whale)". This form is found already in a 10th/11th-century Latin source, as well as in a 13th-century copy of the astrolabe treatise of Pseudo-Messahalla, parts of which are translated from a treatise by the Spanish Arab al-Majrīṭī.[28]
- 9OMIZA (no. 6) from *(al-)ghumayṣā'*, which means something like "the one with eyes impaired by weeping" (compared with Sirius). This form is not attested in the manuscript tradition, which tends to prefer (AL)GOMEI/YZ/SA.[29] #162 has ALGOMIC.
- ON9E (no. 13), a very vernacular derivation from *ʿunq* or *ʿunuq (al-ḥayya)*, "neck (of the serpent)", is not known to be attested elsewhere. The early modern (17th-century) forms are UNUK or UNK, usually followed by ALHAY or variants thereof,

25 *Ibid.*, p. 31.
Western Islamic astrolabes tend to have some 27 star-pointers, following a clearly-defined tradition associated with the astronomer Maslama al-Majrīṭī (Madrid, *ca.* 1000) – see Kunitzsch, *Typen von Sternverzeichnissen*, pp. 17–18 (Typ III); and *idem* & Dekker, "Stars on the Carolingian Astrolabe", pp. 656–658. The earliest known European astrolabe, #3042 from 10th-century Catalonia, has 20 unlabelled pointers – see Kunitzsch & Dekker, *op. cit.*, *passim* – and the astrolabe under discussion has 19 of these but is missing α Ophiuchi (often named *alhawi* on medieval European astrolabes) and has in addition RIGORCA and ON9E. The Catalan astrolabe #162 from *ca.* 1300 has 20 stars, of which three are not found on this astrolabe – see King & Maier, "Catalan Astrolabe", pp. 676–677 and 684–685. A second star-table of al-Majrīṭī contains precisely 21 stars but only 13 of these are found on this piece – see Kunitzsch, *Typen von Sternverzeichnissen*, pp. 15–18. The star-tables associated with al-Zarqāllu – see *idem*, "Two Star Tables from Muslim Spain" – do not have the enigmatic star *kaʿb al-faras* in Pegasus.

26 Information kindly provided by Burkhard Stautz.

27 On Ptolemy see the article in *DSB*. For editions of the Arabic and Latin versions of his star catalogue see Kunitzsch, *Sternkatalog des Almagest*. See also Kunitzsch, *Typen von Sternverzeichnissen*, pp. 15–18 (Typ I), for a 12th-century list in which selected stars are already labelled in Latin.

28 See Kunitzsch, *Arabische Sternnamen*, p. 67, no. 8.

29 *Ibid.*, pp. 73–74, no. 25.

Table 1

No.	Star-name	Arabic additions (bold if original)	Kunitzsch Typ VI	Arabic or Latin name (A/L) (as relevant)	Designation
1	PANTA 9AITOS	baṭn · (illegible letter)	1	A *baṭn qayṭus*	ζ Cet
2	ALGOL	ghūl	2	A *al-ghūl*	β Per
3	D(A)B(A)RAN	**dabrān**	3	A *al-dabarān*	α Tau
4	RIGEL	rijl al-jawzā	6	A *rijl (al-jawzā*ʾ)	β Ori
5	ALABOR	ʿabūr	8	A *al-ʿabūr*	α CMa
6	9OMIZA	**ghumayṣa**	9	A *al-ghumayṣā*ʾ	α CMi
7	9OR LEO(NIS)	asad	12	L *cor leonis*	α Leo
8	RIG<u>OR</u>CA	rijl	10	A *rijl* + L *ursae* (!!)	μ UMa
9	9ORUUS	ghurāb	14	L *corvus*	γ Crv
10	SPI9A	aʿzal	15	L *spica*	α Vir
11	ALRAME(CH)	illegible	17	A (*al-simāk*) *al-rāmiḥ*	α Boo
12	FE9A	fakka	18	A (*al-*)*fakka*	α CrB
13	ON9E	**al-ḥayya**	19	A *ʿunuq a*(*l-ḥayya*)	α Ser
14	9OR S9ORPI(ONIS)	qalb	20	L *cor scorpionis*	α Sco
				A *qalb al-ʿaqrab*	
15	UEGA	wāqiʿ	22	A (*al-nasr*) *al-wāqiʿ*	α Lyr
16	RADF	rid[f]	25	A (*al-*)*ridf*	α Cyg
17	ALTAIR	**al-ṭāyir**	24	A (*al-nasr*) *al-ṭāʾir*	α Aql
18	9AUD:9OARI (error for CAUD(A) CAPRI)	ṭ-r-w (??)	26	L *cauda capricorni*	δ Cap
19	9ABI	–	–	A *kaʿb a*(*l-faras*)	κ Peg
20	OMER9I	mankib faras	27 (β Peg)	L (*h*)*umerus equi*	α Peg
21	D(E)N(E)P 9AITOS:	d(hanab)	29	A *dhanab qayṭus*	ι Cet

which are closer to the Arabic.[30] The phenomenon of short u → o is attested in medieval Spanish Latin[31] and can also be observed in the star-names OMER9I (no. 20) and (RIG)ORCA (no. 8).[32] The final E of ON9E is probably a remnant of the article *al-*.

- UEGA (no. 15) from *(al-)wāqiʿ*, "falling (eagle)". This corresponds to WEGA in some textual sources.[33]
- RADF (no. 16), from *(al-)ridf*, "the person riding on a horse behind the main rider", is not known either from the textual tradition or from any known medieval astrolabe. The more common names are RIDF and REDF, corresponding more closely to the Arabic,[34] and #162 has ALREDAF. With RADF it may be that we are witness to a change i → a that is attested in medieval Spanish Latin.[35]
- 9ABI (no. 19), pronounced KABI, derived from the Arabic *kaʿb al-faras*, "the ankle of the horse (Pegasus)", κ Pegasi. This star is not otherwise known from European astrolabes or star-lists. Indeed it is not a standard astrolabe-star, although it does appear on some of the astrolabes of the prolific Muḥammad ibn Fattūḥ al-Khamāʾirī of Seville *ca.* 1225.[36]
- D(E)N(E)P / D(A)N(A)P (no. 21) from *dhanab*, "tail (of the whale)". The more usual form is DENEB.[37]

Then two forms based on Latin:

- 9AUD:9OARI (no. 18) seems to be an error for CAUD(A) CAPRI(CORNI), with COARI for ACOARI (from Aquarii).[38] The colon is used as a separator (also at the end of no. 21); it was not uncommon for abbreviations such as HU: EQUI (for HUMERUS EQUI – see below) to be used for star-names engraved in restricted space on astrolabe retes. With this erroneous star-name CAUD:COARI it seems that we are dealing with a more developed vernacular form (A)COARI than the ACUARI on the zodiacal scale (with engraving in a different hand), which may be nothing

30 *Ibid.*, p. 217, no. 196. For some reason this star is not listed by Kunitzsch as an astrolabe-star.

31 On the switch u → o see Castro, *Glosarios latino-españoles*, p. xxxiv.

32 Likewise, the form OMBRA on the back of this astrolabe, for Latin UMBRA, reveals Spanish or French vernacular influence. It should be borne in mind that OMBRA is still Latin, albeit with vernacular influence, rather than a true Romance form (such as *sombra* or *ombre*). On the fate of UMBRA in Spanish see Corominas, *Diccionario*, V, pp. 298–300, especially p. 298b. See also the Renaissance Italian astrolabe with OMBRA in Gunther, *Astrolabes*, II, p. 331.

33 *Ibid.*, p. 81, no. 53.

34 *Ibid.*, p. 82, no. 56.

35 See Castro, *Glosarios latino-españoles*, p. xxxvi, especially the form *aladada* from Arabic *al-ʿiḍāda*, meaning "alidade". This form is not listed in Kunitzsch, "Fachausdrücke", pp. 527–528 (pp. 73–74 of the separatum).

36 On al-Khamāʾirī see Mayer, *Islamic Astrolabists*, pp. 64–66. On his astrolabe #130, for example, see Gunther, *Astrolabes*, I, pp. 276–277 (no. 130), and Kunitzsch, *Typen von Sternverzeichnissen*, p. 31, n. 2.

37 *Ibid.*, p. 66, no. 4.

38 Aquarius does not, of course, have a tail (*cauda*), though see Burnett, *Studies*, XVII, p. 120, where he is given a *cadacauda* in a 12th-century source.

other than an orthographical variant (C for Q) rather than a dialectal one.[39] The switch qu $\rightarrow$ c in Spanish Latin is attested.[40]

- OMER9I (no. 20), pronounced OMERQ(U)I / OMERKI, seems to derive from OMER (E)QUI, a vernacular form of the Latin *humerus equi*, "the shoulder of the horse", α Pegasi.

Finally, one very remarkable name which seems to be based on a mixture of Arabic and Latin:

- RIGORCA (no. 8), pronounced RIGORÇA or RIGORSA, appears to be from RIG(EL) ORSA(E), a combination of Arabic and vernacular Latin ("correctly", rijl ursae), μ Ursae maioris.[41] The ORCA, pronounced ORSA, displays Spanish or French influence, or even Italian: compare Spanish *osa*, French *ourse*, and Italian *orsa*. See also ON9E (no. 13) and OMER9I (no. 20). The proper Arabic form would be *rijl al-dubb*, "the (back) leg of the Bear". The precise Latin equivalent is not specifically stated in the Latin translation of Ptolemy's star-catalogue,[42] and medieval European astrolabists seem to have had some problems with the star.[43]

The Arabic names for four stars have been added by Masʿūd secondarily to the European names:

- *dabrān* (no. 3) with a *sukūn* or zero-vowel on the *b* (= *bāʾ*);[44] the standard name is (*al-*)*dabarān*.[45] This form *dabrān* is Spanish Arabic, for there the second vowel of two successive internal short syllables can be suppressed.[46] The usual medieval European forms are Aldebaran/m, although the astrolabe under discussion has DBRAN, perhaps for D(A/E)BRAN, perhaps for D(A/E)B(A)RAN, and ALDEBRAN is attested in Geoffrey Chaucer's treatise on the astrolabe and on one 14th-century English astrolabe (#457).[47]

---

39 An early-13th-century Andalusī astrolabe from Seville with later Northern Spanish inscriptions (#1148) has AQARI, which the maker of #4560 would probably have engraved as A9ARI. Both the astrolabe of Petrus Raimundus of Aragon (#3053), made in Barcelona in 1375, and the medieval Catalan additions to an 11th-century astrolabe from Cordova (#3622) have ACARI. On the latter see Maier, "Astrolab aus Córdoba", pp. 125–126.

40 Castro, *Glosarios latino-españoles*, p. xliv.

41 Kunitzsch, *Arabische Sternnamen in Europa*, pp. 74–75, no. 28.

42 See *idem*, *Sternkatalog des Almagest*, II, p. 38, no. 21.

43 It is not amongst Paul Kunitzsch's list of astrolabe stars (*Arabische Sternnamen in Europa*, pp. 59–96), but is listed as *pes ursi* (*sic*) in one medieval star-table (*idem*, *Typen von Sternverzeichnissen*, p. 32, no. 13). See also *Chicago AP Catalogue*, I, p. 159, where it is not to be found in the list of all stars featured on the astrolabes in the collection.

44 Masʿūd suffered from *sukūn*itis, as is evident from the *sukūn*s, mainly redundant, all over the place in his inscriptions; indeed, he even used *sukūn*s and vowel-signs as decoration in the cartouches on the back of the astrolabe. This phenomenon is attested elsewhere in Andalusī inscriptions and decorative calligraphy, as on, for example, textiles.

45 Kunitzsch, *Sternnomenklatur der Araber*, p. 51, no. 69.

46 See the EI_2 article "al-Andalus, x: Spanish Arabic" by G. S. Colin, especially p. 502b. I did not find any examples of this phenomenon in Corriente, *Spanish Arabic*, but words of the form CaCaCān are not common.

47 Kunitzsch, *Arabische Sternnamen in Europa*, p. 47 and p. 70, no. 18.

- *ghumayṣa* (no. 6), correctly *ghumayṣā'*, from which the Latin form is derived, is the Western Islamic name, the star being called *al-shiʿrā al-sha'āmiya* in the Islamic East.[48] The form *ghumayṣa* is probably to be regarded as a Spanish Arabic simplification of the classical form.
- *al-ḥayya* (no. 13), with a *sukūn* on the *l* (= *lām*), which is correct but superfluous, and a *shadda* on the *y* (= *yā'*), which is appropriate, is abbreviated from *ʿunuq al-ḥayya.*[49]
- *al-ṭāyir* (no. 17), with a *y* (= *yā'*) rather than a carrier for a *hamza*, an acceptable Middle Arabic form of *al-ṭā'ir*.[50]

Note that the stars have been selected because each one is in a different quadrant of the heavens, so that at night at least one is, in theory, always visible. In the case of the second layer of Arabic star names, scratched near their "Latin" equivalents, the hand is totally different from that of Masʿūd. It seems inconceivable anyway that he would have made these messy additions to his own astrolabe. Thus it seems reasonable to postulate a second Arabic hand. These additional names are given in Table 1.[51]

2. An astrolabe from 14th-century Northern France (Picardy)

A Northern French astrolabe (#202) which has been known since the 1930s[52] but only during the past few years become available for study – see Fig. 2 – is remarkable for three main reasons.[53]

- First, the month-names on the back are in the medieval dialect of Picardy,[54] so that its provenance is established. This is confirmed by the additional markings on the plates for latitudes 51° and 52°. A dating to the 14th century seems secure, before or after the Black Death in the middle of that century is uncertain.
- Second, all numbers engraved on the instrument are in monastic ciphers such as were used by some Cistercian communities during the 13th–15th centuries.[55] These ciphers, with which any number between 1 and 9999 is represented by a single geometric cipher, are particularly suitable for incising on metal (or stone of wood) since they are composed solely of straight-line segments.[56]

48 *idem*, *Sternnomenklatur der Araber*, p. 112, no. 290a.

49 On this astrolabe star see n. 30 above.

50 Kunitzsch, *Sternnomenklatur der Araber*, p. 86, no. 194a.

51 Several are difficult to see and not a few difficult to read. No equivalent name has been given for 9ABI (no. 19), doubtless because this was not a standard astrolabe-star and the Latin name seemed so strange; one name (no. 11), scratched around the hole at the base of the star-pointer, is illegible; and another (no. 18: *ṭ-r-w*) makes no sense.

52 The instrument was briefly mentioned in Gunther, *Astrolabes*, II, p. 349 (no. 202). See, more recently, *Christie's London 26.9.1991 Catalogue*, pp. 39–42 (lot 42).

53 It is now studied in detail in King, *Ciphers*, Chapter IV and Appendix L. See already *idem*, "Picard Astrolabe".

54 Ibid., pp. 59–60, and Maier, "Romanische Monatsnamen", A, pp. 240–242.

55 See King, *Ciphers*, and an earlier study addressed to colleagues in Cistercian Studies, listed as *idem*, "Forgotten Cistercian System of Numerical Notation".

56 See King, *Ciphers*, Sections V.5 and VI.8, and Meskens *et al.*, "Wine-Gauging at Damme", on the use of the ciphers in wine-gauging and their engraving on gauging-rods and wine-barrels in Bruges (and Damme) from the 15th to the 18th century.

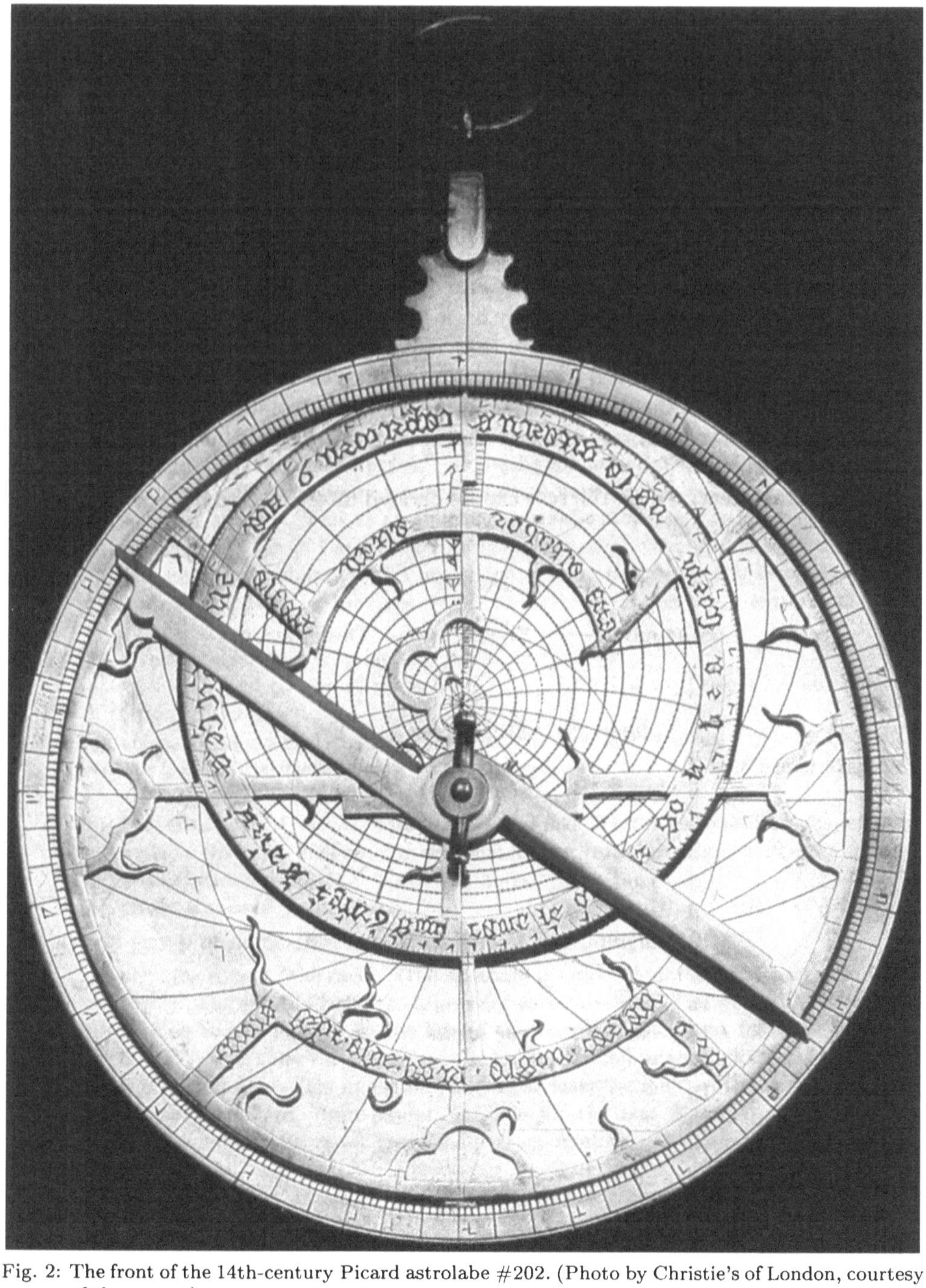

Fig. 2: The front of the 14th-century Picard astrolabe #202. (Photo by Christie's of London, courtesy of the owner.)

- Third, the astrolabe bears a later inscription indicating that its owner, Paschasius Berselius, who was a Benedictine humanist monk of Liège, presented it in 1522 to Hadrianus Amerotius, who was his teacher of Greek and later became professor of Greek at the Collegium Trilingue in Louvain.[57]

The rete bears a half-quatrefoil on the upper solstitial bar as well as an upper bar concentric with the ecliptic, both typically French decorations. An identical rete design is found on another astrolabe (#461), also from 14th-century Northern France.[58] The 29 star-pointers are in the wavy "Fusoris" tradition but, like those on a contemporaneous geared astrolabe from Northern France (#198),[59] clearly predate it. In fact, this kind of star-pointer appears already on a 13th-century astrolabe from Catalonia (#416),[60] and it was probably thence that its use spread to France. On both the Picard astrolabe and the geared astrolabe some of the star-names have been omitted. Whoever labelled 11 of the star-pointers on the Picard astrolabe was using defective sources, and the star-positions confirm the need in the 14th century for new and definitive star-tables.

The star-names featured on the rete can be divided into two groups, those correctly associated with pointers:

Table 2

	Star-name	Signification	Designation
1	–		
2	–		
3	*fluvis* (?)	*finis fluxus*	θ Eri
4	–		
5	*sept*	*septime*	see below
6	–		
7	*alde*	*aldebaran*	α Tau
8	*rigil*	*rigil*	β Ori
9	–		
10	*algon*	*algomeisa*	α CMi
11	–		
12	*cor*^{9}	–	see below
13	*cor leon*	*cor leonis*	α Leo
14	–		
15	–		
16	–		
17	–		
18	–		

57 On these two men see, for example, Bietenholz & Deutscher, eds., *Contemporaries of Erasmus*, I, p. 140 and p. 48.

58 See *Brussels MRAH 1957 Exhibition Catalogue*, p. 8 (no. A12); *Brussels SG 1984 Exhibition Catalogue*, pp. 38–39 (no. 9 – the front is depicted in reverse); also King, *Ciphers*, Appendix K5.

59 See Gunther, *Astrolabes*, II, p. 347 (no. 198); and King, *Ciphers*, Appendix K5.

60 See n. 12 above.

Star-name		Signification	Designation
19	–		
20	*feta*	*feta*	α CrB
21	–		
22	–		
23	*alhabor*	*alhaue* (!)	see below
24	–		
25	*altair*	*altair*	α Aql
26	–		
27	*alfart*	*alfaras* (!)	see below
28	–		
29	–		

Some remarks are in order:

(1) *fluvis*, if my reading is correct,[61] would be for *finis fluvii*, "the end of the river", θ Eridani, although many medieval sources have *finis fluxus*.

(5) *sept*, for *septime*. This is an unusual astrolabe-star (see further below) and is attested, as I far as I know, only on two other medieval astrolabes, both French. The geared astrolabe (#198) has *septime*, and another French astrolabe dating from about 1425 (#163) has *septī* (as well as various other mysterious star-names).

(7) *alde*, for *aldebaran*, standard for α Tauri.

(8) *rigil*, standard (if sometimes spelled *rigel*), referring to the foot (Arabic *rijl*) of Orion, β Orionis.

(10) *algon*, for *algomeisa*, standard for Procyon, α Canis Minoris.

(13) *cor leon* (?), written *leoo* (with the 'e' upside-down) – for *cor leonis*, standard for Regulus, α Leonis.

(20) *feta*, standard (if incorrect from Arabic *al-fakka*) for α Coronæ Borealis, usually written *elfeta* on Fusoris astrolabes, resulting from confusion between a 'c' and 't' in Gothic script.[62]

(25) *altair*, standard, from Arabic (*al-nasr*) *al-ṭā'ir*, "the flying (eagle)", α Aquilæ.

A second group of names are applied to the wrong pointers:

(12) *cor*⁹, for *corvus*, "the raven" (the standard abbreviational suffix -⁹ here stands for -*us*[63]). The star δ Corvi was a popular astrolabe-star. The name has here been applied to the pointer to the left of the right-hand support for the equatorial frame, rather than the one below the half-quatrefoil at the right-hand end of the equinoctial axis. This might be judged the least "serious" of the three errors.

(23) *alhabor*, standard, from Arabic *al-ʿabūr* for the star Sirius, α Canis Maioris. The name is an error for (*ras*)*alhaue* from Arabic (*ra's*) *al-ḥawwā'*, referring to α Ophiuchi.

(27) *alfart*, standard although other renderings are common, from Arabic *al-fard*, "the solitary" star α Hydrae, labelled *ydra* on some Fusoris astrolabes (perhaps

61 The 'uv', if such it is, is written as a single ligature 'ɪɪɪ', which can hardly be an 'x', and the 'is' is secure.

62 See also n. 67 below.

63 See already n. 14.

following French pronounciation) and *alfard* on others. The name is an error for *alferas* or *alferaz* from Arabic *al-faras*, horse, for Pegasus.

Both groups can be compared with star-names featuring on contemporary French astrolabes, in particular with the geared astrolabe (#198). These star-names, with one exception, (2) *sept*, are found in the star-catalogue associated with 'Messahalla', which derives originally from the 11th-century astronomer Maslama al-Majrīṭī (Kunitzsch's Typ VIII).[64] But the engraver of the Picard astrolabe was merely consulting a defective star-catalogue or copying the star-names from another instrument based on such a catalogue.[65] The same holds for the star-positions on the geared astrolabe (#198), although the star-positions on the Ghent astrolabe (#461) correspond better to reality.[66] With the exception of (10) *alfart*, the stars on the Picard astrolabe are a sub-group of those represented on the French astrolabe with similar rete design (#163), which even has *korus* where the Picard astrolabe has *cor*^{9}.[67] The star-names and positions on all four of these French astrolabes (#202, #198, #461 and #163) should be subjected to a detailed analysis.

The positions of the star-pointers are problematic and no explanation is possible for the time being.[68] In view of this, I have resisted the temptation to identify the unnamed star-pointers.

3a. An astrolabe from 14th-century Italy

An astrolabe now preserved in the Museo di Storia della Scienza in Florence (#493) is clearly Italian,[69] and it has a very medieval look about it: most probably it dates from the 14th century (see Fig. 3a).

The rete is distinguished by five short circular arcs inside the southern (upper) ecliptic and two pairs of larger arcs separated by a broad Gothic arch on the lower circumferential frame. A "wine-cup"-shaped frame joins the top of the Gothic arch to the northern (lower) ecliptic. The throne is also distinctive, and the star-pointers most unusual for a European astrolabe. In fact, the design of the rete is "topologi-

64 Kunitzsch, *Typen von Sternverzeichnissen*, especially pp. 51–58 (Typ VIII) and pp. 73–85 (Typ XII). The star-name *septime* is not attested in the medieval star-catalogues that have been studied, although in the Toledan star-table (Typ XII) there is a *septentrionale cornu*. Stars in this region of the heavens caused medieval French astrolabists some problems: in particular, Fusoris put the star *cornu arietis* in this region by assigning the wrong direction to its declination.

65 In the star-catalogue mentioned above (Typ VIII) the star *alfart* is mistakenly associated with the constellation *equus* (for Pegasus): *ibid.*, p. 55 (Typ VIII, no. 19), and the notes on p. 57.

66 On #163 see Gunther, *Astrolabes*, II, pp. 309–311 (no. 163).

67 The letter *k* is attested in Middle French dialects from the North East, and is found elsewhere on #163, as, for example, in *kap(r)icornus* and *akarius*. Also the star-name *altair* is written as *alkair* on this piece (see n. 62 above), and although the form *alkair* is common on German astrolabes, we are dealing here with a Northern French phenomenon.

68 See Stautz, *Mathematisch-astronomische Darstellungen*, p. 102 and fig. 4.5d on p. 261. Stautz compared the positions with the coordinates of the stars from Kunitzsch's Typ VIII (see n. 64) but found no correspondence and no systematic errors. It goes without saying that the pointers correspond only vaguely to the correct positions of the stars for the 14th century.

69 Florence, Museo di Storia della Scienza, inv. no. 1107 (earlier numbering 6785, 1017, 596 and 782); diameter: 99 mm, thickness: 5 mm: unpublished; briefly mentioned in *Florence 1954 Catalogue*, p. 62 (no. 1107), and *Florence 1987–88 Exhibition Catalogue*, p. 14 (no. I.6).

Fig. 3a: The front of the medieval Italian astrolabe #493. (Courtesy of the Museo di Storia della Scienza, Florence.)

cally" related to the rete-design of the prolific Abū Bakr ibn Yūsuf of Marrakesh (*fl.* *ca.* 1200): the concave arcs on the Islamic rete are rendered convex on the Italian one. If this in itself is not convincing, the connection is confirmed by the fact that the distinctive design of both the throne and the star-pointers is also that of Abū Bakr. The basic rete-design of this Italian astrolabe influenced the Vienna school of Peuerbach and Regiomontanus in the mid 15th century,[70] and an identical design was used by Johannes Stöffler *ca.* 1525.[71] See below on other reasons for the exceptional historical interest of this piece.

There are pointers for 23 stars, named as in following list. Regulus (COR LEO) is at *ca.* Leo 20°, and both the selection of stars and the star-positions are related to the star-catalogue compiled in Paris by John of London in 1246 (Kunitzsch Typ VI).[72] The names given in the principal source of that star-catalogue used by Paul Kunitzsch and his numbering and modern identifications are also shown.

Table 3a

The stars on the Italian astrolabe #493, with the corresponding entries in Kunitzsch's Typ VI and identifications

Star-name		Kunitzsch VI		Derivation	Identification
1	BATVL	3	batenkaitoz	*baṭn qayṭus*	ζ Ceti
2	MENCE	7	menkar	*minkhar*	α Ceti
3	ALAV	8	algemb	*al-janb*	α Per
4	ALDBA	10	aldebaran	*al-dabarān*	α Tau
5	BEDEL	13	bedalgeuze	*yad al-jawzā'*	α Ori
6	ALABOR	14	alhaabor	*al-ʿabūr*	α CMa
7	ALGOME	16	algomeisa	*al-ghumayṣā'*	α CMi
8	ALRI	19	–	*al-rijl*	μ UMa
9	ALAB	20	alfart	*al-fard*	α Hya
10	COR LEO	21	cor leonis	*cor leonis*	α Leo
11	ALGORE	24	algorab	*al-ghurāb*	γ Crv
12	ALCHI	25	alchimech	*al-simāk* (*al-aʿzal*)	α Vir
13	ALRAMECH	26	alramech	(*al-simāk*) *al-rāmiḥ*	α Boo
14	COR SCOR	29	cor scorpionis	*cor scorpionis*	α Sco
15	VEDH	28	yed	*yad*	δ Oph
16	RAZT	31	razcaben	*ra's al-tinnīn*	γ Dra
17	BAZALA	30	razalangue	*ra's al-ḥawwā'*	α Oph
18	[V/WEGA]	32	alwega	(*al-nasr*) *al-wāqiʿ*	α Lyr
19	ADDE	36	aldigege	*al-dajāja*	α Cyg
20	DELFI	34	delfin	*dulfīn*	γ Del

70 See King & Turner, "Bessarion's Astrolabe", p. 191.

71 Gunther, *Astrolabes*, pp. 427–431, especially the rete illustrated in fig. 172.

72 Stautz, *Mathematisch-astronomische Darstellungen*, p. 115 and fig. 5.2a, where ALFERAS is interpreted as no. 5 in Typ VI (α Cas), and nos. 35 and 36 are switched.

Star-name		Kunitzsch VI		Derivation	Identification
21	DENELA	35	deneb algedi	*dhanab al-jady*	δ Cap
22	ALFERAS	37	alferaz	*al-faras*	ϵ Peg
23	DENEB	40	denebkaytoz	*dhanab qayṭus*	β Cet

3b. A related Renaissance astrolabe

An astrolabe with a distinct Renaissance appearance (#4506), dated Urbino, 1462, was clearly inspired by one in the tradition of the medieval astrolabe #493. It was alas stolen in the late 1970s from the Musée Départemental in Moulins (Allier), France, and has not been seen since,[73] but fortunately photos of the front and back are available – see Fig. 3b. On the back of the throne is the inscription: "K P",[74] and below the shadow squares on the back is another: "1462 VRBINI K". The precise significance of neither is clear.

The rete is elegantly decorated with five shallow circular arcs inside the southern (upper) part of the ecliptic ring and five more rounded arcs on the lower half of the circumferential ring. This ring has a dragon's head at the left extremity and a tail of sorts at the other end. Of the latter the central one is extended upwards to join a short horizontal collar and "wine-cup"-shaped frame with its two bifurcations attached to the middle of the northern (lower) ecliptic ring. On the ecliptic the name THAVR9 is unusual, and the 'R' for a 'P' in CARRICORN9 is apparently a human error.[75]

The 22 star-pointers are hook- or flame-shaped and are named as in Table 3b. The star-positions are taken from a medieval star-catalogue (Kunitzsch Typ VIII) which is in part derived from that discussed above (Typ VI).[76]

Table 3b
The stars on the 1462 astrolabe #4506, with the corresponding entries in Kunitzsch's Typ VIII and identifications

Star-name		Kunitzsch VIII		Derivation	Identification
1	FINI	5	finis fluxus (*sic*)	*finis fluxus* (*sic*)	θ Eri
2	AREN	7	algenib	*al-janb*	α Per
3	ALDEBERAM	9	aldebaran	*al-dabarān*	α Tau
4	RIGEL	11	rigil	*(al-)rijl*	β Ori
5	ALLABOR	13	alhabor	*al-ʿabūr*	α CMa
6	ALGOMEZA	15	algomeyza	*al-ghumayṣāʾ*	α Cmi
7	ALFARAC	19	alfart	*al-fard*	α Hya

73 It was briefly mentioned in *Paris BN 1973 Catalogue*, p. 75 (no. 176), with illustrations of the front and back on p. 57, and is published in King, "Urbino Astrolabe".

74 Since the two letters are separated I deem it unlikely that the reading "JC P" was intended.

75 'R' for 'B' or 'P' (apparently deliberate) is attested (albeit sporadically) on two medieval Catalan astrolabes – see King & Maier, "Catalan Astrolabe", pp. 688 and 690.

76 See Kunitzsch, *Typen von Sternverzeichnissen*, pp. 51–58 (Typ VIII), and Stautz, *Mathematisch-astronomische Darstellungen*, p. 115 and fig. 5.2b on p. 280, where there is a superfluous no. 24 (β Leo) in the 2nd quadrant.

Star-name		Kunitzsch VIII		Derivation	Identification
8	COR L	20	cor leonis	*cor leonis*	α Leo
9	DVB	23	dubhe	*(al-)dubb*	α UMa
10	CORVVS	22	coruus	–	–
11	ALGER	25	algorab	*al-ghurāb*	γ Crv
12	ALCHIMEC	26	alchimec	*al-simāk*	α Vir
13	ALRAMIC	28	alramech	*al-rāmiḥ*	α Boo
14	FITA	29	alfeca	*al-fakka*	α CrB
15	LAD	31	yed	*yad*	δ Oph
16	COR S	32	cor scorpii	*cor scorpii*	α Sco
17	ALHAVERI	33	alhaue	*al-ḥawwāʾ*	α Oph
18	VVEGA	35	wega	*(al-)wāqiʿ*	α Lyr
19	ALTAIR	36	altair	*al-ṭāʾir*	α Aql
20	DENEP	44	denebalgedi	*dhanab al-jady*	δ Cap
21	SERA	45	sceach (var.: **sera**th)	*sāq*	δ Aqr
22	ALFOR	46	alferaz	*al-faras*	β Peg

Note the confusion in no. 21 caused by the similarity of the Gothic letters for *c*, *e*, *r* and *t* between the already corrupt and curious *sceach* and its variant *serath*, ultimately derived from Arabic *sāq*.

3c. The astrolabe in the Urbino intarsia

Now these two Italian astrolabes are of interest for yet another reason. The astrolabe depicted in one of the intarsias decorating the *studiolo* in the palace of Archduke Federico da Montefeltro in Urbino, built in 1476,[77] was copied from an actual instrument which was very closely related to the 1462 piece. The astrolabe in the intarsia is very carefully represented, so carefully in fact that its component parts are faithfully portrayed in minute detail. Indeed, all of the stars represented by the pointers can be identified; alas, we cannot know how these stars would have been named.

4. Concluding remarks

I find it surprising that the star-names on the sole surviving medieval astrolabe from central Spain (Section 1) should be so distinctively corrupted from their original Arabic forms, not least because they were engraved in a milieu where there were still educated Arabs available to interpret the names correctly. More surprising is the fact that even the Latin names have suffered a similar fate. It is not surprising, on the other hand, that the makers of the Picard astrolabe and the Northern French geared astrolabe (Section 2) fought shy of engraving all of the names on their star-pointers: those they did engrave are problematic enough.[78] There was a clear need in medieval

77 On this see King, "Urbino Astrolabe".

78 Those on #198 would repay detailed investigation. The fact that the stereographic projection is northern rather than the standard southern one means that the positions bear no apparent visual

Fig. 3b: The front of the astrolabe made in Urbino in 1462 (#4506). (Photo courtesy of the Museum of History of Science, Oxford.)

France (and, of course, elsewhere in Europe) for a new star-table and carefully-made instruments based upon it – both were produced *ca.* 1400 by Jean Fusoris in his Paris atelier, albeit still in the medieval tradition.[79] I find it remarkable that the two closely-related Italian astrolabes (Section 3) actually feature two different sets of stars. The maker of the 1462 piece erred severely in thinking that his source for stellar data might be any better than that used for the medieval piece. And when Regiomontanus marked the star-pointers on the astrolabe (#640) which he presented to Cardinal Bessarion in the same year, 1462,[80] he took a monumental step forward as far as European instrumentation was concerned: the stars are all correctly labelled in Latin and their positions correctly indicated.[81] Thus, at least by 1462 one European had finally caught up with the Muslim instrument-makers of the 9th century. Nevertheless, even in the early 16th century, Johannes Stöffler, who used precisely the same rete-design as that of the Urbino astrolabes but yet another star-catalogue type,[82] still used corrupt forms of the Arabic names and was not above confusing the names of Sirius and Procyon, which he affectionately italianized as Algomesia and Alhabro, respectively.

Appendix A: Instruments cited

Note: Each instrument has been assigned a number (here prefixed by #), continuing the tradition started by R. T. Gunther and upheld by D. de Solla Price. The vast majority of the instruments listed below are described in the new Frankfurt catalogue, as yet unpublished. Various instruments mentioned only in passing are not listed here, and for these the reader must have recourse to Price *et al.*, *Checklist.*

#162 Catalan astrolabe from *ca.* 1300 with a rectangular frame and decorative quatrefoil on the rete – London, Society of Antiquaries – See Gunther, *Astrolabes*, II, pp. 306–309 (no. 162), and the detailed study in King & Maier, "Catalan Astrolabe". See also King, "Astrolabes from Medieval Catalonia" (forthcoming).

#163 Medieval Northern French astrolabe – Oxford, Museum of the History of Science, inv. no. LE 2080 – See Gunther, *Astrolabes*, II, pp. 309–311 (no. 163).

#198 Northern French astrolabe from *ca.* 1350 with luni-solar gear mechanism – London, Science Museum, inv. no. 1880.32 – See Gunther, *Astrolabes*, II, p. 347 (no. 198); and King, *Ciphers*, Appendix K5.

#202 Picard astrolabe from *ca.* 1350 with numbers marked in monastic ciphers – Private collection – See Gunther, *Astrolabes*, II, p. 349 (no. 202); *Christie's London 26.9.1991 Catalogue*, pp. 39–42 (lot 42); and the detailed studies in King, "Picard Astrolabe", and *idem*, *Ciphers*, Chapter IV and Appendix L.

---

relationship to those on a standard astrolabe rete. See Stautz, *Mathematisch-astronomische Darstellungen*, pp. 103–104 and fig. 4.5e on p. 262, for a first attempt.

79 See Poulle, *Fusoris*, and the same author's article "Fusoris" in *DSB.*

80 On this see King & Turner, "Bessarion's Astrolabe".

81 See Stautz, *Mathematisch-astronomische Darstellungen*, pp. 111–114 and fig. 5.1a on p. 274. Another astrolabe by Regiomontanus (#452), which probably predates the 1462 piece, still uses Arabic star-names – see King & Turner, "Bessarion's Astrolabe", p. 189 and fig. 14.

82 See n. 71 above.

#416 Catalan astrolabe from *ca.* 1300 with a Y-shaped frame on the rete – Greenwich, National Maritime Museum, inv. no. AST0552 = A21/1936-21C – See King, "Astrolabes from Medieval Catalonia" (in press). Earlier references are: King & Maier, "Catalan Astrolabe", pp. 694–695, n. 60, King, "Earliest European Astrolabe", fig. 14, and *idem*, "14th-Century Spanish Astrolabe", fig. 10.

#452 Astrolabe by the maker of #640 – Private collection – Unpublished; see *Greppin Catalogue*, p. 44 (no. 126); and King & Turner, "Bessarion's Astrolabe", p. 189 and figs. 14–17.

#461 14th(?)-century French astrolabe – Ghent, Bijloke Museum, inv. no. 455 – Unpublished; see *Brussels MRAH 1957 Exhibition Catalogue*, p. 8 (no. A12); *Brussels SG 1984 Exhibition Catalogue*, pp. 38–39 (no. 9 – the front is depicted in reverse); also King, *Ciphers*, Appendix K5.

#493 Medieval Italian astrolabe – Florence, Museo di Storia della Scienza, inv. no. 1107 (earlier numbering 6785, 1017, 596 and 782) – See King, "Urbino Astrolabe" (forthcoming).

#640 Astrolabe dedicated by Regiomontanus to Cardinal Bessarion in 1462 – Private collection – See *Christie's London 28.9.1989 Catalogue*; and the detailed study in King & Turner, "Bessarion's Astrolabe", which also deals with another ten instruments from the same workshop.

#3042 Astrolabe from 10th-century Catalonia – Paris, Institut du Monde Arabe, inv. no. AI 36-31 – See the detailed studies in Destombes, "Astrolabe carolingien"; King, "Earliest European Astrolabe"; Kunitzsch & Dekker, "Stars on the Carolingian Astrolabe"; as well as other papers in Stevens *et al.*, *Oldest Latin Astrolabe*.

#3053 Astrolabe signed by Petrus Raimundus of Aragon, made in Barcelona and dated 1375 – Boston, Mass., Museum of Fine Arts, inv. no. 88.654 – See King, "Astrolabes from Medieval Catalonia" (forthcoming). The front is illustrated in King, "Earliest European Astrolabe", fig. 16, and *idem*, "14th-Century Spanish Astrolabe", fig. 11.

#3622 Unsigned astrolabe from Cordova, dated 446 Hijra [= 1054/55], with later Catalan additions – Cracow, Jagiellonian Museum, inv. no. 4037-35/V – Published in detail in Maier, "Astrolab aus Córdoba" (where earlier studies in Polish are cited).

#4506 Italian astrolabe bearing the initials "KP" and dated Urbino, 1462 – Stolen in the mid 1970s from Musée Départemental d'Alliers, Moulins – Published in King, "Urbino Astrolabe" (forthcoming). See already *Paris BN 1973 Catalogue*, p. 75 (no. 176), with illustrations of the front and back on p. 57; and King & Turner, "Bessarion's Astrolabe", p. 191.

#4560 14th-century astrolabe from Toledo (?) with markings in Hebrew, Latin and Arabic – Private collection – See *Christie's London 15.4.1999 Catalogue*, pp. 98-107 (lot 52), and the detailed study in King, "14th-Century Spanish Astrolabe".

Bibliography and bibliographical abbreviations

AGUIAR AGUILAR, *Tratado de Ibn Mas'ūd*: Maravillas Aguiar Aguilar, Risalat Kayfiyat al-sayb fi l-'amal bi-l-jayb *de 'Izz al-Dīn b. Mas'ūd: Un tratado del siglo XIV acerca de los usos del cuadrante de senos*, unpublished doctoral thesis, Universidad de La Laguna (Tenerife), 1995.

–, "'Abd al-'Azīz ibn Mas'ūd": *eadem*, "En torno a 'Abd al-'Azīz b. Mas'ūd", *Boletín de la Asociación Española de Orientalistas* (Madrid) 32 (1996), pp. 187–196.

AIOS: Fuat Sezgin *et al.*, ed., *Arabische Instrumente in orientalistischen Studien*, 6 vols., Frankfurt: Institut für Geschichte der Arabisch-Islamischen Wissenschaften, 1990–91.

BEAUJOUAN, "Rotation des chiffres": Guy Beaujouan, "Étude paléographique sur la rotation des chiffres et l'emploi des apices du Xe au XIIe siècle", *Revue d'Histoire des Sciences* 1 (1947), pp. 301–311, reprinted in *idem*, *Studies*, A–IX.

–, *Studies*, A–B: *idem*, *Par raison de nombres – L'art du calcul et les savoirs scientifiques médiévaux* (A), and *Science médiévale d'Espagne et d'alentour* (B), Aldershot: Variorum, 1990 and 1992.

BENOIT *et al.*, eds., *Histoire des fractions*: P. Benoit, K. Chemla and J. Ritter, eds., *Histoire des fractions: Fractions d'histoire*, Basle: Birkhäuser, 1992.

BIETENHOLZ, Peter G. and Thomas B. DEUTSCHER, eds., *Contemporaries of Erasmus – A Biographical Register of the Renaissance and Reformation*, 3 vols., Toronto: University of Toronto Press, 1985–87.

BISCHOFF, Bernhard, *Latin Palaeography – Antiquity and the Middle Ages*, Cambridge, *etc.*: Cambridge University Press, 1990.

Brussels MRAH 1957 Exhibition Catalogue: [Henri Michel *et al.*], *Exposition – Art, Histoire et Sciences: Catalogue*, Brussels: Musées royaux d'Art et d'Histoire, 1957.

Brussels SG 1984 Exhibition Catalogue: Georges Baptiste *et al.*, *La mesure du temps dans les collections belges*, Brussels: Société Générale de Banque, 1984. [Catalogue of an exhibition held at the Bank in Brussels during 26.1.–7.4.1984.]

Cairo ENL Survey: David A. King, *A Survey of the Scientific Manuscripts in the Egyptian National Library*, (Catalog vol. 5 of the American Research Center in Egypt), Winona Lake, Ind.: Eisenbrauns, 1986.

CAJORI, *Mathematical Notations*: Florian Cajori, *A History of Mathematical Notations*, 2 vols., Chicago, Ill.: The Open Court Publishing Company, 1928–29, reprinted La Salle, Ill.: The Open Court Publishing Company, 1951–52.

CAPPELLI, *Lexicon abbreviaturarum*: Adriano Cappelli, *Lexicon abbreviaturarum – Dizionario di abbreviature latine ed italiane*, Milan: Ulrico Hoepli, 1990 (reprint of the 6th edition, 1961).

CASTRO, Américo, *Glosarios latino-españoles de la edad media*, *Revista de filología española* (Madrid) 22 (1936).

CHABÀS, Josep, and Daniel BOSCH, *L'astrolabi pla – Guia per a la construcció i utilització*, Barcelona: Institut de Ciències de l'Educació de la Universitat Politècnica de Catalunya, 1987.

Chicago AP Catalogue, I–II: Bruce Chandler with Sara Schechter Genuth, eds., *Historic Scientific Instruments of the Adler Planetarium & Astronomy Museum*, vol. I:

Western Astrolabes, by Roderick and Marjorie Webster, Chicago, Ill.: Adler Planetarium & Astronomy Museum, 1998, and vol. II: *Eastern Astrolabes*, in press.

Christie's London 28.9.1989 Catalogue: Gerard L'E. Turner, *The Regiomontanus Astrolabe dated 1462*, supplement to Fine Scientific and Philosophical Instruments, 28 September, 1989, London: Christie's South Kensington, 1989.

Christie's London 26.9.1991 Catalogue: *Engineering and Scientific Works of Art, Instruments and Models – London, Thursday, 26 September, 1991*, London: Christie's South Kensington, 1991.

Christie's London 15.4.1999 Catalogue: *Exceptional Scientific and Engineering Works of Art, Instruments and Models, Thursday, 15 April 1999*, London: Christie's South Kensington, 1999.

COROMINAS, Joan, with assistance from José A. Pascual, *Diccionario crítico etimológico castellano e hispánico*, revised from the 1954 edn., 5 vols., Madrid: Gredos, 1983.

CORRIENTE, *Spanish Arabic*: Federico Corriente, *A Grammatical Sketch of the Spanish-Arabic Dialect Bundle*, Madrid: Instituto Hispano-Árabe de Cultura, 1977.

DESTOMBES, Marcel, "Un astrolabe carolingien et l'origine de nos chiffres arabes", *Archives Internationales d'Histoire des Sciences* 58–59 (1962), pp. 1–45, reprinted in G. Schilder, P. van der Krogt, S. de Clercq, eds., *M. Destombes (1905–1983): Selected Contributions to the History of Cartography and Scientific Instruments*, Utrecht: Hes, & Paris: A. G. Nizet, 1987, pp. 153–200.

DMA: *Dictionary of the Middle Ages*, 13 vols., New York, N. Y.: Charles Scribner's Sons, 1982–89.

DSB: *Dictionary of Scientific Biography*, 16 vols., New York, N. Y.: Charles Scribner's Sons, 1970–80.

EI_2: *Encyclopaedia of Islam*, new (2nd) edn., 10 vols., Leiden: E. J. Brill, 1960–2000.

Florence MSS 1954 Catalogue: *Catalogo degli strumenti del Museo di Storia della Scienza*, Florence: Leo S. Olschki, 1954.

Florence MSS 1987–88 Exhibition Catalogue: Mara Miniati, ed., *L'Età di Galileo – Il secolo d'oro della scienza in Toscana*, Florence: Istituto e Museo di Storia della Scienza, 1987. [A catalogue of an exhibition held at the Museum from 1.7.1987 to 9.1.1988.]

Frankfurt IGN Festband: Anton von Gotstedter, ed., *Ad radices – Festband zum fünfzigjährigen Bestehen des Instituts für Geschichte der Naturwissenschaften Frankfurt am Main*, Stuttgart: Franz Steiner, 1994.

Greppin Catalogue: *La Collection Greppin, instruments scientifiques*, Zurich: Galerie Koller, 1975. [Catalogue of an auction on 17.11.1975.]

GUNTHER, *Astrolabes*: Robert T. Gunther, *The Astrolabes of the World* ..., 2 vols., Oxford: University Press, 1932, reprinted in 1 vol., London: The Holland Press, 1976.

HARTNER, "Astrolabe", A–B: Willy Hartner, "The Principle and Use of the Astrolabe" (A), a chapter in Pope, ed., *Survey of Persian Art*, reprinted without the original plates in *Willy Hartner – Oriens–Occidens – Ausgewählte Schriften zur Wissenschafts- und Kulturgeschichte – Festschrift zum 60. Geburtstag*, Hildesheim: Georg Olms, 1968, pp. 287–311, and again with them in *Astrolabica* (Paris: Société

Internationale de l'Astrolabe), 1 (1978), as well as *idem*, article "Asturlab" (B) in *Enc. Islam*, 2nd edn., reprinted in *idem*, *Oriens–Occidens*, 1968, pp. 312–318.

HARTNER, *Oriens–Occidens*, I–II: *Willy Hartner – Oriens–Occidens – Ausgewählte Schriften zur Wissenschafts- und Kulturgeschichte – Festschrift zum 60. Geburtstag*, Hildesheim: Georg Olms, 1968 (I); and Yasukatsu Maeyama, ed., *Oriens–Occidens – Ausgewählte Schriften zur Wissenschafts- und Kulturgeschichte – Band II*, Hildesheim, etc.: Georg Olms, 1984 (II).

HILL, *Arabic Numerals*: George Francis Hill, *The Development of Arabic Numerals in Europe Exhibited in Sixty-Four Tables*, Oxford: Clarendon Press, 1915.

IFRAH, *Histoire des chiffres*: Georges Ifrah, *Histoire universelle des chiffres*, 2nd edn., Paris: Robert Laffont, 1994. [The first edition is available in English as *From One to Zero – A Universal History of Numbers*, New York, N. Y., etc.: Penguin Books, 1988.]

KENNEDY & KENNEDY, *Islamic Geographical Coordinates*: Edward Stewart Kennedy and Mary Helen Kennedy, *Geographical Coordinates of Localities from Islamic Sources*, Frankfurt am Main: Institut für Geschichte der Arabisch-Islamischen Wissenschaften, 1987.

KENNEDY *et al.*, *Studies*: E. S. Kennedy, Colleagues and Former Students, *Studies in the Islamic Exact Sciences*, ed. by Mary Helen Kennedy and David A. King, Beirut: American University of Beirut, 1983.

Kennedy Festschrift: David A. King and George Saliba, eds., *From Deferent to Equant – A Volume of Studies in the History of Science in the Ancient and Medieval Near East in Honor of E. S. Kennedy*, *Annals of the New York Academy of Sciences* 500 (1987).

KING, David A., "Astronomical Instruments between East and West", in *Krems 1992 Conference Proceedings*, pp. 143–198.

–, *The Ciphers of the Monks: A Forgotten Number Notation of the Middle Ages*, to appear in the series *Boethius: Texte und Abhandlungen zur Geschichte der Mathematik und der Naturwissenschaften*, published by Franz Steiner, Stuttgart.

–, "The Earliest Known European Astrolabe in the Light of Other Early Astrolabes", in Stevens *et al.*, eds., *The Oldest Latin Astrolabe*, pp. 359–404.

–, "A Forgotten Cistercian System of Numerical Notation", *Cîteaux – Commentarii Cistercienses* 46:3–4 (1995), pp. 183–217.

–, "14th-Century Spanish Astrolabe: *idem*, "An Astrolabe from 14th-Century Christian Spain with Inscriptions in Latin, Hebrew and Arabic – A Unique Testimonial to an Intercultural Encounter", to appear in *Medieval Encounters – Jewish, Christian and Muslim Culture in Confluence and Dialogue* in 2000.

–, "Geography of Early Astrolabes": *idem*, "Bringing Astronomical Instruments back to Earth: The Geographical Data on Early Astrolabes (to *ca.* 1100)", in *North Festschrift*, pp. 3–53.

–, "Instrument Catalogue", A–B: *idem*, "Medieval Astronomical Instruments: A Catalogue in Preparation", *Bulletin of the Scientific Instrument Society* (London) 31 (Dec., 1991), 3–7, also *ibid.*, 36 (March, 1993), pp. 17–18, and 44 (March, 1995), pp. 5–12 (A); and *idem*, "Der Frankfurter Katalog mittelalterlicher astronomischer Instrumente", in Gerhard Endress and Remke Kruk, eds., *The Ancient Tradition in*

Christian and Islamic Hellenism – Studies on the Transmission of Greek Philosophy and Sciences dedicated to H. J. Drossaart Lulofs on his Ninetieth Birthday, Leiden: Research School CNWS, School of Asian, African, and Amerindian Studies, 1997, pp. 145–164 (C). (B).

–, "Picard Astrolabe": *idem*, "Rewriting History through Instruments: The Secrets of a Medieval Astrolabe from Picardy", in *Turner Festschrift*, pp. 42–62.

–, "Urbino Astrolabe": *idem*, "The Astrolabe in the Studiolo in the Ducal Palace in Urbino", in F. Vetrano, ed., *Science in Urbino*, Urbino: Università di Urbino, in press.

KING & MAIER, "Catalan Astrolabe": *idem* and Kurt Maier, "The Medieval Catalan Astrolabe in the Society of Antiquaries, London", in *Vernet Festschrift*, II, pp. 673–718.

KING & TURNER, " Bessarion's Astrolabe": *idem* and Gerard L'E. Turner, "The Astrolabe Presented by Regiomontanus to Cardinal Bessarion in 1462", *Nuncius – Annali di Storia della Scienza* (Florence) 9 (1994), pp. 165–206.

Krems 1992 Conference Proceedings: Harry Kühnel, ed., *Kommunikation zwischen Orient und Okzident*, (*Sitzungsberichte der Philosophisch-Historischen Klasse der Österreichischen Akademie der Wissenschaften*, Band 619, *Veröffentlichungen des Instituts für Realienkunde des Mittelalters und der Frühen Neuzeit* (Krems), Nr. 16), Vienna: Österreichische Akademie der Wissenschaften, 1994.

KUNITZSCH, *Arabische Sternnamen in Europa*: Paul Kunitzsch, *Arabische Sternnamen in Europa*, Wiesbaden: Otto Harrassowitz, 1959.

–, "Astrolabe Stars": *idem*, "al-Ṣūfī and the Astrolabe Stars", *Zeitschrift für Geschichte der arabisch-islamischen Wissenschaften* (Frankfurt am Main) 6 (1990), pp. 151–166.

–, "Fachausdrücke": *idem*, "Glossar der arabischen Fachausdrücke in der mittelalterlichen europäischen Astrolabliteratur", *Nachrichten der Akademie der Wissenschaften in Göttingen*, I. Philologisch-Historische Klasse, Jahrgang 1982, Nr. 11, pp. 455–571. [The separatum is paginated 1–117.]

–, "Hindu-Arabic Numerals": *idem*, "The Transmission of Hindu-Arabic Numerals Reconsidered", *Perspectives on Science in Medieval Islam*, ed. J. Hogendijk and A. Sabra, Cambridge, Mass. (in press).

–, *Sternkatalog des Almagest*: *idem*, *Claudius Ptolemäus – Der Sternkatalog des Almagest – Die arabisch-mittelalterliche Tradition*, 3 vols., Wiesbaden: Otto Harrassowitz, 1986–1991.

–, *Sternnomenklatur der Araber*: *idem*, *Untersuchungen zur Sternnomenklatur der Araber*, Wiesbaden: Otto Harrassowitz, 1961.

–, *Studies*: *idem*, *The Arabs and the Stars*, Northampton: Variorum, 1989.

–, "Two Star Tables from Muslim Spain", *Journal for the History of Astronomy* 11 (1980), pp. 192–201, reprinted in *idem*, *Studies*, IV.

–, *Typen von Sternverzeichnissen in astronomischen Handschriften des zehnten bis vierzehnten Jahrhunderts*, Wiesbaden: Otto Harrassowitz, 1966.

– and Elly Dekker, "The Stars on the Rete of the So-Called 'Carolingian Astrolabe'", in *Vernet Festschrift*, II, pp. 655–672, reprinted in Stevens *et al.*, *The Oldest Latin Astrolabe*, new edn.

LEMAY, "Arabic Numerals": Richard Lemay, article "Arabic Numerals" in *DMA*.

LORCH, "Rudolf of Bruges on the Astrolabe": Richard P. Lorch, "The Treatise of Rudolf of Bruges on the Astrolabe", in *North Festschrift*, pp. 55–100.

MAIER, "Astrolab aus Córdoba": Kurt Maier, "Ein arabisches Astrolab aus Córdoba (11. Jh.) mit späteren altkatalanischen Inschriften", in *Saltzer Festschrift*, pp. 119–133.

–, "Romanische Monatsnamen", A–B: *idem*, "Bemerkungen zu romanischen Monatsnamen auf mittelalterlichen Astrolabien", in *Frankfurt IGN Festband*, pp. 237–254 (A); and *idem*, "Zeugen des Mehrsprachigkeit: mittelalterliche romanische Monatsnamen auf islamischen astronomischen Instrumenten", in Jens Lüdtke, ed., *Romania arabica. Festschrift für Reinhold Kontzi zum 70. Geburtstag*, Tübingen: Gunther Narr, 1996, pp. 251–270 (B).

MAIER: see also KING & MAIER.

MAYER, Leo A., *Islamic Astrolabists and Their Works*, Geneva: Albert Kundig, 1956.

MESKENS, Ad, Germain BONTE, Jacques de GROOTE, Mieke de JONGHE, and David A. KING, "Wine-Gauging at Damme: The Evidence of a Late Medieval Manuscript", *Histoire et mesure* (Paris: C. R. H.–C. N. R. S.) 14 (1999), pp. 51–77.

Munich Astrolabe Catalogue: Burkhard Stautz, *Die Astrolabiensammlungen des Deutschen Museums und des Bayerischen Nationalmuseums*, Munich: Oldenbourg, 1999. [See already the summary "Mit dem Himmel in der Hand – Astrolabien und die Astrolabiensammlung des Deutschen Museums", *Kultur & Technik* (Munich: Deutsches Museum) 2 (1997), pp. 38–43.]

NORTH, "Astrolabe": John D. North, "The Astrolabe", *Scientific American* 230 (1974), pp. 96–106, reprinted in *idem*, *Studies*, I, no. 14.

–, *Studies*, I: *idem*, *Stars, Minds and Fate – Essays in Ancient and Medieval Cosmology*, London & Ronceverte, W. V.: The Hambledon Press, 1989.

North Festschrift: Arjo Vanderjagt and Lodi Nauta, eds., *Between Demonstration and Imagination: Essays in the History of Science and Philosophy Presented to John D. North*, Leiden: E. J. Brill, 1999.

Nuremberg GNM 1992–93 Exhibition Catalogue: Gerhard Bott, ed., *Focus Behaim-Globus*, 2 vols., Nuremberg: Germanisches Nationalmuseum, 1992. [Catalogue of an exhibition held at the Museum during 2.12.1992–28.2.1993.]

Paris BN 1973 Exhibition Catalogue: *Nicolas Copernic ou La révolution astronomique*, Paris: Bibliothèque Nationale, 1973.

POULLE, "L'astrolabe": Emmanuel Poulle, "L'astrolabe", *La Recherche* (Paris) 178 (juin, 1986), pp. 756–765.

–, *Fusoris*: *idem*, *Un constructeur d'instruments astronomiques au 15e siècle – Jean Fusoris*, Paris: Honoré Champion, 1963.

–, *Instruments du Moyen Age*: *idem*, *Les instruments astronomiques du Moyen Age*, (first published in *Le Rouban Rouge* (Paris), no. 32 (mars 1967), pp. 18–29), revised edn., *Astrolabica* (Société internationale de l'astrolabe), no. 3, Paris: Alain Brieux, 1983.

PRICE *et al.*, *Checklist*: Derek J. de Solla Price *et al.*, *A Computerized Checklist of Astrolabes*, New Haven, Conn.: Yale University, Department of History of Science and Medicine, 1973.

Saltzer Festschrift: Peter Eisenhardt, Frank Linhard and Kaisar Petanides, eds., *Der Weg der Wahrheit – Aufsätze zur Einheit der Wissenschaftsgeschichte – Festgabe zum 60. Geburtstag von Walter G. Saltzer*, Hildesheim: Olms, 1999.

SARRUS, "Astrolabe": Frédéric Sarrus, "Description d'un astrolabe construit à Maroc en l'an 1208", *Mémoires de la Société du Musée d'Histoire Naturelle de Strasbourg* 4:2–3 (1853), pp. 1–32 and 6 pls. (separatum dated 1852), reprinted in *AIOS*, I, pp. 193–230.

SARTON, *IHS*: George Sarton, *An Introduction to the History of Science*, 5 parts in 3 vols., Baltimore, Md.: Wilkins & Wilkins, for the Carnegie Institution of Washington, 1927–1948, reprinted Huntington, N. Y.: Robert E. Krieger, 1975.

STAUTZ, Burkhard, *Untersuchungen von mathematisch-astronomischen Darstellungen auf mittelalterlichen Instrumenten islamischer und europäischer Herkunft*, Bassum: Verlag für Geschichte der Naturwissenschaften und der Technik, 1997.

STAUTZ: see also *Munich Astrolabe Catalogue.*

STEVENS, Wesley, Guy BEAUJOUAN, and Anthony J. TURNER, eds., *The Oldest Latin Astrolabe*, *Physis* (Rome) 32:2–3 (1995) [published 1996], new edition, with additions, Florence: Olschki, 1998.

TOMBA, "Astrolabi", A–D: Tullio Tomba, "Due astrolabi latini del XIV secolo conservati a Milano", *Physis* 8 (1966), pp. 295–306 (A); *idem*, "Nuove osservazioni sui due astrolabi latini del XIV secolo conservati a Milano", *Physis* 10 (1968), pp. 119–122 (B); *idem*, "Un astrolabio del XIV secolo di probabile origine italiana", *Physis* 12 (1970), pp. 82–87 (C); and *idem*, "Tre astrolabi latini, del XIV secolo, conservati in Italia", *Rassegna di Studi e di Notizie* (Milan), 20 (Anno XXII [= 1996]) (the offprint is unpaginated) (D).

TRENTO, Paolo, *L'astrolabio – storia, funzioni, costruzione*, Rome: Biblioteca del Vascello, 1989.

TURNER: see *Christie's London 28.9.1989 Catalogue*, and King & Turner.

Turner Festschrift: Robert G. W. Anderson, James A. Bennett and William F. Ryan, eds., *Making Instruments Count – Essays on Historical Scientific Instruments Presented to Gerard L'Estrange Turner*, Aldershot: Variorum, 1993.

Vernet Festschrift: Josep Casulleras and Julio Samsó, eds., *From Baghdad to Barcelona. Studies in the Islamic Exact Sciences in Honour of Professor Juan Vernet*, 2 vols., Barcelona: Instituto "Millás Vallicrosa" de Historia de la Ciencia Árabe, 1996.

Washington NMAH Catalogue: Sharon Gibbs with George Saliba, *Planispheric Astrolabes from the National Museum of American History*, Washington, D. C.: Smithsonian Institution Press, 1984.

ZINNER, *Astronomische Instrumente*: Ernst Zinner, *Deutsche und niederländische astronomische Instrumente des 11.–18. Jahrhunderts*, 2nd rev. edn., Munich: C. H. Beck, 1972.

VIII

A *VETUSTISSIMUS* ARABIC TREATISE ON THE *QUADRANS VETUS*

In memory of Frans Bruin (1922–2001)

1. *Introduction*

The universal horary quadrant is an ingenious mathematical device that was used in one form or another for close to a thousand years,[1] yet its first mention in the primary sources is not generally known, and the function for which it was intended has sometimes been misunderstood in the modern secondary literature.[2] Its function was, for any reasonable terrestrial latitude and at any time of the year, to find the time of day in seasonal hours[3] from the altitude of the sun given its meridian altitude; it is based on an approximate formula that works better than one would expect at first encounter. Because of its simplicity and efficacy, the universal horary quadrant in its different manifestations became, after the astrolabe, and insofar as one can separate it from the astrolabe,[4] the second most widely-used instrument in the Middle Ages, at least in Europe if not in the Islamic world. In the history of medieval instrumentation it went hand-in-hand with a shadow square.

This paper is abridged from part of a history of the universal quadrant for over a millennium, currently in press.[5] I use the term "universal" to describe any astronomical instrument or table that works for essentially all terrestrial latitudes.[6] I use the expression "universal horary quadrant" because the markings are universal, the purpose solely to measure the time of day in (seasonal) hours, and the form usually a quadrant.[7] Other modern expressions, such as "unequal-hour lines", "unequal-hour diagram", and "horary quadrant for planetary hours", tend to obscure the function and the scope of the markings.

Hitherto, the first descriptions of the *quadrans vetus*, that is, the universal horary quadrant with movable solar longitude cursor, also provided with a shadow square, were known only from medieval Latin and Middle Castillian sources:[8] in the present study, an Arabic text from ninth- or tenth-century Baghdad on the very same instrument is presented. I leave it to others to investigate the relationship of this "new" treatise to the later European texts; certainly the European tradition was derived from early Islamic sources, not least because the horary markings in both traditions are associated with shadow squares. A late thirteenth-century European development of the *quadrans vetus*, namely, the *quadrans novus*, included some stereographic markings for the ecliptic and the horizons. With this rather unhappy combination of approximate and accurate markings one could find the time of day approximately, and solve horizon-related problems such as finding the length of daylight accurately, but to determine the time accurately one had to have recourse to the trigonometric quadrant sometimes, but not always, included on the back of the instrument.[9] The modern literature on the quadrant is cluttered with a confusion of the *quadrans novus* with the astrolabic quadrant, an Islamic invention quite different in conception.

The universal horary quadrant without a solar scale, which, like the shadow scales, is found occasionally on the backs of astrolabes, has been misunderstood by some modern authors as essentially non-functional without an additional solar scale:[10] in fact, however, the basic markings even without the solar scale serve perfectly well the purpose for which they were intended. The correct procedure, which is beautifully simple, is very seldom (only once?) mentioned in the modern literature on the astrolabe.[11] It is, however, mentioned in several medieval and Renaissance treatises.[12]

It is often overlooked that the universal horary markings provide a graphical representation of the altitude of the sun at the seasonal hours which is *approximate*. There is, however, no better way of representing graphically the solar altitude at the seasonal hours for all latitudes. If one wants to display graphically the solar altitude at the seasonal or equinoctial hours throughout the year for a specific latitude, one needs a different set of markings. Different ways of graphically representing the altitudes at either kind of hours for a fixed latitude were used by astronomers from the ninth century onwards.[13]

2. *The Baghdad Treatise and its Author*

Various forms of the universal horary quadrant are described in an Arabic treatise from ninth-century or, at the latest, tenth-century Baghdad. The text survives in a manuscript preserved in the Egyptian National Library in Cairo; this is a very late copy, prepared *c.* 1800, probably in the holy city of Meshed in northeastern Iran (see Figure 1).[14]

The treatise is anonymous, and its author has not been determined. I have considered whether he might have been the celebrated ninth-century astronomer Abū Ja'far al-Khwārizmī,[15] who also authored the earliest known treatises on the horary quadrant for a specific latitude and the sine quadrant.[16] The attribution of this treatise to the ninth–tenth centuries is suggested first by the content, for it clearly predates the entire Latin *quadrans vetus* tradition. Furthermore, due attention is paid by the author to the determination of the times of the *ẓuhr* and the *'aṣr* prayers, associated here with the ends of the 6th and 9th seasonal hours (that is, midday and the middle of the afternoon), and the treatise is of prime importance since it represents a stage in the development of the Islamic prayer ritual *before* the standard definitions, in terms of shadow increases, were introduced; this process seems to have been achieved by the ninth century.[17] The association with al-Khwārizmī is suggested not least by the mention of the value 33° for the local latitude, which, in the light of our present knowledge of early Islamic astronomy, can only serve Baghdad. Also it was used only in the ninth and tenth centuries.[18] Furthermore, in the book on astronomical instruments for timekeeping by Abū 'Alī al-Marrākushī (*fl.* Cairo, *c.* 1280), the sine quadrant is called *al-jayb al-Khwārizmī*, "the sine (quadrant) of al-Khwārizmī", and

FIG. 1. The beginning and end of the text in the Cairo manuscript, courtesy of the Egyptian National Library, Cairo.

the shadow square *al-ẓill al-Khwārizmī*, "the shadow (square) of al-Khwārizmī".[19]

However, if al-Khwārizmī was the author, then this must be a later work of his, and he must surely have learned a great deal of trigonometry during his active life, for at least one of his early writings was heavy indeed, to the point of being somewhat absurd.[20] And it must be a later work of his than, say, his treatises on the construction and use of the astrolabe.[21] But he must also have learned a lot about the potential of Arabic as the new language of science, for the style is quite different from that of his astrolabe treatises.[22] Indeed, the language of the text is crisp and elegant, and no words are wasted. It is almost as if the work was, say, a tenth-, eleventh-, twelfth- or thirteenth-century recension of some ninth-century original.

Against the hypothesis of an origin in, say, the eleventh–thirteenth centuries, we note, besides the various points mentioned above, the fact that all of the other works copied in the same manuscript are from the ninth and tenth centuries.[23] Certainly some version of the treatise ended up in al-Andalus before the thirteenth century, for it was surely thence that the *quadrans vetus* came to the attention of Europeans in the twelfth century. But most of the astronomical materials transmitted from the Islamic East to al-Andalus were of ninth- and tenth-century origin.[24] al-Marrākushī in Cairo in the late thirteenth century seems to have found his sine quadrant with movable cursor in an Andalusī source. Furthermore, the Baghdad treatise was apparently unknown to Egyptian and Syrian astronomers of the fourteenth century, who were *the* specialists in different varieties of quadrants.

Other possibilities for authorship are the very innovative Ḥabash al-Ḥāsib (*fl.* Baghdad, *c.* 830–880),[25] who was particularly interested in trigonometry, or the well-known astronomer Thābit ibn Qurra (*fl.* Baghdad, *c.* 880),[26] but I consider both unlikely.

In any case our anonymous author, who makes no claim to have invented the instrument, was fully in control of his subject, not least because he proposed three different equivalent forms of the instrument, and he assumed a certain basic knowledge on the part of his readers. The surviving copy of his text is partially defective, but it is not difficult to reconstruct the author's intentions. In the Appendix to this paper I present the Arabic text, a translation, and a commentary.

The markings of the universal horary quadrant described in the Baghdad treatise are in three basic forms (see Figure 2) with an optional movable or fixed calendrical/declination scale fitted in a groove on the rim (Figure 3). Alas the description of the cursor has dropped out of the text as it now stands, but from references elsewhere in the treatise it seems that the argument was the date in a specific solar calendar.[27] Essentially the markings in each of the three kinds are linked to the 15°-divisions of the outer scale, either by arcs of circles (type 1), a set of parallels (type 2), or a set of radii (type 3). Superposed on the horary markings is a shadow square with scales on each of the two non-axial sides. Special markings on the curves for the third and sixth hours may have been intended, in order to show their relationship to the times of prayer; these inevitably disappeared in the Latin and Castillian tradition.[28]

3. *The Approximate Formula for Timekeeping*

The main markings serve the solution of an approximate formula for timekeeping adopted by Muslim astronomers (along with the exact formula) from Indian sources.[29] In modern notation we may render this as:

$$T = 1/15 \text{ arc sin } \{\sin h / \sin H\},$$

where h is the instantaneous solar altitude, H is the meridian altitude, and T is the time since sunrise or before sunset in seasonal hours ($0 < T < 6$). By setting the movable bead on the radial thread in a certain way (different for each variety) to H, entered on the outer scale, and then aligning one side of the quadrant towards the sun with altitude h on the same scale, the time of day T can be read from the position relative to the hour markings of the bead on the now vertical thread (see Figure 4).

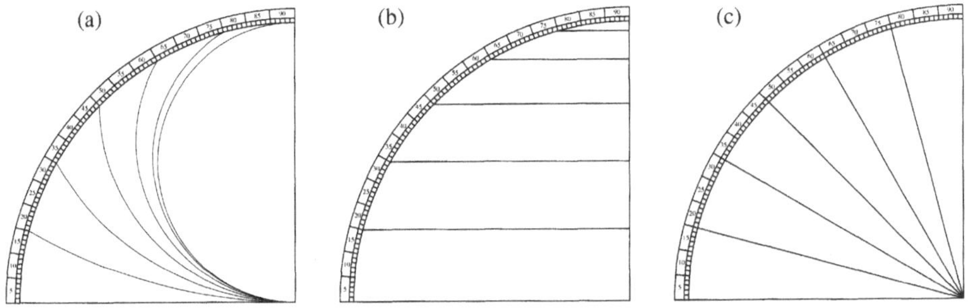

FIG. 2. The three different kinds of markings (types 1–3) mentioned in the Baghdad treatise.

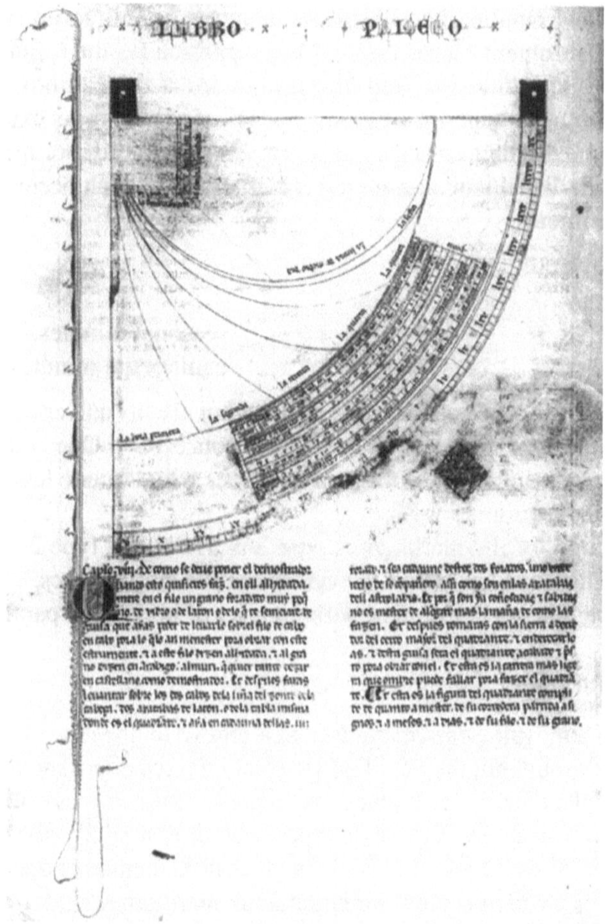

FIG. 3. The universal horary quadrant illustrated in the *Libros del saber del astronomía* which is the first of the three described in the Baghdad treatise. From the facsimile of MS Madrid Biblioteca Nacional L97 (fol. 173v) published as vol. i of the 2-volume Barcelona, 1999, reprint of Madrid, 1863–67, edition of M. Rico y Sinobas, courtesy of Professor Julio Samsó, Barcelona.

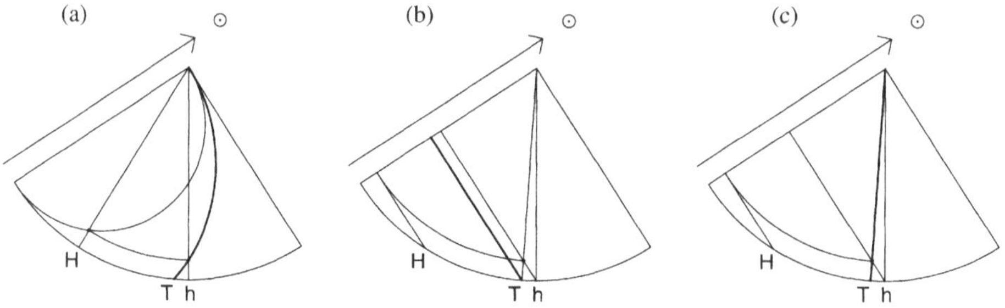

FIG. 4. The use of the three kinds of universal quadrants to find T(h, H). Our Baghdad astronomer simply wrote: "the use of all of these is the same", and left the details to the imagination of his readers.

Elsewhere I have studied the substantial literature on this formula, and the various tables and instruments based on it.[30] The inspiration for the formula was clearly the fact that both at latitude 0° and for all latitudes at the equinoxes it is correct. The formula is approximate for other solar longitudes but works extremely well in Mesopotamian and Mediterranean latitudes but less well in, say, northern Europe. The errors are latitude-dependent and the maximum errors (which occur at the summer solstice)[31] for latitudes:

	30°	40°	50°	
are of the order:				
	6	9	14	seasonal minutes,
that is,	7	11	19	equinoctial minutes.

The virtue of the formula is not only its simplicity but also its universality and efficacy (at least within the regions in which it was first conceived). One could add that for most times of day during most of the year in places with latitude less than, say, 40°, the formula works extremely well.

The construction of the markings of type 3 is trivial, for type 2 less so, and for type 1 simple when one has found the centres of the circular arcs.[32] The use of the markings of type 3 is not convenient without a set of horizontal parallels.

4. *The Optional Cursor*

The addition of a movable cursor attached in a groove on the outer scale enables the user to enter H throughout the year knowing only the local latitude $\phi°$ and the solar longitude $\lambda°$ or the date if a calendar scale is present: it is not essential to the basic operation of the instrument. The scale on the cursor essentially serves to show the solar declination $\delta°$ as a function of $\lambda°$,[33] so that if the centre of the cursor is set at argument $90° - \phi°$, which is the solar equinoctial meridian altitude, the mark for $\lambda°$ on the cursor will be at argument $90° - \phi° + \delta°$, which is H, on the outer scale. It was not too much to expect that the medieval astronomer or user in general might know the solar meridian altitude on a given day: this function was tabulated in medieval ephemerides,[34] and medieval tables displaying time of day as a function of solar altitude sometimes had the solar meridian altitude as the other argument (otherwise it would be the solar longitude).[35]

The inclusion of a fixed cursor immediately limits the universal application of the cursor itself, but not the potential universal application of the horary markings. A solar scale anywhere other than on the outer rim of the quadrant, such as on the radii of the quadrant or on a radial rule, would also compromise the spirit of the markings, because the scale itself is necessarily latitude-dependent.

For more information on the history of the universal horary quadrant I refer the reader to my more detailed study mentioned at the beginning of this paper.

APPENDIX: THE BAGHDAD TREATISE
MS Cairo DM 969,4 (fols. 8v–9v)

Text

المصدر: مخطوطة دار الكتب المصرية ميقات ٩٦٩ (ق ٨ظ - ٩ظ)

ذكر العمل بربع الدائرة وهذه أبوابه

[١] في معرفة خطوطه وأسمائها اعلم أنّ الربع ربع دائرة تامّة والأوّل من خطوطه الخطّان اللّذان يتقاطعان على المركز أحدهما يسمّى خطّ المشرق والآخر خطّ نهاية الارتفاع وفيما بين هذين الخطّين[1] في محيط الربع أجزاء مقسومة متساوية وهي تسعون جزءا < . . . > تسمّى المجرة[2] وتعمل في بعض الأرباع متحرّكة وفي بعضها ثابتة ثم دونها ستّة خطوط مقوّسة[3] وهي قطع دوائر تسمّى الساعات الزمانية وقد تعمل هذه الخطوط في بعض الأرباع مستقيمة وفي بعضها أقطارًا تخرج[4] من مركز الربع والعمل بالكلّ سواء ثم مربّع الظلّ وكلّ ضلع من المربّع مقسوم باثنى عشر قسمًا يسمّى الضلع القائم على خطّ نهاية الارتفاع الظلّ المبسوط والآخر الظلّ[5] المنكوس وفي رأس الربع زيادتان فيهما ثقبان يسمّيان الدفّتين ويخرج من مركز الربع خيط فيه شاقول وفي الخيط عقدة متحرّكة تسمّى المري فهذا ما في الربع من الخطوط فاعلمه

[٢] في أخذ الارتفاع أيّ وقت شئت تمسك الربع بيدك وتحاذي بمنكبك وتحرف الربع القائم جرم الشمس ثم لا تزال تحرّك الربع صاعدًا ونازلًا حتّى ينفذ شعاع[6] الشمس من ثقب الدفّة العليا إلى ثقب الدفّة الأخرى وليكن إرسال الشاقول مع سطح الربع فإذا سكن الشاقول انظر كم قطع الخيط من درج الارتفاع فعدّ من خطّ المشرق والمغرب إليه فما كان فهو الارتفاع في ذلك الوقت

[٣] في معرفة ميل الشمس إذا أردت ذلك فاعرف اليوم الذي تريد ميل الشمس فيه من أيّ شهر يكون فضع الخيط على مثل ذلك اليوم من ذلك الشهر ثم انظر ما بين الخيط وخطّ نصف المجرة وهو خطّ الاعتدال من الدرج فما كان فهو ميل الشمس في ذلك اليوم فإن كان يومك فيما بين خطّ الاعتدال وخطّ المشرق والمغرب فالميل جنوبي

---

1 الخطّين] وفيما بين هذين *bis* MS   2 المجرة] المحرة MS   3 مقوّسة] معسو مقوسة MS   4 تخرج] يخرج MS   5 الظلّ] ظل MS   6 شعاع] الشعاع MS   7 خطّ] *bis* MS

وإن كان فيما بين خطّ[7] الاعتدال وخطّ نهاية الارتفاع فهو شمالي

[٤] في معرفة الماضي من النهار من الساعات الزمانية والظهر والعصر في كلّ زمان إذا أردت ذلك فضع الخيط على يومك في المجرة ثم حرّك المري حتّى تضعه على خطّ نصف النهار وهذا الخطّ الثالث فإذا عدّلته حدد[8] ارتفاع الشمس كما تقدّم ثم انظر أين وقع المري من الساعات فما كان فهي الساعات المارّة من النهار إن كانت[9] الشمس مشرقة وإن كانت مغربة فهي الساعات الباقية من النهار فتنقصها من اثنى عشر تبقى الساعات الماضية من النهار كذلك تعرف وقت الظهر والعصر بأن تأخذ الارتفاع فإذا وقع [9r] المري على خطّ نصف النهار فهو أوّل الظهر وكذلك إذا وقع على خطّ العصر والشمس مغربة فهو أوّل وقت العصر

[٥] في معرفة الظلّ المبسوط والمنكوس إذا أردت ذلك فاعرف ارتفاع الشمس كما تقدّم ثم انظر ما يقع عليه الخيط من الظلّ المبسوط إن أردته أو المنكوس فما كان فهو الظلّ لذلك الوقت فإن أردت الظلّ المبسوط ووقع الخيط في الظلّ المنكوس فاقسم عليه أبدًا مائة وأربعة وأربعين يخرج من القسمة الظلّ المبسوط فاعلم ذلك

[٦] في معرفة ارتفاع نصف النهار وارتفاع أوّل العصر في كلّ فصل إذا أردت ارتفاع أوّل الظهر فضع الخيط على اليوم الذي تريد فما قطعه الخيط من درج الارتفاع فهو ارتفاع نصف نهار ذلك اليوم وأمّا الارتفاع أوّل العصر فإنّك تعدّل المري على خطّ نصف النهار على ما تقدّم ثم تضعه أعني المري على خطّ العصر فما يقع عليه الخيط من الارتفاع فهو ارتفاع أوّل العصر

[٧] في معرفة الظلّ المبسوط لأوّل وقت الظهر وأوّل وقت العصر إذا أردت فاعرف ارتفاع أيّهما[10] أردت على ما تقدّم وضع الخيط في الأصابع فما كان فهو الظلّ لذلك الوقت فإن وقع الخيط على الظلّ المنكوس لأوّل الظهر والعصر < ... > فاعلم ذلك

[٨] في معرفة الأقدام إذا كانت معك أصابع وأردت أن تردّها أقداما فاضربها أبدًا في

---

[8] حدد] حد MS [9] إن كانت] *bis* MS [10] أيّهما] ايتهما MS

سبعة فما اجتمع فاقسمه على اثنى عشر تخرج من القسمة الأقدام لتلك الأصابع وكذلك إذا كانت معك أقدام وأردت أن تردّها أصابع فاضرب ما معك من الأقدام في اثنى عشر واقسمها على سبعة تخرج أصابع تلك الأقدام

[٩] في معرفة عرض البلد إذا دخلت بلدًا وأردت معرفة عرضه فخذ ارتفاع الشمس أرفع ما يكون بالرصد واحفظه[11] ثم اعرف ميل الشمس كما أعلمتك فإن كان جنوبيًا نقصته من ارتفاع الشمس نصف النهار وإن كان شماليًا زدته فما بلغ أو بقي فهو ارتفاع المعدّل فانقصه أبدًا من تسعين فما بقي فهو عرض ذلك الموضع واعلم أنّ المجرة المتحرّكة لا يعمل بها حتّى تعمل على مثل عرض البلد فأمّا الثابتة[12] فهي موضوعة[13] على عرض مخصوص فإذا أردت معرفة العرض الذي وضعت المجرة الثابتة عليه فانظر إلى خطّ نصف المجرة وهو خطّ الاعتدال على كم وقع من الارتفاع فما كان انقصه من تسعين فما بقي فهو عرض البلد الذي عملت له المجرة الثابتة وعرض هذه المجرة لجَ

[١٠] في معرفة كيفيّة العمل بالمجرة المتحرّكة أو الثابتة في جميع البلاد فإذا[14] دخلت بلدًا وأردت أن تعمل بالربع فاعرف عرض ذلك البلد على ما تقدّم ثم ضع خطّ نصف المجرة وهو خطّ الاعتدال على مثل ذلك العرض وهو أن تعدّ من خطّ نهاية الارتفاع مثل ذلك العرض وتضع المجرة على منتهى العدد ثم اعمل بها ما تقدّم من الأعمال فإن لم تكن[15] المجرة على عرض البلد لا يصحّ معك العمل فأمّا المجرة الثابتة فتعرف عرض البلد الذي عملت له المجرة الثابتة وتحفظه فإذا دخلت مدينة فاعرف عرضها فإن كانا متساويين فاعمل بالمجرة الثابتة على ثباتها بلا زيادة ونقصان وإن كان عرض البلد الذي دخلته أزيد من عرض البلد الذي أثبتت[16] فيه المجرة فاحفظ فضل ما بينهما وحرّك الخيط من يومك بمقدار هذه الفضلة إلى جهة خطّ المشرق والمغرب فيقع الخيط على يوم المعدّل لذلك البلد وإن كان عرض البلد الذي دخلته[17] انقص من عرض البلد الذي أثبتت[18] فيه المجرة فحرّك الخيط من يومك بقدر فضل ما بين العرضين إلى جهة خطّ نهاية الارتفاع فيقع الخيط على اليوم المعدّل لذلك البلد فاعرف ذلك وفّقك الله وهذه صورته

---

[11] واحفظه] MS واحفظ [12] الثابتة] MS الثانية [13] موضوعة] MS موضعة [14] فإذا] MS اذا [15] تكن] MS يكن [16] أثبتت] MS اثبت [17] دخلته] MS دخلت [18] أثبتت] MS اثبت

Translation

Notes: Arabic words in parentheses are the corresponding technical terms in the original. English words in parentheses are inserted to ensure the flow of the translation. Text in square brackets corresponds to lacunae in the Arabic text. Remarks in curly brackets are my comments.

(A Treatise on the Use of the Quadrant)

A statement concerning the use of the quadrant (*rubʿ al-dā'ira*). These are the ways it can be used (*abwābuhu*):

[**1**] On its markings (*khuṭūṭ*) and their names. The quadrant is a quarter of a complete circle. The first of its markings are the two lines which intersect at the centre, one of which is called the east(–west) line (*khaṭṭ al-mashriq*) and the other the meridian line (*khaṭṭ nihāyat al-irtifāʿ*). In between these two lines around the circumference of the quadrant there are equal divisions which make up ninety degrees. ... ... {LACUNA}; this is called the cursor (*al-majarra, sic* for *al-majrā* ?) and it is made movable in some quadrants and fixed in others (*tuʿmal fī baʿḍihā mutaharrika wa-fī baʿḍihā thābita*). Below (the cursor) there are six curved lines which are segments of circles called the seasonal hours (*al-sāʿāt al-zamāniyya*). On some quadrants these markings are made straight and on others (they are made) radii (*aqṭār*) radiating from the centre of the quadrant. The use of all of these is the same.

Then (there is) the shadow square (*murabbaʿ al-ẓill*), each side of which is divided into twelve parts. The side standing on the meridian line is called the horizontal shadow (*al-ẓill al-mabsūṭ*), and the other the vertical shadow (*al-ẓill al-mankūs*). On the upper side (*ra's*, lit., head) of the quadrant there are two parts which stick out (*ziyādatāni*) with two holes {i.e., one each} called the sights (*daffa*). A thread bearing a plummet (*shāqūl*) extends from the centre of the quadrant, also bearing a movable bead (*ʿuqda*, lit., knot!) called the marker (*al-murī*). These are the markings on the quadrant, so know this.

[**2**] On measuring the altitude at any time you wish. You hold the quadrant in your hand, line (it) up with your shoulder, and turn the now vertical quadrant towards the sun. Then you keep moving the quadrant up and down until the solar rays pass through the hole on the upper sight to (fall on) the hole on the other sight. (Also) the plummet should hang properly (*irsāl*) on the surface of the quadrant. When the plummet is still, see how many degrees of altitude are marked off by the thread, then count from the east–west line to that (point), and the result will be the altitude at that time.

[**3**] On finding the declination of the sun. If you want to do this, find the day on which you want the declination of the sun, whatever month it is in, and place the thread on that day of the month. Then look at how many degrees are between the thread and the line (joining the centre to) the middle of the cursor, which is the

equinox line. The result will be the declination of the sun on that day. If your day is between the equinox line and the east–west line, then the declination is southerly; if it is between the equinox line and the line of the limit of the altitude, then (the declination) is northerly.

[**4a**, corrupt] On finding the time of day elapsed in seasonal hours, and the (times of) the *ẓuhr* and the *ʿaṣr* (prayers) at any time (of the year). If you want this, place the thread on your day on the cursor and then move the marker until you place it on the meridian line, which is the third line (?). Then if you adjust it, fix the altitude of the sun as (described) previously. Then look where the marker has fallen on the hours, and this will be the hours of daylight passed if the sun is in the east; if (the sun) is the west, it will be the hours of daylight remaining, so you subtract it from twelve and the remainder will be the hours of daylight passed.

[**4b**, corrupt] Likewise you find the time of the *ẓuhr* and the *ʿaṣr* by taking the altitude and if the marker falls on the meridian line, it will be the beginning of the *ẓuhr*. Similarly, if it falls on the *ʿaṣr* line and the sun is in the west, it will be the beginning of the time of the *ʿaṣr*.

[**4a**, reconstructed] On finding the time of day elapsed in seasonal hours, and the (times of) the *ẓuhr* and the *ʿaṣr* (prayers) at any time (of the year). If you want this, place the thread on your day on the cursor and then move the marker until you place it on the meridian line. Then if you adjust it, fix the altitude of the sun as (described) previously. Then look where the marker has fallen on the hours, and this will be the hours of daylight passed if the sun is in the east; if (the sun) is the west, it will be the hours of daylight remaining, so you subtract it from twelve and the remainder will be the hours of daylight passed.

[**4b**, reconstructed] Likewise you find the time of the *ẓuhr* and the *ʿaṣr* by (first) taking the altitude when the marker falls on the meridian line: this will be the beginning of the *ẓuhr*. Similarly, if it falls on the *ʿaṣr* line, which is the third hour line, and the sun is in the west, it will be the beginning of the time of the *ʿaṣr*.

[**5**] On finding the horizontal and vertical shadows. If you want this, find the altitude of the sun as (described) previously and then look where the thread falls on the horizontal shadow (scale) if you want that or on the vertical shadow (scale) (if you want that). The result will be the shadow at that time. If you want the horizontal shadow and the thread has fallen on the vertical shadow, divide it into one hundred and forty-four and the resulting quotient will be the horizontal shadow, so know this.

[**6**] On finding the altitude at midday and the altitude at the beginning of the *ʿaṣr* at any time of the year. If you want the altitude at the beginning of the *ẓuhr*, put the thread on the day you want (on the cursor) and the amount of altitude degrees the thread cuts off will be the altitude at midday on that day. The altitude at the beginning of the *ʿaṣr* (is found as follows). You adjust the marker on the meridian as (described) previously,

then you place it, that is, the marker, on the *ʿaṣr* line. Where the thread falls on the altitude (scale) will be the altitude of the beginning of the time of the *ʿaṣr*.

[**7**] On finding the horizontal shadow at the beginning of the times of the *ẓuhr* and *ʿaṣr*. If you want, find the altitude of whichever one of the two you wish as (described) previously, and place the thread on the digits (of the shadow scale) and the result will be the shadow at that time. If the thread falls on the vertical shadow (scale) at the beginning of the *ẓuhr* and the *ʿaṣr*, ... {LACUNA}, so know that.

[**8**] On finding (the shadow in) feet if you know (the shadow) in digits and you want to convert it to feet. Multiply (the digits) by seven and divide the product by twelve: the quotient will be the feet corresponding to those digits. Likewise, if you have (the shadow in) feet and you want to convert it to digits; multiply the feet you have by twelve, divide (the product) by seven, and the result will be the digits corresponding to those feet.

[**9**] On finding the local latitude. If you come to a locality and you want to know its latitude, observe the altitude of the sun when it is at its highest and keep it in mind. Then find the declination of the sun as I have told you, and if it is southerly subtract it from the altitude of the sun at midday, and if it is northerly add it. The result of addition or subtraction will be the altitude of the celestial equator: subtract it from ninety and the remainder will be the latitude of that place. Know that the movable cursor is not to be used until it has been placed on the equivalent of the local latitude (on the altitude scale). The fixed cursor is set at a particular latitude. If you want to know the latitude for which the fixed cursor is set, look at the line (from the centre to) the middle of the cursor, which is the line of the equator, (and see) where it falls on the altitude (scale). Subtract this from ninety and the remainder will be the local latitude for which the fixed cursor was made. The latitude of this cursor {in the illustration that follows?} is 33°.

[**10**] On the method of using the movable or fixed cursor in all localities. If you come to a locality and you want to work with the quadrant, find the latitude of that locality as (described) previously. Then place the line (joining the centre) to the middle of the cursor, which is the equinox line at that latitude, by counting the latitude from the meridian line and placing the cursor at the argument which you reach. Then perform with it the operations that have been described previously. If the cursor is not set at the local latitude, then the operations are not correct. With regard to the fixed cursor, find the local latitude for which it was made and keep it in mind. If you enter any town determine its latitude, and if the two (latitudes) are the same, work with the fixed cursor as it is without modification. If the latitude of the locality you entered is greater than the latitude for which the cursor is fixed, keep in mind the difference between them and move the thread from your day by the amount of this difference in the direction of the east–west-line so that the thread falls on the day adjusted for that locality. If the latitude of the locality you entered is less than the

latitude of the locality for which the cursor was fixed, move the thread from your day by the difference between the two latitudes in the direction of the meridian line, and the thread will fall on the day adjusted for that locality. Know this and may God grant you success. This is an illustration of (the quadrant): {BLANK}.

Commentary

Introduction: The term *bāb*, pl. *abwāb*, is often used in Arabic technical texts in the sense of "method", rather than the standard meaning "chapter" (literally "door"). In the manuscript there are empty spaces which normally might be filled with the chapter numbers (*al-bāb al-awwal*, "the first chapter", *etc.*) in red ink. However, I doubt that the chapters were numbered in the ninth-century text; perhaps all that is missing each time is the word *bāb*, "chapter".

Chap. 1: The terminology is eminently sensible and in its time set new standards for the field.[36] Here and elsewhere for the movable cursor in its groove the manuscript has *al-m-j-r-h*, perhaps to be read *al-majarra* (root *j-r-r*, "to pull"), and meaning "a place or path where something is pulled"; this is perhaps an error for *al-majrā* (written *m-j-r-y*, root *j-r-y*, "to run"), which literally means "a place for running" and would be singularly appropriate if applied to the groove.[37] The word *al-majarra* is used for the Milky Way in Arabic.[38] One could argue that a Persian-speaker rather than an Arabic-speaker would be more likely to mistakenly write the word pronounced *majrā* as *m-j-r-h*. On the other hand, *m-j-r-h* is treated consistently (and correctly) as feminine in the manuscript, whereas *majrā* is masculine. In al-Marrākushī's treatise, the term used for the cursor on the sine quadrant on the back of the *zarqāllī* quadrant is *m-j-r-h*, sometimes with a *shadda* on the "r" (to double the "r").[39]

Chap. 4: See Figure 3(a) for the basic procedures to find T(h, H). The text is corrupt and I have restored the meaning in the translation. If my restoration of this chapter is correct (see below *ad* Chap. 7), our author takes the beginning of the *ẓuhr* as midday and the beginning of the *ʿaṣr* as the middle of the afternoon, using the "third line". No other medieval Arabic astronomical texts proposing this definition for the *ʿaṣr* are known to me, although it is precisely this definition that underlies and inspired the standard definition in terms of the shadow increase that was used from the ninth century onwards. Certainly our author is not using those "new" definitions, because in Chap. 7 he proposes finding the shadows at the prayer-times from the corresponding altitudes.[40] An early work of al-Khwārizmī is a table displaying the shadows to one digit (base 12) at the two prayer-times for the latitude of Baghdad (33°) and each 6° of solar longitude. Each set of values was independently computed, showing that he was not using the "new" definitions.[41]

Chap. 5: The horizontal and vertical shadows, s and s′, for a gnomon length n = 12 when the solar altitude is h are related by:

$$s \times s' = n \cot h \times n \tan h = n^2 = 144.$$

In the Latin tradition a distinction is made between solar altitudes less than and greater than 45°.[42]

Chap. 7: The fact that the shadow-lengths need to be calculated confirms my restoration of Chap. 4.

Chap. 8: The two most common values for the length of the gnomon in Islamic astronomy from the eighth century onwards were 12 digits and 7 feet, others being $6\frac{1}{2}$ and $6\frac{2}{3}$.[43]

Chap. 9: We note the mention of latitude 33°. This is an early value for Baghdad used by al-Khwārizmī,[44] but it was still used in the tenth century: thus, for example, it is found on the astrolabes of Ḥāmid ibn ʿAlī al-Wāsiṭī and Ḥāmid ibn Khiḍr al-Khujandī.[45] Better values had been determined already in the early ninth century. The value 33;9° for the latitude of Baghdad is also associated with al-Khwārizmī, although this may be result from a scribal error for the better value 33;21°, which, however, is attested first in the tenth century.[46]

Chap. 10: Perhaps the copyist in Meshed thought that a quadrant with a cursor set for latitude 33° was not worth copying. Or maybe it was already missing from his original (for the same reason).

Acknowledgements

It is a pleasure to thank François Charette for his critical reading of the penultimate version of this paper.

REFERENCES

1. See my article "Rubʿ" [= quadrant] in *The encyclopaedia of Islam*, new edn (10 vols to date, Leiden, 1960–), viii (1995), 574–5 and 4 plates, and David A. King, *Studies in astronomical timekeeping in medieval Islam* (hereafter *SATMI*), 12 parts (Leiden, in press), VIII: *Astronomical instrumentation in the Islamic world*, Section 6, for overviews of the early history of the different kinds of quadrant in the Middle Ages. Peter Schmalzl, *Zur Geschichte des Quadranten bei den Arabern* (Munich, 1929), is still useful.

   The Baghdad treatise was announced already in King, "Early Islamic astronomy", *Journal for the history of astronomy*, xii (1981), 55–59, p. 59. The findings of this paper are summarised in *idem*, "The neglected astrolabe", in *Mathematische Probleme in Mittelalter: Der lateinische und arabische Sprachbereich*, ed. by M. Folkerts (Wiesbaden, 1996), 45–55, pp. 48 and 51–52, and *idem*, "Astronomical instruments between East and West", in *Kommunikation zwischen Orient und Okzident: Alltag und Sachkultur*, ed. by Harry Kühnel (Vienna, 1994), 143–98, p. 165.
2. See ref. 10 below. In the modern literature the best overview of the markings and their history is M. Archinard, "The diagram of unequal hours", *Annals of science*, xlvii (1990), 173–90.
3. The seasonal day hours of Antiquity and the Middle Ages are the one-twelfth divisions of the length of daylight; they are hence dependent on terrestrial latitude and solar longitude.
4. On the astrolabe see John D. North, "The astrolabe", *Scientific American*, ccxxx (1974), 96–106, repr. in *idem*, *Stars, minds and fate: Essays in ancient and medieval cosmology* (London and

Ronceverte, W.V.), 1989, no. 14, and King, "Astrolabe" (ref. 1), the latter dealing, in particular, with the origin of some of the components.

5. King, *SATMI* (ref. 1), IXa: *On the universal horary quadrant for timekeeping by the Sun.* This includes studies of the development of the shadow-square and the universal sundial on astrolabe alidades.
6. On universal instruments and tables see further King, *SATMI* (ref. 1), VI: *Universal solutions in Islamic astronomy.* Alas the universal horary quadrant was omitted from the earlier studies (first published in the 1980s and repr. in King, *Astronomy in the service of Islam* (Aldershot, 1993), VI and VII) on which the new versions are based. On the concept of universality in relation to the standard astrolabe see *idem*, "Bringing astronomical instruments back to Earth: The geographical data on medieval astrolabes (to *ca.* 1100)", in *Between demonstration and imagination: Essays in the history of science and philosophy presented to John D. North*, ed. by A. Vanderjagt and L. Nauta (Leiden, 1999), 3–53, pp. 6–11 and 14–17.
7. Likewise I use the expression "universal horary dial" to refer to the kind of markings found, *i.a.*, on the medieval instrument known as the *navicula* or the *Uhrtäfelchen* of Regiomontanus: they too are universal, they serve the determination of the (equinoctial) hours using the exact formula, and they consist of a set of fixed horary markings to be used in conjunction with an ingenious movable device with thread and movable bead attached with which one can enter the local latitude and the solar declination. This seems preferable to various other expressions in the modern literature, some with good historical backgrounds, such as "Uhrtäfelchen" or "cadran solaire rectiligne" or "rectilinear dial" or "universal sundial". The correct use of the universal horary dial on the *navicula* is nowhere described in the modern literature. See further the parallel study listed as King, "14th-century England or 9th-century Baghdad? New insights on the origins of the elusive astronomical instrument called *navicula de Venetiis*", to appear in a special issue of *Centaurus* in honour of Bernard R. Goldstein, and a more detailed investigation in *idem, SATMI* (ref. 1), IXb: *On universal horary dials for determining time by the sun and stars.*

   There is a sense in which the quadrant and the dial serve the same purpose: both serve essentially only timekeeping by the sun; both are universal, but one is approximate and the other is accurate; one is very easy to use and the other requires some dexterity, especially to arrive at the exact solution.

   A more complex mathematical device is required for timekeeping for any latitude by the stars, since this also involves the sun and various kinds of stellar data must be accessible. Such a device was invented in Baghdad in the ninth century: see King, *World-maps for finding the direction and distance to Mecca: Innovation and tradition in Islamic science* (Leiden and London, 1999), 354–9, for some first reflections; F. Charette and P. G. Schmidl, "A universal plate for timekeeping with the stars by Ḥabash al-Ḥāsib: Text, translation and preliminary commentary", *Suhayl — Journal for the history of the exact and natural sciences in Islamic civilisation* (Barcelona), ii (2001), 107–59, for the text and an annotated translation; and King, "*Navicula*", and a more detailed investigation in *idem, SATMI*, IXb-13, on the *modus operandi.* The universal horary quadrant can of course be used to find the time of night from the instantaneous and culminating altitudes of any star, but the time is then given in "stellar hours" related to the arc of visibility of the star in question. Only one medieval astronomer who used such "hours" is known to me (*SATMI*, I-4.6.1).
8. Of the substantial literature on the Latin *quadrans vetus* I cite only W. R. Knorr, "The Latin sources of *quadrans vetus*, and what they imply for its authorship and date", in E. Sylla and M. McVaugh (eds), *Texts and contexts in ancient and medieval science: Studies on the occasion of John E. Murdoch's seventieth birthday* (Leiden, 1997), 23–67, and *idem*, "Sacrobosco's *quadrans*: Date and sources", *Journal for the history of astronomy*, xxviii (1997), 187–222, where the reader will find references to earlier works. On the Middle Castillian treatise in the *Libros del saber* (ed. Rico y Sinobas (Madrid, 1873), v, 285–316), see J.-M. Millás Vallicrosa, "La introducción del cuadrante con cursor en Europa", *Isis*, xvii (1932), 218–58, repr. in *idem, Estudios sobre historia de la ciencia española* (Barcelona, 1949; repr. Madrid, 1987), 65–110, pp. 66–67, etc.
9. Likewise on the *quadrans novus* I cite only E. Poulle, "Le quadrant nouveau médiéval", *Journal des*

savants, 1964, 148–67 and 182–214. On the surviving instruments see E. Dekker, "An unrecorded medieval astrolabe quadrant from *c*. 1300", *Annals of science*, lii (1995), 1–47.

10. Thus John North, after finding these markings on about 40 of some 130 astrolabes in the Museum of the History of Science, Oxford, wrote ("Astrolabes and the hour-line ritual", *Journal of history of Arabic science* (Aleppo), v (1981), 113–14, repr. in *idem*, *Essays* (ref. 4), no. 15):

    "At best, the lines (on the back of an astrolabe) can give the unequal hour with an accuracy only about half as great [!] as that given by the conventional astrolabe itself. At worst, the lines are carelessly drawn, unnumbered, very small indeed, and — worst of all — not associated with an auxiliary scale of solar positions.... Not a single medieval (astrolabe) has survived in a form which would suggest that the unequal-hour lines were used meaningfully."

    The correct use of these markings — with a mark on the alidade to represent the solar meridian altitude — is described in medieval and Renaissance treatises, but it is not found in the standard modern literature on the astrolabe. See further King, *SATMI*, IXa-9.

11. An exception is S. García Franco, *Catálogo crítico de los astrolabios existentes en España* (Madrid, 1945), 218–21, p. 219, where the application of ink to mark the meridian position on the alidade is mentioned.

12. For example, in the popular Latin astrolabe treatise attributed to Messahalla but actually based on the Arabic of Maslama al-Majrīṭī (*c*. 1000) (R. T. Gunther, *Early science in Oxford*, v: *Chaucer and Messahalla on the astrolabe* (Oxford, 1929), 221 and 174); the astrolabe treatise of Jean Fusoris, compiled in Paris *c*. 1400 (E. Poulle, *Un constructeur d'instruments astronomiques au 15e siècle — Jean Fusoris* (Paris, 1963), 116); the early sixteenth-century astrolabe treatise of Johannes Stöffler (*Elucidatio*, 1524 edn, Oppenheim, fol. 28r); and an English astrolabe treatise partly based on Stöffler, *The travailer's ioy and felicitie*, by Robert Tanner (facsimile in R. T. Gunther, *The astrolabes of the world* (2 vols, Oxford, 1932; repr. in 1 vol., London, 1976), ii, 511).

13. See ref. 16 on the earliest text, and M. Viladrich, "Medieval Islamic horary quadrants for specific latitudes and their influence on the European tradition", *Suhayl — Journal for the history of the exact and natural sciences in Islamic civilisation*, i (2000), 273–355, for an overview. New insights are in F. Charette, "Mathematical instrumentation in 14th-century Egypt and Syria", doctoral thesis, Johann Wolfgang Goethe University, Institute for History of Science, 2002, Section 3.1.

14. MS Cairo DM 969,4, fols. 8v–9v, copied in an orderly but inelegant Persian hand. The manuscript, and also MS DM 970 by the same copyist *c*. 1800, contains a series of important treatises on instruments from the ninth and tenth century, some, like this treatise on the universal horary quadrant, unique, and all unpublished. For details on the quadrant text see further D. A. King, *A catalogue of the scientific manuscripts in the Egyptian National Library* (in Arabic) (2 vols, Cairo, 1981 and 1986), i, 168–9, and ii, 540–1 (*ad* 4.6.12), and *idem*, *A survey of the scientific manuscripts in the Egyptian National Library* (in English) (American Research Center in Egypt, Catalogs, v; Winona Lake, Ind., 1987), p. 53 (no. B105) and pl. LIVa (caption on p. 207).

15. On al-Khwārizmī see article by Gerald Toomer in *Dictionary of scientific biography*. In D. A. King, "al-Khwārizmī and new trends in mathematical astronomy in the ninth century", *Occasional papers on the Near East* (New York University, Hagop Kevorkian Center for Near Eastern Studies), ii (1983), 43 pp., a study of various treatises attributed to him discovered in the 1970s, mainly dealing with instruments, this anonymous treatise is not mentioned.

16. On his writings on the horary quadrant for a fixed latitude see King, "al-Khwārizmī" (ref. 15), 30–31. On his writings on the sine quadrant and the purpose for which he invented it, namely, to solve the formula underlying the universal horary quadrant, see *ibid.*, 27–29, and 7 and 10–11. For the text and a new translation, see F. Charette (with P. Schmidl), "Scientific initiative in 9th-century Baghdad: al-Khwārizmī on astronomical instruments", to appear, text J.

    Of course, the universal horary quadrant of type 2 — see Fig. 2(b) — is nothing other than one variety of the sine quadrant. R. Lorch ("A note on the horary quadrant", *Journal of history of Arabic science*, v (1981), 115–20 (repr. in *idem*, *Arabic mathematical sciences: Instruments, texts, transmission* (Aldershot, 1995), XVII), and "Some early applications of the sine quadrant",

Suhayl — Journal for the history of the exact and natural sciences in Islamic civilisation, i (2000), 251–72) has stressed the origin of the sine quadrant as a device for solving the approximate formula for timekeeping. See also J. P. Hogendijk, "Al-Khwārizmī's tables of the 'sine of the hours' and the underlying sine table", *Historia scientiarum*, xlii (1991), 1–12.

17. On the times of prayer in Islam see the article "Mīḳāt, ii" [= astronomical timekeeping and the regulation of the times of prayer] in *Encyclopaedia of Islam* (ref. 1). See further refs 21 and 28 and the commentary to Ch. 4 of the Baghdad treatise.

18. This parameter, which is not particularly accurate, is used in various other treatises associated with him: see King, "al-Khwārizmī" (ref. 15), 2, and also *idem*, *Astronomy in the service of Islam* (ref. 6), XIV, 129. See further the commentary to Chap. 9 of the Baghdad treatise.

19. On al-Marrākushī see the article in *Encyclopaedia of Islam* (ref. 1), and J.-J. Sédillot, *Traité des instruments astronomiques des Arabes composé au treizième siècle par Aboul Hhassan Ali de Maroc* ... (2 vols, Paris, 1834–35, repr. in 1 vol., Frankfurt am Main, 1985), and L. A. Sédillot, "Mémoire sur les instruments astronomiques des Arabes", *Mémoires de l'Académie Royale des Inscriptions et Belles-lettres de l'Institut de France*, i (1844), 1–229, (repr. Frankfurt am Main, 1989). On some of these references see Charette, "Mathematical instrumentation" (ref. 13), Section 5.1.

20. See King, "al-Khwārizmī" (ref. 15), 10–11, on some mysterious tables by him entitled *jayb al-sā'āt*, "sine of the hours", and the rather surprising explanation in Hogendijk, "Al-Khwārizmī's sine of the hours" (ref. 16).

21. A German translation of the treatise on the *use* of the instrument was published by J. Frank in 1922. Brief remarks on the treatise on the *construction* of the instrument are in King, "al-Khwārizmī" (ref. 15), 22–27. See now the edition of both, with translation and commentary, in Charette, "al-Khwārizmī on instruments" (ref. 16).

    Two reasons why this would have to be a late work of his are: (1) In the treatise on the *use* of the astrolabe he is still experimenting with new definitions of the times of the *ẓuhr* and *'aṣr* prayers: see the commentary to Chap. 4. (2) In his treatise on the *construction* of the astrolabe al-Khwārizmī proposes a simple linear horizontal shadow-scale, 12 units below the horizontal diameter, and in this treatise on the universal horary quadrant we are dealing with a more developed shadow-square.

22. Compare, for example, the Arabic in the texts edited in Charette, "al-Khwārizmī on instruments" (ref. 16).

23. See ref. 14 above.

24. D. A. King and J. Samsó, with a contribution by B. R. Goldstein, "Astronomical handbooks and tables from the Islamic world (750–1900): An interim report", *Suhayl — Journal for the history of the exact and natural sciences in Islamic civilisation*, ii (2001), 9–105, pp. 56–59. (A summary of this paper is in the article "Zīdj" [= astronomical handbooks with tables] in *Encyclopaedia of Islam* (ref. 1).)

25. Most of Ḥabash's treatises on instruments use 34° for the latitude of the Abbasid capital, Samarra: see King, *Mecca-centred world-maps* (ref. 7), 40–41 and 345–59.

26. J. Drecker (*Die Theorie der Sonnenuhren*, vol. 1E of *Die Geschichte der Zeitmessung und der Uhren*, ed. by E. von Bassermann-Jordan (Berlin and Leipzig, 1925), 86) mentions copies of Sacrobosco's treatise in the Bayerische Staatsbibliothek in Munich in which the text is associated with "Thabit" (Clm 10661) and "Albatenius" (Clm 14583). No such treatise was known to Francis Carmody, who in 1960 surveyed various Arabic and Latin treatises associated with Thābit ibn Qurra (*Dictionary of scientific biography*, *s.v.*). Likewise, no such treatise was known to Régis Morelon, who in 1987 published various original Arabic texts by Thābit. "Albatenius", the early tenth-century Syrian astronomer al-Battānī (*ibid.*, *s.v.*), is not known to have written on instruments other than those in the *Almagest* (the section on sundials in the unique copy of his astronomical handbook is spurious).

27. On the calendars used by Muslim astronomers see King and Samsó, "Islamic astronomical handbooks and tables" (ref. 24), 19–20, and the article "Ta'rīkh, iv" [= calendars] by B. van Dalen in

Encyclopaedia of Islam (ref. 1).

28. In Christian Europe, all of the specifically Islamic materials in Islamic astronomical works, such as the treatment of the times of the *ẓuhr* and the *ʿaṣr* prayers, were invariably suppressed in new Latin versions. This was perhaps unfortunate in the case of the Baghdad treatise, for the procedures advocated here (see the commentary to Chap. 4) would have been very useful for determining the *sexts*, as well as the *terces* and *nones*, but these were surely of less interest at the university scene in Paris than in monasteries.
29. See already, for example, J.-B. Delambre, *Histoire de l'astronomie du moyen âge* (Paris, 1819; repr. New York and London, 1965), 243–7; Drecker, *Theorie der Sonnenuhren* (ref. 26), 86, for a derivation of the formula; and Lorch, "Horary quadrant" (ref. 16), where the hour-angle is used instead of the time of day.
30. See King, *SATMI*, VII: *An approximate formula for timekeeping from 750 to 1900*, and also Charette, "Mathematical instrumentation" (ref. 13), Section 3.2–3.
31. Compare North, "Hour-line ritual", 113, quoted in ref. 10 above. The errors using the formula are investigated more fully in King, *SATMI*, VII-2.3 (ref. 30).
32. See the discussions in Lorch, "Horary quadrant" (ref. 16), 116–17; King, *SATMI*, VII-2.2; and Charette, "Mathematical instrumentation" (ref. 13), Section 5.2.
33. Similar scales for converting λ to δ or sin δ or tan δ are found on other early medieval instruments. An example for δ is found on a universal instrument for timekeeping by the stars by Habash al-Hāsib, a contemporary of al-Khwārizmī (see Charette and Schmidl, "Ḥabash's universal plate" (ref. 7), 154). An example with which one could find sin δ and cos δ is on the universal horary dial on the sole surviving *albion* in the Osservatorio Astronomico, Rome (illustrated in J. D. North, *Richard of Wallingford: An edition of his writings with introductions, English translation and commentary* (3 vols, Oxford, 1976), iii, pl. XXIII). The declination scale on the bottom of the universal horary dial on the medieval English *navicula*s serves to determine δ and that at the side tan δ. See further King, "*Navicula*" (ref. 7), and *idem*, *SATMI*, IXb-7-8.
34. As, for example, in the fourteenth-century tables for the Yemen and Oxford — see the article "Taḳwīm" [= ephemeris] in *Encyclopaedia of Islam* (ref. 1), and S. Eisner and G. MacEoin, *The Kalendarium of Nicholas of Lynn* (Athens, Ga., 1980), respectively. In the tables of solar altitude for the equinoctial hours at the latitude of Baez compiled *c.* 1435 by pseudo-Enrique de Villena (*Tratado de astrología atribuido a Enrique de Villena*, ed. by P. M. Cátedra and Julio Samsó (Madrid, 1980), 171–6), one argument is the dates for which the solar meridian altitude can be rounded to a given degree.
35. See King, *SATMI*, I: *A survey of Islamic tables for reckoning time by the sun and stars*, Section 2.
36. On some of the terms see P. Kunitzsch, "Glossar der arabischen Fachausdrücke in der mittelalterlichen europäischen Astrolabliteratur", *Nachrichten der Akademie der Wissenschaften in Göttingen*, Philologisch-historische Klasse, 1982, no. 11, 455–571 (separatum paginated 1–117).
37. Ḥabash uses *shaẓiyya* and *majrā* in his treatise on the universal horary dial for timekeeping by the stars to describe first the movable cursor and then the hollowed groove in which it fits on a diametrical rule: see Charette and Schmidl, "Ḥabash's universal plate" (ref. 7), 152–3.
38. See the article "Madjarra" [= Milky Way] by P. Kunitzsch in *Encyclopaedia of Islam* (ref. 1).
39. Sédillot-*fils*, "Mémoire" (ref. 19), 104–5: *mujerrih*; also Schmalzl, *Geschichte des Quadranten* (ref. 1), 124–6.
40. Using the new definitions in terms of shadow increases, the shadows would be given, and one would need to calculate the altitudes.
41. See further King, "al-Khwārizmī" (ref. 15), 7–9; *idem*, *SATMI*, II: *A survey of Islamic tables for regulating the times of prayer*, Section 3.1; and IV: *On the times of prayer in Islam*, Section 4.5. The apparent divergence between these definitions — nowhere more apparent than on medieval sundials or astrolabe plates showing both the seasonal hours and the times of the *ẓuhr* and *ʿaṣr* prayers — is the result of the approximate nature of the Indian formula that was used to relate the time of day to shadow increases.
42. On shadow scales see ref. 5 above.

43. See E. S. Kennedy, *The exhaustive treatise on shadows by Abū al-Rayḥān Muḥammad b. Aḥmad al-Bīrūnī* (2 vols, Aleppo, 1976), i, 68–79, and ii, 25–31; *idem et al.*, *Studies in the Islamic exact sciences* (Beirut, 1983), 23–25; and King, *SATMI*, III: *A survey of Islamic arithmetical shadow-schemes for simple time-reckoning*, Section 1.2.
44. King, "al-Khwārizmī" (ref. 15), 2. See also E. S. Kennedy and M. H. Kennedy, *Geographical coordinates of localities from Islamic sources* (Frankfurt am Main, 1987), 55.
45. Illustrated in King, *Mecca-centred world-maps* (ref. 7), 18–19 and 356.
46. See further Kennedy and Kennedy, *Islamic geographical coordinates* (ref. 44), 55–56; King, "Too many cooks ... — A newly-rediscovered account of the first Muslim geodetic measurements", *Suhayl — Journal for the history of the exact and natural sciences in Islamic civilisation*, i (2000), 207–41, pp. 226–7; and *idem*, "Geography of astrolabes" (ref. 6), 13–14.

IX

14th-Century England or 9th-Century Baghdad? New Insights on the Elusive Astronomical Instrument Called *Navicula de Venetiis*

To the best thesis advisor a guy ever had.

1. Introduction

The medieval English instrument for timekeeping by the sun known as *navicula de Venetiis*, "the little ship of the Venetians", has had a very chequered career in the modern literature; indeed, one could argue that no single historical instrument was so misunderstood by several generations of scholars in the 20th century. Yet there are good reasons why this should have been the case, and why only now we can begin to see some light at the end of the tunnel.[1] The most important component of the *navicula* is a universal horary dial with which one can find the time of day more or less accurately[2] from the solar altitude and the solar declination for any terrestrial latitude. A series of modern scholars have accepted that the device functions only approximately, but without considering the construction of the various scales.[3] However, the mode of construction and use described in the available medieval texts on the *navicula* does indeed yield an approximate solution for the general case. Furthermore, until recently, only two medieval *navicula*s were known, the first carelessly made and now preserved in Oxford, and the second not in the form of a ship and now preserved in Florence.[4] However, just in the last 15 years, three other 14th-century examples of English *navicula*s – which I label Greenwich, Colchester and Geneva – have been rediscovered, so the time is ripe for a new look at the *navicula*. The reader will soon see that many questions surrounding this remarkable device cannot be answered yet.

In the 1920s when Robert Gunther first turned his attention to the *navicula*, only the crudely-fashioned Oxford piece was known; this he maintained was German, although, in fact, it is 14th- or 15th-century English. Gunther also published a medieval English text in Latin on the construction and use of the instrument.[5] In the 1950s Ernst Zinner identified a medieval German tradition of the same instrument in a different manifestation, that is, not in the form of a ship.[6] The universal horary dial thus circulated in more than one form, one variety having a rotatable rectilinear latitude scale, which on the *navicula* was developed into the mast. The other had a *brachiolus* and fixed latitude scale, later adapted as the well-known *Uhrtäfelchen* by Regiomontanus, who mentions in his 1434 treatise that he found the device in the writings of an earlier writer (*antiquus compositor*), alas unidentified.[7] In 1960 Derek Price published a Middle English treatise on the *navicula*, clearly derived from the Latin one published by Gunther.[8] In 1980 Giuseppe Brusa published an overview of *navicula*s and identified the other medieval piece in Florence, now known to be of English provenance and probably 14th century,[9] as well as several Renaissance examples. Around 1990 Jan Kragten investigated for the first time both the construction and use of the *navicula* in the light of these texts, confirming that the instrument there described indeed functioned only approximately, but perhaps better than others had thought, and also that the available *navicula*s (notably the Oxford and Greenwich pieces) were not particularly well made anyway.[10] In brief, he demonstrated how the latitude scale on the mast, and each of the two declination scales, were constructed according to singularly complicated procedures in order to achieve an approximate solution for the general case. Apparently the motivation for this was to facilitate the use of the device, although, as we shall see, much simpler constructions can yield an instrument that will yield accurate solutions for the general case at the expense of having to make a slight adjustment during the course of setting up the instrument for use.

The first of the new set of medieval *navicula*s that have recently come to light is now in Greenwich; it is not carefully made, but at least is better than

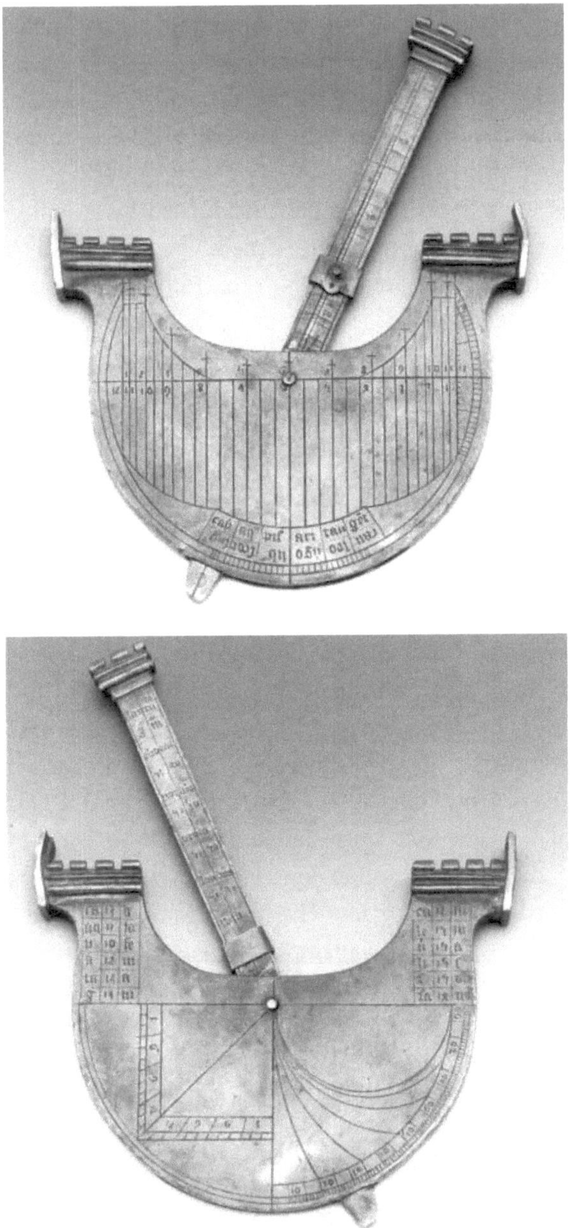

Fig. 1. The front and back of the Geneva *navicula.* This is the only one of the medieval English pieces that is made according to the accurate ("Regiomontanus") method. In addition it is the only one that has a set of markings on the right-hand 12th-hour line for the tangent of the declination. (Photos courtesy of Sotheby's of London.)

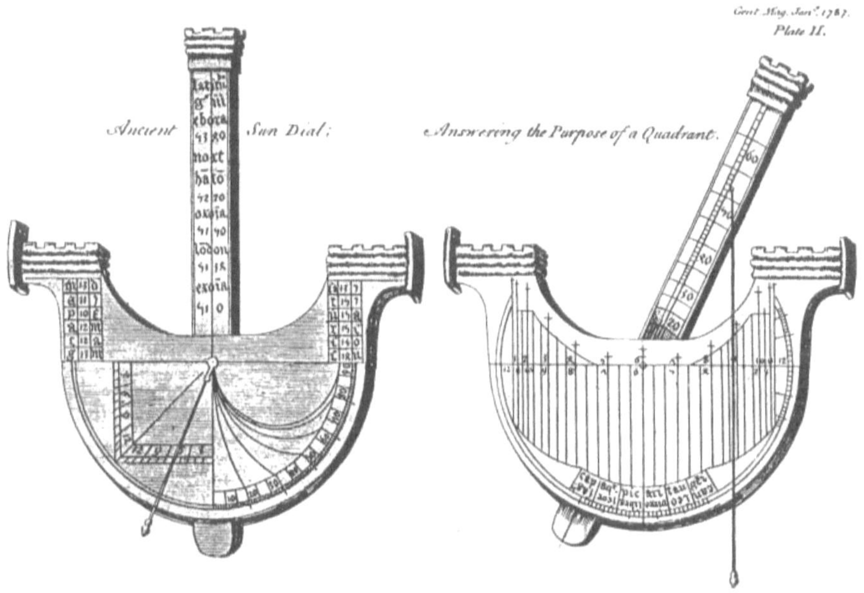

Fig. 2. A sketch of the Colchester *navicula* published in 1787 but forgotten until 1993. The sketch is carefully drawn and seems to faithfully represent the original, alas lost. This instrument represents the "modified" version. (From Delehar, "Illustrations", p. 387.)

the Oxford piece. (The problems associated with the scales on these pieces were identified by Kragten.) The second is now in Geneva (see Fig. 1.); the third, once in Colchester, is known only from an 18th-century illustration (see Fig. 2.).[11] It has not been previously stressed that at least these last two "new" *navicula*s are 14th-century English (the Greenwich piece may be 14th or 15th century English), and that they possibly hail from the same workshop, if not from the same craftsman. Indeed, there can be no doubt that the *navicula de Venetiis* as we know it (probably identical with the instrument called simply *navis*) hails from 14th-century England.[12] The calendrical data on these three new pieces, also found in the English treatise on the *navicula* in Latin (the Middle English version is missing this part), can be derived from 14th-century English astronomical tables; also, the latitudes of various cities in England engraved on the backs of the masts of the same three *navicula*s are taken from earlier English geographical tables.[13]

It has also not been previously noted that the Geneva *navicula* is based on a different construction procedure from that underlying the Greenwich and

Colchester pieces.[14] Indeed, the Geneva piece is apparently based on simpler, more direct procedures, and is complete with all of the components necessary to yield an accurate solution. Henceforth I shall be referring to the Geneva instrument when talking about the "standard" variety of the *navicula.* Also I shall be at liberty to reconstruct its *modus operandi*, since this is necessarily not the same as that described in the medieval English texts. These present a "modified" manifestation of the same instrument, represented by the Oxford, Greenwich and Colchester pieces, and also the Florence dial. Nobody could have conceived the "modified" version without being fully cognizant of, or himself having invented, the "standard" version. As we shall see, the "*standard*" version requires a little juggling to produce an *accurate* solution for the general case; in the "*modified*" version there is no need for such juggling, but the solution is *approximate.*

The horary dial on the *navicula* is "universal" in the sense that it works for essentially all terrestrial latitudes.[15] If used properly, at least the "standard" variety provides a complicated, but accurate means of measuring the time in *equinoctial* hours. Now on the back of all *navicula*s we find a universal horary quadrant and a shadow-square superposed thereon. The universal horary quadrant provides a simple alternative to measuring the time of day in any latitude, this time approximately in *seasonal* hours.[16] It is now well established that both these devices, and the combination of the two, fitted with a movable declination cursor – better known from the medieval Latin tradition of the *quadrans vetus* – have their origin in 9th-century Baghdad.[17] In the light of this alone, it is not unreasonable to seek an early Islamic origin for the universal horary dial on the *navicula.* In addition, however, it is now known that Habash al-Hāsib,[18] perhaps the most innovative astronomer of the 9th century, devised a yet more complicated universal instrument for timekeeping by the stars.

This new appreciation of the universal horary dial raises substantial problems for the history of medieval astronomy, in particular for the history of the transmission of scientific innovations from the Islamic world to medieval Europe. It also demonstrates the importance of looking at instruments in the light of available textual material and *vice versa.* Alas my study contributes but little to our understanding of the way in which the universal horary dial was transmitted to Europe, for it was surely there that one variety of the instrument assumed the form of a Venetian ship.

In the sequel I shall first explain the theory underlying the universal horary dial on the *navicula* in terms of modern mathematical notation, albeit using functions that are of prime importance in medieval (Islamic) timekeeping.[19] The following notation is used freely:

- a — azimuth (measured from the east- or west-points)
- $\mathcal{B}$ — the auxiliary function $\cos\delta \cos\phi$
- $\mathcal{C}$ — the auxiliary function $\sin\delta \sin\phi$
- d — the excess of the half-arc of daylight over 90° (=D−90°)
- D — the half-arc of daylight (the time from sunrise to midday or from midday to sunset)
- $\mathcal{E}$ — the auxiliary function $\tan\delta \tan\phi$ (=sin d)
- $\mathcal{G}$ — the auxiliary function $\sec\delta \sec\phi$
- h — instantaneous solar altitude
- H — the solar meridian altitude ($=90°-\phi+\delta$)
- $_{\text{-H}}$ — as subscript, refers to the horoscopus: see λ_H below
- $\mathcal{K}$ — the auxiliary function $\sin h \tan\phi$ (for azimuth calculations)
- $\mathcal{L}$ — the auxiliary function $\sin\delta/\cos\phi$ (*ditto*)
- t — hour-angle, or time before or after midday, measured in degrees
- t* — hour-angle for stars, measured before or after midnight
- T — the time of day measured from sunrise before midday or until sunset thereafter (=D−t)
- x,y — independent variables
- α — right ascensions[20]
- α_ϕ — oblique ascensions for latitude ϕ
- α' — normed right ascensions (measured from ♑ 0°, so that $\alpha'=\alpha+90°$)
- δ — solar declination[21]
- Δ — stellar declination
- ε — obliquity of the ecliptic[22]
- ϕ — local latitude
- λ — solar longitude
- λ_H — longitude of the horoscopus or ascendant, that is, the point of the ecliptic instantaneously rising over the horizon
- θ — independent variable (arc or angle).

2. *On Aspects of Islamic Astronomical Timekeeping*

The three-dimensional problem of finding t or T from h, δ and ϕ can be reduced to two dimensions using a procedure called the *analemma* and known from Greek mathematical astronomy.[23] But Greek astronomy was not geared to astronomical timekeeping. In fact, Muslim astronomers learned the procedures for deriving the formulae for $t(h,\delta,\phi)$ or $T(h,\delta,\phi)$ from Indian sources, which in turn relied on early Greek procedures. And it was the Muslims who in the 9th, 10th and 11th centuries developed such projection methods into an art in itself. Indeed, Muslim astronomers already in the 9th century but continuing into the 16th century developed sophisticated techniques for and aids to timekeeping, as well as varieties of astronomical instruments, which were in the main unknown to European astronomers in the Middle Ages and the Renaissance. In the 1970s some 500-odd Islamic manuscripts relating to astronomical timekeeping were studied for the first time and the results are only now in press.[24]

Before investigating the possible Islamic origin of the universal horary dial it is necessary for us to appreciate the sophistication of spherical astronomy as practiced by Muslim astronomers. Most medieval Muslim authors of astronomical handbooks with tables (*zījes*) treated the problem of the determination of $t(h,\delta,\phi)$ or $T(h,\delta,\phi)$ and their procedures are well documented.[25] All of these exact procedures are, of course, mathematically equivalent to the modern formula:

$$\begin{aligned}\cos t &= [\sin h - \sin\delta \sin\phi]/[\cos\delta \cos\phi] \\ &= \sin h \sec\delta \sec\phi - \tan\delta \tan\phi\,.\end{aligned}$$

Muslim astronomers over the centuries compiled sets of tables of auxiliary functions for facilitating their calculations (in the following discussion the trigonometric functions are reduced to base 1). From the 9th and 10th centuries we have tables of the following functions from the localities noted:[26]

$\sin\delta(\lambda)$ (Baghdad), $\cos\delta(90°-\lambda)$ (Baghdad, Khwarizm),
$\tan\delta(\lambda)$ (Baghdad, Cairo, Iran),
$\sin\alpha(\lambda) = \tan\delta(\lambda)/\tan\varepsilon$ (Baghdad, Cairo, Khwarizm),
$B(\lambda,\phi) = \cos\delta(\lambda)\cos\phi$ (Rayy),
$\mathcal{K}(h,\phi) = \sin h \tan\phi$ (Baghdad, Cairo).

From later centuries we have tables of the following functions (the digits between curly brackets indicate the number of such tables for different latitudes that have been identified):

$\mathcal{B}(\lambda) = \cos\delta(\lambda)\cos\phi\ \{20\}$, $\mathcal{C}(\lambda) = \sin\delta(\lambda)\sin\phi\ \{17\}$,
$\mathcal{G}(\lambda) = \sec\delta(\lambda)\sec\phi\ \{4\}$,
$\mathcal{K}(h) = \sin h \tan\phi\ \{8\}$, $\mathcal{L}(\lambda) = \sin\delta(\lambda)/\cos\phi\ \{4\}$,

and of the following for all latitudes (the associated digits indicate the number of such sets known):

$$\mathcal{B}(\lambda,\phi)\ \{2\}\ ,\ \mathcal{B}(\Delta,\phi)\ \{1\}\ ,\ \mathcal{C}(\lambda,\phi)\ \{0\}\ ,\ \mathcal{C}(\Delta,\phi)\ \{1\}\ ,$$
$$\mathcal{G}(\lambda,\phi)\ \{2\}\ ,\ \mathcal{K}(h,\phi)\ \{1\}\ ,\ \mathcal{L}(\lambda,\phi)\ \{1\}\ ,\ \mathcal{L}(\Delta,\phi)\ \{1\}\ .$$

Some astronomers compiled tables for a specific latitude of the functions:[27]

$$t(h,\lambda)\ \{14\}\ ,\ T(h,H)\ \{7\}\ ,\ a(h,\lambda)\ \{4\}\ ,$$

and one, Najm al-Dīn al-Misrī (Cairo, *ca.* 1325), even went to the trouble of compiling a table of T(h,H,D) for each degree of each of the three arguments, which served both the sun and stars and worked for all latitudes, at the cost of computing over 440,000 entries.[28]

In addition, a dozen or so Muslim astronomers compiled tables of auxiliary functions for solving problems of spherical astronomy.[29] Those of Habash essentially serve the conversion of ecliptic and equatorial coordinates:

$$F_1(\lambda) = \delta_2(\lambda) = \text{arc tan}\ \{\tan \varepsilon \sin \lambda\}\ ;$$
$$F_2(\lambda) = \cos \delta(90° - \lambda) = \cos\ \{\text{arc sin}\ [\cos \lambda \sin \varepsilon]\}\ ;$$
$$F_3(\lambda) = \cos \lambda / F_2(\lambda) = \cos \lambda / \cos \delta(90° - \lambda)\ ;$$
$$F_{4a}(\theta) = \tan \theta \sin \varepsilon;\ \text{and}\ F_{4b}(\theta) = \tan \theta\ .$$

Habash was clearly completely at home with such auxiliary functions. We can see the development of this kind of tables through to those of Najm al-Dīn, who advocated the use of his monumental three-argument table of F(x,y,z) for solving all of the standard problems of spherical astronomy for any latitude, and al-Khalīlī (Damascus, *ca.* 1360),[30] who tabulated the very useful functions:

$$f_\phi(\theta) = \sin \theta / \cos \phi = \mathcal{L}(\vartheta,\pi)\ ;\ g_\phi(\theta) = \sin \theta \tan \phi = \mathcal{K}(\vartheta,\pi)\ ;$$
$$F(x,y) = \text{arc cos}\ \{x / \cos y\}\ ,$$

with which:

$$t(h,\Delta,\phi) = F\{[f_\phi(h) - g_\phi(\Delta)], \Delta\}\ \text{and}\ a(h,\Delta,\phi) = F\{[g_\phi(h) - f_\phi(\Delta)], h\}\ .$$

It was not only strict timekeeping that was served by some of these tables, because knowledge of the time of day or night is a prerequisite for the practice of mathematical astrology. Another function tabulated by Muslim astronomers, albeit infrequently, was the longitude of the horoscopus, λ_H, as a function of the altitudes of the sun or specific fixed stars.[31]

In medieval European astronomical treatises not directly derived from Islamic sources such as al-Khwārizmī and al-Battānī the problem of reckoning time from solar or stellar altitude by an accurate formula is discussed only rarely. The only one of the above auxiliary tables that appears to have been known in Europe is the table of tan $\delta(\lambda)$ by al-Khwārizmī that ended up in the *Toledan Tables.*[32] European tables of the altitude of the sun at the hours for each zodiacal sign – mainly intended for marking vertical sundials – are well known,[33] but less well known are the extensive tables displaying the solar altitude as a function of time which were compiled by medieval European astronomers,[34] thereby continuing, perhaps unwittingly, the tradition of the more sophisticated (and more useful) Islamic tables for time as a function of solar and/or stellar altitude.[35] Only in the late 15th century did European astronomers begin to compile extensive auxiliary tables, essentially for the same purposes as the Islamic ones.[36]

3. The Use of the Universal Horary Dial on the Standard Navicula

The universal horary dial on both kinds of *navicula* has four scales. The *first scale* is on the mast, along which a cursor with a thread and bead can move to enter the latitude. Then there are *two declination scales*, the lower one for setting the mast according to the solar declination, and the one at the side for setting the bead in position on the thread again according to the declination. One feeds in the solar longitude on a non-uniform scale on a kind of fixed cursor in the form of a circular arc of about one-half of a quarter circle, and the declination is shown on an attached uniform scale. Both declination scales on the *navicula* are identical to the scales on the universal horary quadrant first described in 9th-century Baghdad,[37] although the one at the side of the main markings has an additional scale on the chord joining its ends, which is the scale that actually has to be used. This second declination scale defines the 12th-hour line (midday), and its distance from the central vertical (6th hour) defines the base for the underlying trigonometry. On the *fourth scale* the hour-lines are drawn and numbered from midnight on the left to midday on the right.

We first set the cursor on the mast to the latitude. Then we rotate the mast to the solar declination on the lower scale. Next we set the bead on the thread at the solar declination on the side scale (see below for details). When the instrument is tilted so that the rays of the sun pass through the hole on the upper sight and fall on the corresponding part of the lower sight, the inclination of the horizontal axis of the instrument measures the solar altitude. The bead then falls on the appropriate marking of the hour-scale. The markings can also be used for timekeeping by stars whose declination is less than the obliquity of the ecliptic.

Now the way in which these scales are used is critical for the correct application of the "standard" instrument. *When the latitude-cursor is set on the mast the bead must be set so as to rest at the base of the mast.* Then when the mast is turned so that its inclination to the vertical axis of the instrument corresponds to the solar declination, the bead will hang slightly below the horizontal axis. *The latitude-cursor must then be moved slightly on the mast so that the bead falls on the horizontal axis again.* Next, when the thread and bead are moved to the side declination scale, *the bead must not be laid on the declination scale itself, but rather on the scale on the graduated vertical line segment (whose divisions are defined by the declination scale).*

The instructions in the Latin and Middle English treatises on the use of the device do not call for this fine tuning that I have proposed since they are dealing with a different, "modified", manifestation of the instrument. In that, each of the latitude and two declination scales are modified in such a way that one places the bead on the side scale itself and does not have to use any "fine tuning" of the kind I have suggested: inevitably only an approximate solution can be obtained for the general case.[38] The errors are of course minuscule on an instrument that one can hold in the palm of one's hand.

4. The Mathematics Underlying the Universal Horary Dial on the Navicula

The application of the accurate formula for $t(h,\delta,\phi)$ involves a mathematical device for generating mechanically two auxiliary functions of medieval (Islamic) timekeeping, $\mathcal{E} = \tan\phi \tan\delta$ and $\mathcal{G} = \sec\phi \sec\delta$. Once this has been done, with a minimum of effort one can solve mechanically what has been reduced to a very simple relation between h and t, namely:

$$t(h,\delta,\phi) = \text{arc cos}\,\{\sin h \cdot \mathcal{G} - \mathcal{E}\}\ .$$

The inventor of the universal horary dial has exploited a lemma (see Fig. 3a):

> Given a triangle OAB right-angled at O, a point C on the perpendicular to OA at A, and a triangle OCD right-angled at O constructed on OC, in such a way that BD be parallel to OA, then triangle OCD will be similar to triangle OAB.

I know of no reference to this lemma in the ancient or medieval literature. The proof is left as an exercise for the reader. Now in a right-angled triangle OAB with horizontal base OA=1, O = 90°, and A = ϕ, we have OB = tan ϕ and AB = sec ϕ (see Fig. 3b). If we then rotate OA by an angle of δ (>0) and extend it to OC so that AC is vertical, then AC (on the vertical declination scale) measures tan δ, and OC = sec δ; furthermore, with the triangle OCD similar to OAB we have BD = tan ϕ tan δ = $\mathcal{E}$ and DC = sec ϕ sec δ = $\mathcal{G}$. In this way we have generated the auxiliary quantities $\mathcal{E}$ and $\mathcal{G}$. The procedure is elegant as well as simple and efficacious.

Now on the universal horary dial on the *navicula*, we have set the cursor on the latitude scale to ϕ in position B, with the bead of the vertical thread set at O, and when we rotate the mast by δ (>0) to ON the cursor moves from B to E and the bead hangs below the horizontal OA (see Fig. 3c). In order to bring the cursor to D we simply have to move it on the latitude scale so that the bead is again precisely on the horizontal OA. We now move the bead to the point C on the vertical scale AC measuring tan δ. Then the cursor is at distance BD = $\mathcal{E}$ from the "vertical" on the *navicula* (that is, the central position of the mast) and the thread now measures DC = $\mathcal{G}$ from the cursor at D on the latitude scale to the appropriate point C on the vertical tan $\delta(\lambda)$ scale. We now tilt the *navicula* till OA is aligned with the sun (see Fig. 3d), and the thread with the bead indicating $\mathcal{G}$ will fall at an angle h to the "vertical" axis of the *navicula*. The bead will then be at $\mathcal{G} \cdot \sin h$ from this "vertical". The algebraic difference between this and $\mathcal{E}$ equals cos t, so the bead will indicate the appropriate time of day on the horary markings.

When we understand the simplicity, elegance and efficacity of this construction, we can only wonder at the complicated procedures proposed in the available medieval English texts. A detailed new investigation of those procedures should be conducted, for it would be interesting to know *why* anyone should have proposed "modifications" of the three scales in order to yield an *approximate* solution without adjusting the bead, when with the regular scales and the "fine tuning" that I have suggested for the bead one

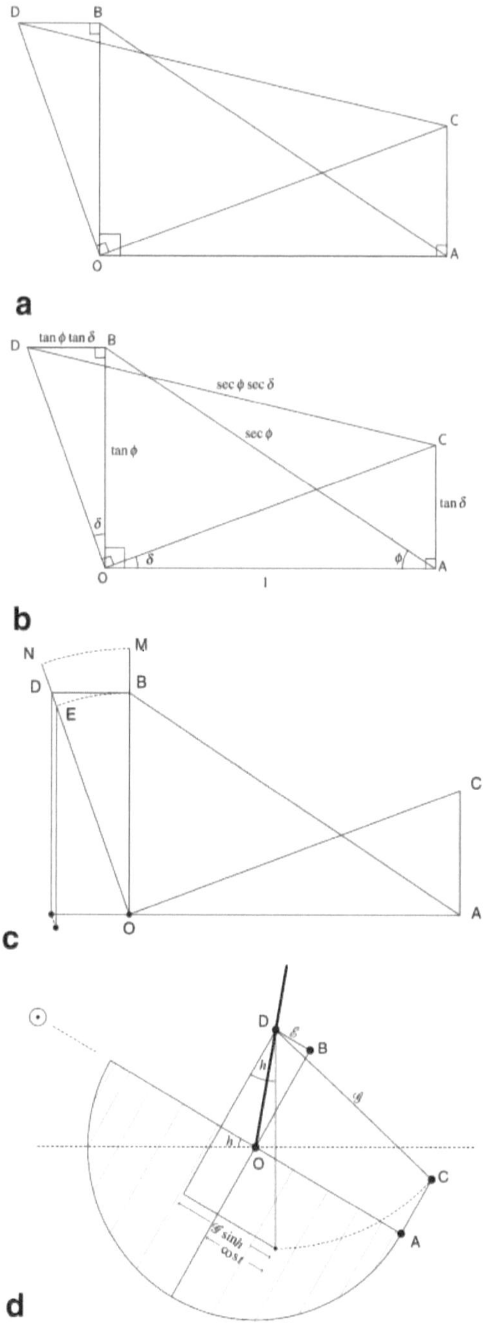
D
B
C
O
A
a
D
tan φ tan δ
B
sec φ sec δ
sec φ
tan φ
C
tan δ
δ
δ
φ
O
A
1
b
N
M
D
B
E
C
O
A
c
D
B
C
O
A
h
d

Fig. 3.

can derive an *accurate* solution. And one may well wonder *who* in medieval England was addressing these problems. Or were both the standard and the modified versions of the universal horary dial transmitted together?

5. Habash's Universal Instrument for Timekeeping by the Stars

Now Habash devised a highly complex universal computing device for time-keeping by the stars (see Fig. 4). Whilst we have no information that would lead us to suppose that this instrument was ever made by any later astronomers, Muslim or European, it is clear that the text of the unique 13th-century manuscript now in the Bodleian Library in Oxford has suffered at the hands of successive copyists. Habash's treatise on this instrument has now been published with an English translation,[39] with commentary restricted to the construction of the instrument, for its use is not mentioned in the available text and is by no means self-evident. Furthermore, it seems that even the text on the construction may not be complete, so that we may be missing a description of a key part of the instrument.

It is, however, clear that the instrument serves to find for all latitudes first the time from an observed stellar altitude, and also the azimuth of the star (which, of course, could also be observed). But, as we shall see, it also achieves something else. Thus the instrument is inevitably more complicated than the universal horary dial for timekeeping by the sun and, necessarily, has a different morphology. Nevertheless Habash's device has in common with the early European universal hour dials the following features:

(1) a latitude scale based on tan ϕ;
(2) a movable cursor on a scale;
(3) a scale showing $\delta(\lambda)$; and
(4) parallel markings for the sine function (actually three sets!).

In addition we note the presence of:

(5) a mechanical device to insert the auxiliary function max $\mathcal{E}$ = sin max $d(\phi) = \tan \varepsilon \tan \phi$.

To find the time t* before or after midnight from the altitude of a star with declination Δ and right ascension α observed at altitude h in a locality with

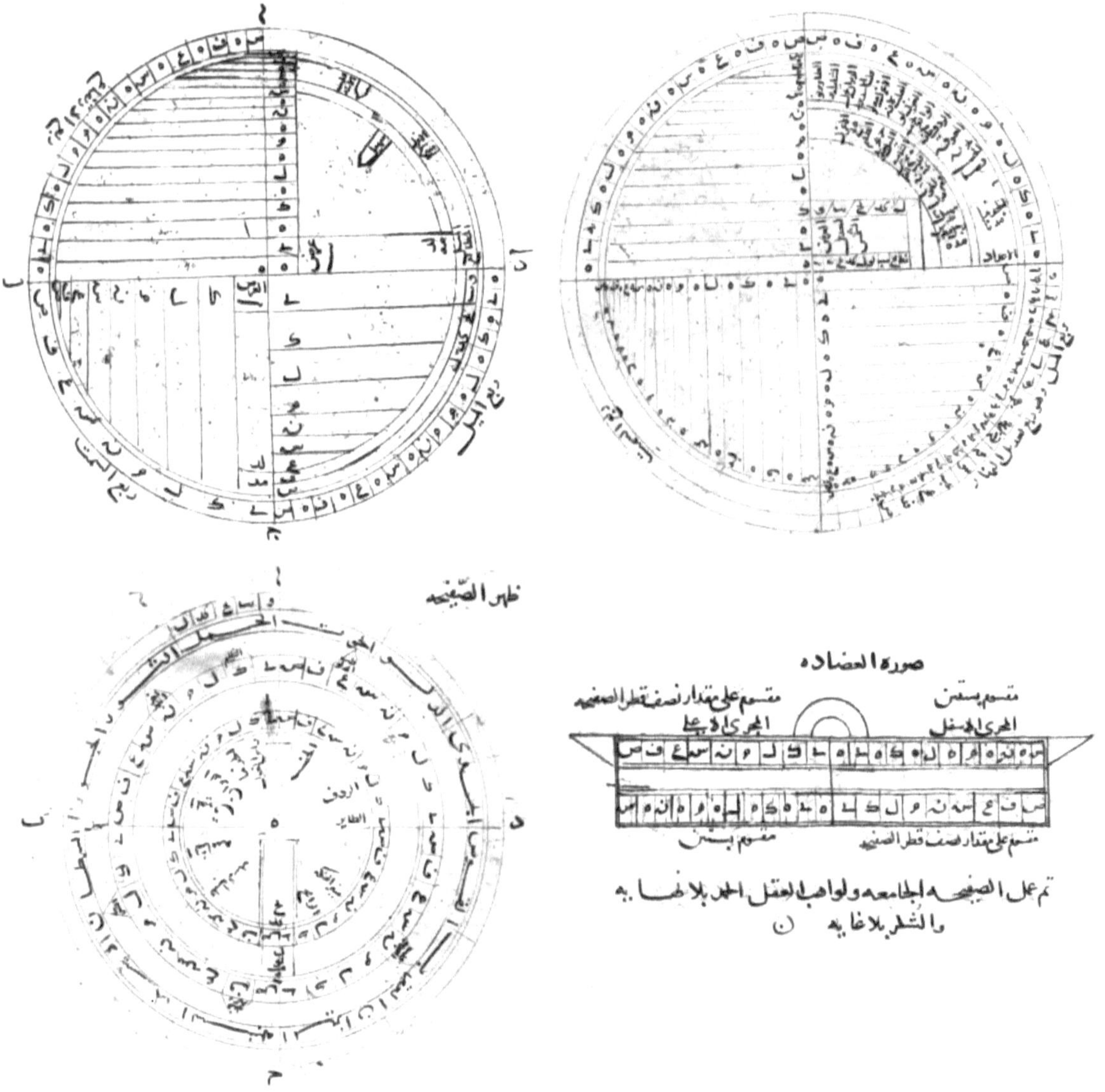

Fig. 4. The main components of Habash's universal dial for operations with the stars. The top two extracts show the trigonometric markings on the front of the instrument (illustrated twice to show all the markings). On the lower left is shown the back of the instrument with the ecliptic and equatorial scales on the outside and the small movable plate that can be displaced from the centre in order to provide a device for showing oblique ascensions for any latitude. On the lower right is one of the two alidades, this one having both a uniform 0–90 scale and a 0–60 sine scale. (From MS Oxford Bodleian Marsh 663, pp. 243, 247, 245 and 246, courtesy of the Bodleian Library.)

latitude ϕ when the solar longitude is λ, one needs the equivalent of the formula in Section 2 with the stellar declination Δ substituted for the solar declination δ. Once t* has been found, the azimuth a can be found by one of several procedures. But what was important in Islamic timekeeping, now in the service of astrology rather than religion, was not only to know the time but also to know the longitude of the horoscopus, λ_H.[40] To find this, one needs to know the hour-angle t* and the longitude of the sun λ, and to have some facility with right and oblique ascensions.

Any attempt to reconstruct the use of the various parts of this instrument must be provisional failing the original text, and the ultimate function of the instrument remains a mystery. I shall also assume that the reader is familiar with the basic operations with the sine quadrant, as explained recently by Richard Lorch on the basis of a 9th-century Arabic text:[41] these include the calculation of sines and cosines and the corresponding arcs, as well as the calculation of products and quotients of sines and (co)sines. For simplicity, I again keep the base for trigonometric functions as unity and ignore signs.

Given that one has measured the altitude of a star h with declination Δ on a day when the solar longitude is λ and the solar declination δ, a hypothetical *modus operandi* would be the following. First, we use a sine quadrant on the front of the instrument to compute $\sin h \cdot \mathcal{G}(\Delta,\phi)$ and $\mathcal{E}(\Delta,\phi)$. Then their algebraic difference is $\cos t(h,\Delta,\phi)$, whence from the same quadrant we can find t* (measured from midnight). Second, we use a sine quadrant to compute $\mathcal{K}(h,\phi)$ and $\mathcal{L}(\Delta,\phi)$. Then compute $[\mathcal{K}\sim\mathcal{L}] / \cos h$, which is $\sin a(h,\Delta,\phi)$, whence also a.[42] Third, move the centre of the small plate on the back of the instrument from the centre of the larger plate beneath it by the amount $\max \mathcal{E} = \sin \max d(\phi)$, using the scale provided, to produce some means, still obscure to me, of displaying oblique ascensions $\alpha_\phi(\lambda)$ for any latitude.[43] I would like to see the instrument as now using the solar longitude λ and the star at t* to read off the longitude of mid-heaven, λ_M. By virtue of the well-known relation:

$$\alpha_\phi(\lambda_H) = \alpha'(\lambda_M)\,,$$

one could then find λ_H.[44] But, alas, the final operation with the instrument still defies explanation.

6. *Early Islamic predecessors to the European universal horary dial?*

The universal hour dial on the "standard" *navicula* has the following features:

(1) a linear latitude scale based on tan ϕ, as on Habash's instrument;
(2) a movable cursor on that scale (on Habash's instrument a cursor moves on a different linear scale);
(3) two scales showing $\delta(\lambda)$ within an arc of length 2ε, as on the 9th-century "*quadrans vetus*" (Habash's instrument has the declination scale over an entire quadrant of arc); and
(4) a double set of parallel markings for the (co)sine function (as opposed to the three single sets on Habash's instrument).

A subtle innovation is the use of the side declination scale giving tan $\delta(\lambda)$ rather than $\delta(\lambda)$. The mechanical device on Habash's instrument for inserting max $\mathcal{E}$ = tan ε tan ϕ so as to find the oblique ascensions for any ϕ and λ is now replaced by:

(5) a yet more ingenious procedure for inserting both $\mathcal{E}$ = tan δ tan ϕ and $\mathcal{G}$ = sec δ sec ϕ.

I strongly suspect that Habash was the first to devise such markings because he invented a yet more complex instrument for timekeeping by the stars, which has similar principal components. But his hypothetical dial may have served the stars as well as the sun, that is, its declination scale may have gone beyond the obliquity of the ecliptic. This feature it would have had in common with his other device. And, of course, there is no reason to suppose that the hypothetical instrument was shaped like a ship. However, whilst I would expect that Habash would have devised an horary dial like that on the Geneva *navicula*, he was an adept at proposing "modified" trigonometric functions to facilitate complicated procedures, such as are represented on the scales of the Greenwich and Colchester *navicula*s. These, however, were not to produce an approximate solution when he already knew of the accurate solution.[45] On the other hand, the constructions outlined in the medieval English texts, in their attempt to balance the effects of one modified declination scale against those of another, reveal an extraordinary mastery of the techniques of the analemma,[46] and perhaps these too derive from the same Arabic source.

Alas for my hypothesis, there is not a trace of the universal horary dial in the well-documented medieval literature on astronomical instrumentation in al-Andalus[47] or Syria and Egypt.[48] An 18th-century Ottoman illustration of such a dial, even though it bears traces of earlier Islamic influence, is not conclusive, and perhaps not even convincing, evidence for a very early Islamic origin of the universal horary dial.[49]

Now there are about 10,000 surviving manuscripts in Arabic (also Persian and Turkish) dealing with astronomy, most of which were copied after, say, 1500. Most of the early manuscripts have not survived the vicissitudes of time. What we know about *early* Islamic astronomy is very much a matter of chance, for it has, in the main, to be reconstructed from materials in later works or in later manuscripts. Some aspects of late Islamic science can only be understood when their inspiration is found in early Islamic texts.[50] Some aspects of medieval European astronomy, on the other hand – and here I am thinking particularly of the universal horary quadrant and shadow square – can only be fully appreciated from early Islamic treatises which just happen to be preserved only in late manuscripts.[51]

The history of universal horary dials from its beginnings to the late Renaissance calls out for a detailed investigation. In the study from which this short paper is condensed I hope to have provided the initial stimulus for such an endeavour.[52] Any accepted conclusions or over-enthusiastic hypotheses could be blown away by new evidence. In brief, we are perhaps talking about a future doctoral dissertation for someone who can handle the trigonometry of spherical astronomy and primary sources in medieval Arabic, medieval Latin, Middle English and Middle German, as well as secondary sources in English, German and Italian. That future researcher also needs to understand the language of instruments, medieval and Renaissance, which is quite different from the language of any texts. There are relevant English manuscripts that have never been studied, and nobody since Zinner has looked at the texts from the German tradition of the universal horary dial. Due attention should be paid to the texts on the *construction* of the various dials, which I have neglected, and to the details of the individual surviving instruments.[53] The two different medieval English manifestations and various German manifestations of universal horary dials need to be clearly identified, and their underlying mathematics understood. The history of the dial after *ca.* 1550 should also be investigated afresh. The same person will also need some good luck to discover some new sources – instruments, texts, diagrams or tables – that are relevant to the topic.[54] But this is not too much to hope for:

the three *navicula*s from 14th-century England all came to light only during the past 15 years!

Acknowledgements

It is a pleasure to thank François Charette and Sven Ruhberg for their constructive criticism and corrections to earlier drafts of this paper, without in any way implicating them in any problems that may still lurk in this version.

BIBLIOGRAPHY

Archinard, Margarida
1988: "Les cadrans solaires rectilignes", *Nuncius* (Florence) 3:2, pp. 149–181.
1995: "*Navicula de Venetiis*: une acquisition prestigieuse du Musée d'histoire des sciences [Genève]", Geneva (Musée d'histoire des sciences).
Brusa, Giuseppe
1980: "Le navicelle orarie di Venezia", *Annali dell'Istituto e Museo di Storia della Scienza di Firenze* 5 (1980), pp. 51–59.
Charette, François
1998: "A Monumental Medieval Table for Solving the Problems of Spherical Astronomy for all Latitudes", *Archives internationales d'histoire des sciences* 48, pp. 11–64.
2003: *Mathematical Instrumentation in Fourteenth-Century Egypt and Syria – The Illustrated Treatise of Najm al-Dīn al-Misrī*, Leiden (E. J. Brill).
Charette, François & Petra Schmidl
2001: "A Universal Plate for Timekeeping with the Stars by Habash al-Hāsib: Text, Translation and Preliminary Commentary", *Suhayl* 2, pp. 107–159.
Davidian, Marie-Louise
1960: "al-Bīrūnī on the Time of Day from Shadow Lengths", *Journal of the American Oriental Society* 80, pp. 330–335, repr. in Kennedy *et al.* 1983, pp. 274–279.
Debarnot, Marie-Thérèse
1985: *al-Bīrūnī: Kitāb Maqālīd 'ilm al-hay'a. La trigonométrie sphérique chez les Arabes de l'est à la fin du Xe siècle*, Damascus (Institut Français de Damas, publ. no. 114).
1987: "The *Zīj* of Habash al-Hāsib: A Survey of MS Istanbul Yeni Cami 784/2", in *Kennedy Festschrift*, pp. 35–69.
Delambre, Jean-Baptiste
1819: *Histoire de l'astronomie du moyen âge*, Paris, 1819, repr. New York & London (Johnson), 1965.
Delehar, Peter
1993: "Illustrations of Scientific Instruments in the *Gentleman's Magazine* 1746–1796", in *Turner Festschrift*, pp. 383–394.

EI$_2$=*The Encyclopaedia of Islam*, new edn., 11 vols. and various supplements to date, Leiden (E. J. Brill), 1960 to present. [Also available in French.]

Fuller, A. W.
1957: "Universal Rectilinear Dials", *The Mathematical Gazette* (London) 41, pp. 9–24.

Jones, Malcolm
2000: "The Names Given to Ships in Fourteenth- and Fifteenth-Century England", *Nomina* 23, pp. 23–36.

Kennedy Festschrift=*From Deferent to Equant – A Volume of Studies in the History of Science in the Ancient and Medieval Near East in Honor of E. S. Kennedy*, D. A. King and G. Saliba, eds., *Annals of the New York Academy of Sciences* 500 (1987).

Kennedy, E. S. & Paul Kunitzsch & Richard P. Lorch
1999: *The Melon-Shaped Astrolabe in Arabic Astronomy* ..., (Boethius 43), Stuttgart (Franz Steiner).

Kennedy, E. S. *et al.*
1983: *Studies in the Islamic Exact Sciences*, D. A. King and M. H. Kennedy, eds., Beirut (AUB).

King, David A.
1986: *Islamic Mathematical Astronomy*, Aldershot (Variorum), rev. edn. 1993.
1993: *Astronomy in the Service of Islam*, Aldershot (Variorum).
1994: "Astronomical Instruments between East and West", in *Kommunikation zwischen Orient und Okzident*, H. Kühnel, ed., Vienna (Öst. Akad. d. Wiss., Phil.-Hist. Kl., SB 619), pp. 143–198.
1999: *World-Maps for Finding the Direction and Distance to Mecca* ..., Leiden (E. J. Brill) & London (Al-Furqan).
2002: "A *Vetustissimus* Arabic Treatise on the *Quadrans Vetus*", *Journal for the History of Astronomy* 33, pp. 1–19. [A more extensive version is in *SATMI*, XIIa.]

King: see also *SATMI*.

King, David A. & Julio Samsó, with Bernard R. Goldstein
2001: "Astronomical Handbooks and Tables from the Islamic World (750–1900): An Interim Report", *Suhayl* 2, pp. 9–106. [A summary is in the article "Zīdj" in *EI*$_2$.]

Kragten, Jan
1989: *The Little Ship of Venice [sic] (Navicula de Venetiis)*, Eindhoven (NL), published privately.
1992: *The Venetian Ship of Sibton Abbey*, Eindhoven (NL), published privately.

Kunitzsch, Paul & Richard Lorch
1994: "Abū Nasr and Habash on *matāli*^{c} al-samt", *Zeitschrift für Geschichte der arabisch-islamischen Wissenschaften* (Frankfurt am Main) 9, pp. 43–82.

Kunitzsch: see also Kennedy & Kunitzsch & Lorch.

Lippincott, Kristen
1992: "The Navicula Sundial", *Bulletin of the Scientific Instrument Society*, no. 35, p. 22.

Lorch, Richard
2000: "Some Early Applications of the Sine Quadrant", *Suhayl* 1, pp. 251–272.

Lorch: see also Kennedy & Kunitzsch & Lorch and Kunitzsch & Lorch.

Luckey, Paul
1927: "Das Analemma von Ptolemäus", *Astronomische Nachrichten*, Bd. 230, Nr. 5498, pp. 18–46.

Madrid 1992 Catalogue=*El legado científico Andalusí, Museo Arqueológico Nacional, Madrid, Abril–Junio 1992*, Madrid, 1992.

Martin, Lilian Ray
2001: *The Art and Archaeology of Venetian Ships and Boats*, Texas A&M University Press.
North, John
1976: *Richard of Wallingford ...*, 3 vols., Oxford.
1988: *Chaucer's Universe*, Oxford.
Price, Derek J. de Solla
1960: "The Little Ship of Venice, a Middle English Instrument Tract", *Journal of the History of Medicine and Allied Sciences* 15, pp. 399–407.
Samsó: see King & Samsó.
SATMI = D. A. King, *Studies in Astronomical Timekeeping in Medieval Islamic Civilization*, 15 pts., Leiden (E. J. Brill), in press. Parts I–IX are to appear first, under the title *The Call of the Muezzin*.
Sédillot, J.-J. (*père*)
1834–35: *Traité des instruments astronomiques des Arabes composé au treizième siècle par Aboul Hhassan Ali de Maroc ...*, 2 vols., Paris, repr. in 1 vol., Frankfurt am Main (IGAIW), 1985.
Sédillot, L.-A. (*fils*)
1844: "Mémoire sur les instruments astronomiques des Arabes," *Mémoires de l'Académie Royale des Inscriptions et Belles-Lettres de l'Institut de France* 1, pp. 1–229, repr. Frankfurt am Main (IGAIW), 1989.
Stebbins, Frederick A.
1961: "A Medieval Portable Sun-Dial", *Journal of the Royal Astronomical Society of Canada* (Toronto) 55:2, pp. 49–56.
Suhayl= *Suhayl – Journal for the History of the Exact and Natural Sciences in Islamic Civilisation* (Barcelona).
Turner, Gerard L'E.
ca. 1994: *The Late-Mediæval Navicula*, London (Trevor Philip & Sons Ltd.), n.d. [On the *navicula* now in Geneva.]
Turner Festschrift= *Making Instruments Count – Essays on Historical Scientific Instruments presented to Gerard L'Estrange Turner*, R. G. Anderson, J. A. Bennett, and W. F. Ryan, eds., Aldershot (Variorum), 1993.
Zinner, Ernst
1956: *Deutsche und niederländische astronomische Instrumente des 11.-18. Jahrhunderts*, Munich, 1956, repr. 1972.

NOTES

1. This paper is condensed from a more detailed study to appear in *SATMI*, XIIb: *On Universal Horary Dials for Timekeeping by the Sun and Stars*. I have already hinted at some of the findings in King 1994, p. 151, and King 1999, pp. 351–359.

   My interest in the *navicula* was aroused during the long-term Frankfurt-based project to catalogue all medieval astronomical instruments (both Islamic and European to *ca.* 1550). Funding for this project (1992–96, Islamic, and 1999–2002, European) was generously provided by the Deutsche Forschungsgemeinschaft. The catalogue is still far from being complete. On some of the results of the project see King 1994. For a provisional table of contents of the catalogue see www.uni-frankfurt.de/fb13/ign/instrument-cata-

logue.html. All medieval instruments have been assigned an International Instrument Checklist number (here prefixed by #), continuing a tradition inaugurated by Derek de Solla Price and his colleagues.
2. We are dealing with instruments small enough to be held in the palm of one's hand, on which the divisions on the various scales are rather crudely marked. Nevertheless we shall find that it is possible to distinguish between universal horary dials that either (1) have the theoretical potential to yield exact solutions, or (2) are so constructed that they can yield only an approximate solution. The *modus operandi* of the two types is different. Since the instruments are so small, this does not necessarily mean that properly-executed examples of the two kinds would in practice yield noticeably different results. These remarks refer only to the *universal horary dial* on the *navicula*: the results that one can obtain from the *universal horary quadrant* on the back of the *navicula* are different in essence and noticeably less accurate, especially for European latitudes.
3. Fuller (1957) started the tradition, and Price (1960) saw no reason to fault his findings.
4. The only *naviculas* from before 1500 known before 15 years ago were: (1) #8531 (Museum of the History of Science, Oxford) – see Gunther 1923, pp. 40–41; Zinner 1956, pp. 111 and 116; Price 1960, p. 401; and Brusa 1980, p. 55 and figs. 3–4. (2) #8532 (Museo di Storia della Scienza, Florence) – see Brusa 1980.
5. Gunther 1923, pp. 375–379 and 38–41. On the Oxford *navicula* see already n. 4.
6. Zinner 1956, pp. 112–113 and 116.
7. See Zinner 1956, pp. 110–117. On the underlying mathematics see Delambre 1819, pp. 326–333; Fuller 1957, pp. 9–15; Stebbins 1961; and Archinard 1988.
8. Price 1960.
9. See already n. 4. The superfluous fixed solar scale on the universal horary quadrant is for latitude 52°, which would serve London. This having been said, the piece is not shaped like a ship and looks more like the illustrations of universal horary dials in manuscripts from the Vienna school.
10. Kragten 1989 and 1992, neither, alas, published or easily accessible. See also n. 38 below.
11. These are: (3) #8536 (National Maritime Museum, Greenwich) – unpublished, though see Lippencott 1992 and Kragten 1992. (4) #8537 (Musée d'Histoire des Sciences, Geneva) – see Turner 1994 and Archinard 1995; illustrated in King 1999, p. 354, fig. 9.5.2, where it is falsely identified as the Greenwich piece. (5) #8538 (formerly Colchester), known only from an illustration in the English *Gentleman's Magazine* of 1787, reproduced as Fig. 1 – see Delehar 1993, pp. 386–388. Kragten 1989 deals with the medieval treatises and the Oxford instrument; Kragten 1992 on the Greenwich instrument was written before the Colchester and Geneva pieces came to light.
12. On the attribution of the available texts to English monks see North 1976, III, pp. 113–114, and *SATMI*, XIIb-6. On the use of the expression "de Venetiis" see *ibid.*, XIIb-6d. A new study of "real" Venetian ships is Martin 2001.
13. *Ibid.*, IXb–9d–e.
14. This was first pointed out to me by Sven Ruhberg. At least the mast scale is in order: the markings correspond to the tangent of the latitude multiplied by the distance from the centre to the 12th hour line. (On the other pieces the mast scale is unhappily based on the radius of the circle bounding the horary markings, so that it contains a factor depending on the obliquity.) The declination scales are not particularly carefully drawn and merit further examination. (At least they are not as different from each other as they are on the other pieces.) Furthermore, only on the Geneva instrument is the chord

subtended by the side declination scale, which chord is the 12th hour line, properly marked as a scale for the tangent of the declination. This explains why in Archinard 1995 no recourse is made to the complicated constructions proposed in the available medieval texts: they relate to a different manifestation of the *navicula*.

15. On universal instruments and tables see further *SATMI*, VI: *Universal Solutions in Islamic Astronomy*. Alas the universal horary quadrant was omitted from the earlier studies reprinted in King 1993, VI–VII, on which the new versions are based.
16. For the same reason, the universal horary quadrant was put on the back of astrolabes. The main markings on the front provide an accurate solution; the quadrant on the back provides a quick alternative approximate solution, sometimes misinterpreted in the standard literature on the astrolabe. See *SATMI*, XIIa-9 (n. 17 below) on the way in which it was intended that these markings be used.
17. King 2002, and the more extensive version in *SATMI*, XIIa: *On the Universal Horary Quadrant for Timekeeping by the Sun*. See also n. 37.
18. On Habash see King 1999, pp. 40–41, 61–64, 345–349, and the references there cited, and also n. 25 below.
19. The same notation is used in *SATMI*, I–II (see n. 24).
20. See the article "Matāli^{c}" [=right and oblique ascensions] in EI_2.
21. Article "Mayl" [=declination] in EI_2.
22. Article "Mintakat al-burūdj" [=ecliptic, also featuring obliquity] by Paul Kunitzsch in EI_2.
23. On the analemma see Luckey 1927, and the more recent literature cited in King 1999, p. 15, n. 29, *etc.* (see p. 431 of the index).
24. See *SATMI*, I: *A Survey of Tables for Timekeeping by the Sun and Stars*, and II: *A Survey of Tables for Regulating the Times of Prayer*, and the brief summary in the article "Mīkāt" [=astronomical timekeeping] in EI_2. On instruments see *SATMI*, X: *Astronomical Instrumentation in the Islamic World* (updated from the first version which was published in Italian in 1991), and King 1994.
25. On the topics treated in the typical Islamic *zīj*, of which we have over 225 from the period 750–1900, see King & Samsó 2001. For the treatment of Habash see Debarnot 1987, p. 48, and also n. 44 below. For more detailed treatments of his spherical astronomy see Debarnot 1985, pp. 29–33; Kennedy & Kunitzsch & Lorch 1999; and Kunitzsch & Lorch 1994. For treatments by projection methods and spherical trigonometry compare Davidian 1960 and Debarnot 1985.
26. *SATMI*, I-6–7–8.
27. *Ibid.*, I-2, 5.
28. On this table, which can also be used as a universal auxiliary table, see the detailed analysis in Charette 1998.
29. *SATMI*, I-9.
30. See King 1986, XI, and *SATMI*, I-9.5 and II-10.7.
31. See the articles "Tāli^{c}" [=ascendant] and "Matāli^{c}" [=ascensions] in EI_2. On the tables see *SATMI*, I-3.2.1, 3.2.2, 3.1.1, 6.16.3, and II-14.2 (also IV-5.3).
32. *Ibid.*, I-7.1.
33. Zinner 1956, pp. 50–51 and 159; North 1988, pp. 104–130; and *SATMI*, I-4, and I-10.1.
34. *SATMI*, I-10.1–2.
35. All such tables are analysed *ibid.*, I-1–3.
36. For references see *ibid.*, I-10.2.
37. See n. 17 above.

38. Compare Gunther 1923, pp. 378–379, with Price 1960, pp. 406–407, and the commentary in Kragten 1989.
39. Charette & Schmidl 2001.
40. See n. 31 above. Habash in his *Zīj* treats consecutively the determination of time from solar and stellar altitude and the determination of the ascendant from the time: see Debarnot 1987, pp. 48 and 56.
41. See Lorch 2000.
42. In spite of the inclusion of a procedure for determining the azimuth, there is apparently no intention to solve a favourite problem of Habash, namely, to determine the time since rising for the sun or a star to attain a particular azimuth. On this see Kunitzsch & Lorch 1994.
43. On ascensions see n. 20.
44. See n. 31.
45. See King 1999, pp. 61 and 63–64, on Habash's modified functions for finding the direction of Mecca from any locality.
46. See King 1999, pp. 62–63, on Habash's analemma for the same purpose.
47. See *Madrid 1992 Catalogue*, and the numerous other recent works of the Barcelona School.
48. See especially Sédillot-*père* 1834–35 and Sédillot-*fils* 1844, and now Charette 2003.
49. King 1999, pp. 352–353 and 355 (fig. 9.5.3), and now *SATMI*, XIIb-12.
50. In King 1999 I hypothesized an early Islamic origin for the inspiration behind the highly sophisticated mathematical grids on two 17th-century world-maps preserving direction and (the sine of the) distance to Mecca at the centre. In 2001 Jan Hogendijk located two previously unknown texts – one from 10th-century Baghdad and the other from 11th-century Isfahan – in which the determination of the direction of Mecca by means of conic sections is discussed. The latitude curves on the 17th-century grids are arcs of circles approximating segments of ellipses.
51. The 9th-century Arabic treatise on the "*quadrans vetus*" (see n. 17) is preserved in a manuscript copied in Meshed *ca.* 1800.
52. In the longer version I devote attention to the yet more elusive Florence "*navicula*", the problematic universal horary dial on the sole surviving albion, from 15th-century Vienna (North 1976, III, pl. XXIII), and a remarkable composite instrument made in 1527 by one Roger Brechte (on which see already Gunther 1923, pp. 135–140 (no. 56)).
53. Kragten 1992 established that the declination scales on each of the Oxford and Greenwich instruments were reversed, when considered in the light of the constructions proposed in the medieval English texts.
54. See *SATMI*, XIIb-11, and Delehar 1993, pp. 388–389.

X

The Astrolabe Depicted in the Intarsia of the *Studiolo* of Archduke Federico in Urbino*

In one of the intarsias decorating the studiolo in the palace of Archduke Federico of Urbino, built in 1476, there are depicted two astronomical instruments, an armillary sphere and an astrolabe of distinctive design – see Fig. 1. The instruments are carefully represented, so carefully in fact that their component parts are faithfully portrayed in minute detail. The purpose of this article is to present a detailed description of the two illustrations, and to show how much they can teach us. As we shall see, it is possible to identify an immediate Italian source and an ultimate Islamic source for the inspiration behind the design of at least the astrolabe, and to outline what influence this design had in later European instrumentation.[1] I shall assume that readers have an idea what these instruments are and how they function.[2]

The armillary sphere depicted in the intarsia is typical of a Renaissance instrument of this kind. The celestial axis has a sphere representing the earth in the middle and through the poles there are four vertical rings: (a) an outer ring apparently representing the limits of the universe; (b) a middle ring representing the solstitial colure; (c) a second middle ring perpendicular to the first, representing the equinoctial colure; and (d) an inner ring. Perpendicular to (b) and (c) and attached to both we see six horizontal rings, representing (e) the north polar circle, (f) the north tropic circle, (g) the equator, (h) the south tropic circle, and (i) the south polar circle. Inclined to the tropic rings is a band for the ecliptic (j), with divisions apparently for individual degrees but without reference to zodiacal signs. Attached

* This paper is a shortened version of my contribution to *La scienza del Ducato di Urbino – The Science of the Dukedom of Urbino*, ed. Flavio Vetrano, Urbino: Accademia Raffaello, 2001 (pp. 101-139), which appeared with English and Italian versions side-by-side. Reprinting the English version on its own would not have been satisfactory. The introduction to the underlying astronomical concepts and to the astrolabe in general has been suppressed here.

---

1 Each medieval and Renaissance instrument has been assigned a number, here preceded by the symbol #. Research on medieval instruments in Frankfurt has been supported by grants from the Deutsche Forschungsgemeinschaft.

2 For two introductions to medieval instruments see Poulle, 'Instruments astronomiques', and King, 'Astronomical Instruments'. A useful study of the astrolabe in Italian is Trento, *Astrolabio*, which supplements the standard works of J. North and W. Hartner. On Islamic instruments see King, 'Strumentazione', and *In Synchrony with the Heavens*, vol. 2.

Fig. 1: The armillary sphere and the astrolabe depicted in the Urbino intarsia.

to the inner vertical ring (d) is another ring (k) inclined at about 18° to the equator. This inclination is arbitrary because the ring serves to represent the horizon, here shown for an approximate latitude of 90°–18°, that is, 72°. The east and west points are here intended to be on either side of the ensemble, but the two rings (d) and (k) can be rotated around the axis and the inclination of the latter to the former can be adjusted so that (k) represents the horizon of any locality. The fact that the ecliptic (j) and the equator (g) do not meet on the equinoctial colure (c) could be interpreted as an attempt to represent the result of the trepidation of the equinoxes, which effect is sometimes incorporated on Renaissance armillary spheres.[3] On the other hand it may result from artistic licence. The apparently successful representation in perspective of the various rings merits a study in its own right.[4]

The astrolabe depicted in the intarsia is shown with its rete 'upside down'. In Fig. 2 the rete is shown in the standard position so that now the astrolabic markings on the plate beneath are upside down. The astrolabe is depicted with great care and with remarkable attention to detail. Alas there are no inscriptions. The piece is rather small when compared with the armillary sphere and the books below. The throne consists of a raised central circular part, to which is attached a swivel and circular suspension ring, with two shallow circular arcs concave outwards on each side. The outer scale is divided into 72 5°-divisions and an effort has been made to show individual 1°-divisions. The ecliptic is divided into 12 zodiacal signs each subdivided into six 5°-intervals. The horizontal diameter of the rete is counter-changed three times on each side of the rather large central disc: inside the ecliptic, at the ecliptic and again outside the ecliptic. A distinctive and aesthetically pleasing rete-pattern has five circular arcs around the inside of the southern (upper) ecliptic and five similar arcs around the inside of the lower half of the outer frame. Of the latter the middle one has been distended upwards along the vertical axis to a short horizontal collar and a small bifurcated frame joined to the lower ecliptic. Various star-pointers are depicted, and both their number and their positions are highly significant – see below.

We may well ask: What was the inspiration for this elegant rete? Is there perhaps an astrolabe still existing with such a distinctive rete-pattern? In fact, there are two, and the reader will imagine my delight upon identifying them. The first, hereafter denoted by **X**, is now preserved in the Museo di Storia della Scienza in Florence (#493): see Figs. 3–4. It is clearly Italian, and it has a very medieval look about it: it probably dates from the 14th century. The second, **Y**, was alas stolen in the summer of 1977 during visiting hours from the former Musée Départemental, now the Musée Anne de Beaujeu, in Moulins (Allier), France, and has not been seen since (#4506): see Figs. 5–6.[5] It is Renaissance Italian, and is dated Urbino, 1462. The design of the

---

3 See, for example, *Schweinfurt 1993 Exhibition Catalogue*, pp. 362–5.

4 On the context see Edgerton, Linear Perspective.

5 The piece was briefly mentioned in *Paris BN 1973 Catalogue*, p. 75 (no. 176), with illustrations of the front and back on p. 57.

Fig. 2: The astrolabe in the intarsia, now with the rete in standard position and the astrolabic markings on the plate upside down.

Fig. 3: The front of the medieval Italian astrolabe #493.
[Courtesy of the Museo di Storia della Scienza, Florence.]

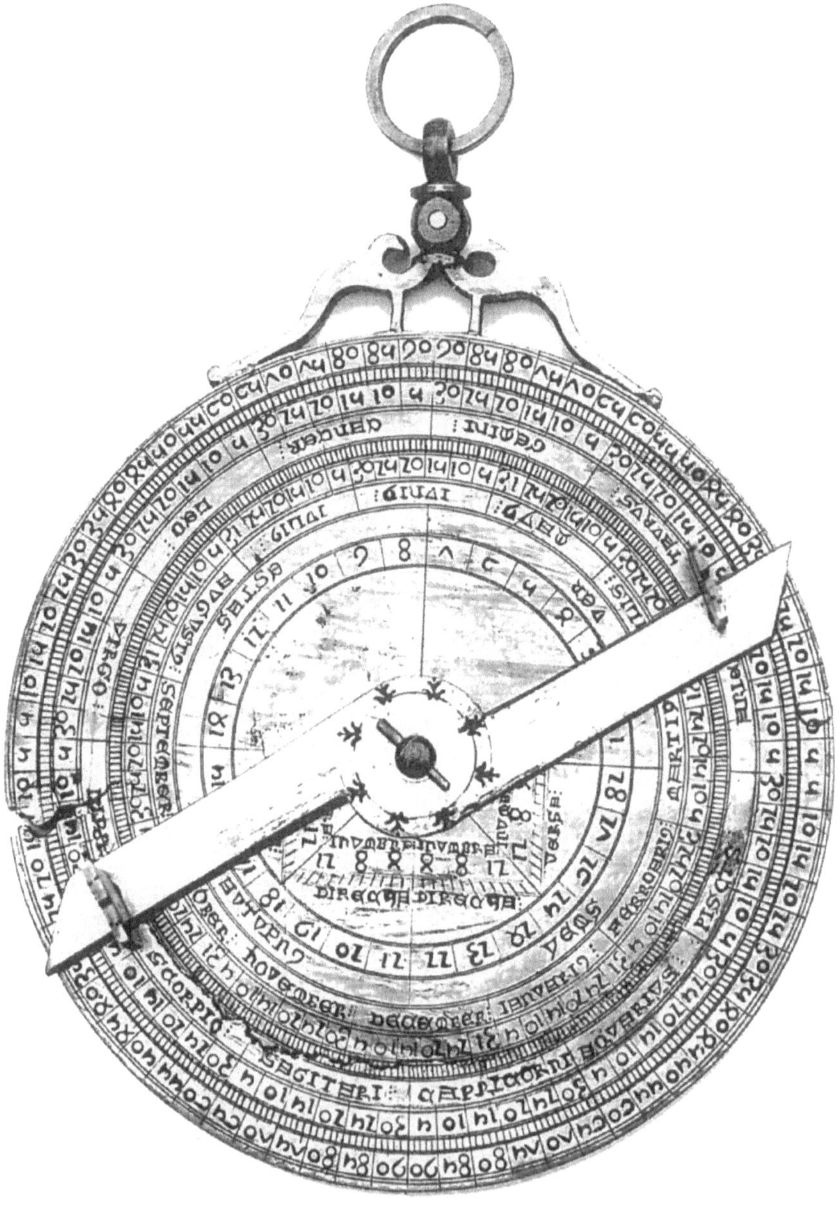

Fig. 4: The back of the medieval Italian astrolabe #493.
[Courtesy of the Museo di Storia della Scienza, Florence.]

Fig. 5: The front of the 1462 Urbino astrolabe #4506.
[Courtesy of the Museum of History of Science, Oxford.]

Fig. 6: The back of the 1462 Urbino astrolabe #4506.
[Courtesy of the Museum of History of Science, Oxford.]

medieval astrolabe **X** clearly inspired the design of the Renaissance piece **Y**, which bears such strong resemblance to the astrolabe depicted in the intarsia, hereafter denoted by **U**, that it is tempting to suspect that **Y** was the model for **U**. Both **X** and **Y** are small (diameter 10 cm and 13 cm, respectively) when compared with other medieval astrolabes, and this may account for the relative smallness of **U**. However, there are a few subtle differences, and we may suppose that the actual model for **U** was a closely-related astrolabe.

In Appendixes A and B I present detailed descriptions of astrolabes **X** and **Y**, neither of which has been published previously. I shall steer clear of any linguistic commentary on the various inscriptions, many of which – in particular, the star-names – are highly unusual.[6] It suffices here to point to the similarity of **Y** and **U**, although certain aspects of the design of these can only be understood with reference to **X**. For example, on **X** each of the five circular arcs within the ecliptic bears a single star-pointer, and each of the four circular arcs on the lower frame bears two star-pointers. Between each of the two pairs of these arcs there is a Gothic arch framing a single pointer, connected to the ecliptic by a frame shaped like the outline of a wine-glass. The other star-pointers on **X** reflect a delicious but slightly deceptive symmetry rarely found on European retes. We are here looking at an *original* design, which has been modified in **Y** in two ways (apart from the new Renaissance look instead of the medieval one). First, the Gothic arch has been brought into line with the other arcs, although the distinctive 'wine-cup' connector (or simply 'bifurcation') has been preserved, the bowl now 'rectangular' rather than 'semi-circular'. Second, some liberties have been taken with the star-pointers and the symmetry has been sacrificed (in fact, we are dealing with a different set of stars). Thus, for example, the second arc on the right inside the ecliptic has a star-pointer on the *concave* side (whereas the other four have pointers on the convex side), and a *double* star-pointer is found on the second arc from the right on the outer frame. Precisely these features are found on **U**! In fact, it is possible to identify the stars which are featured on **U** – see Appendix C. Alas we cannot know how these stars would have been named; some of the star-names used on **X** and **Y** are very corrupt renderings of Arabic originals, attested neither in the known manuscript tradition nor on other surviving astrolabes.

The astrolabic markings on the plate partially visible underneath the rete on **U** are also executed with great care – see Fig. 7. We can detect the circles for Cancer and the equinoxes, the horizon and altitude circles for each 10° up to the zenith, and the 10 curves which together with the two sides of the horizon define the 12 seasonal hours below the horizon. The markings are standard, and we note that **X** has altitude circles for each 6°, and Y has them for each 5°, with additional azimuth curves for each 10° of argument. One could be forgiven for expecting that on an

---

[6] See the various studies of Kurt Maier listed in the bibliography, in which such considerations have lead to the identification of the provenance of an instrument and – in the case of instruments with more than one layer of inscriptions – the location of its subsequent mutations.

Fig. 7: Computer-generated markings for latitude 42° superposed on the astrolabe depicted in the Urbino intarsia. [Computer graphics by Dr François Charette.]

astrolabe illustrated with such technical competence in an intarsia in Urbino we might find markings for the latitude of that city, but in fact this plate serves a latitude of about 42°, which would serve Urbino. It is not modern latitudes that are relevant here, but rather those which were used in the Middle Ages and Renaissance. The state of documentation of early European geographical tables leaves much to be desired.[7] For this study I have consulted only two 15th-century geographical tables, one Italian and one English.[8]

The latitudes represented on **X** are 41° and 42°, which could serve Rome and Pisa / Urbino, respectively.[9] The later sets of markings on another plate are for 43° and 44°, which could serve, say, Ancona / Florence / Perugia and Bologna, respectively.[10] But there are no markings for any lower latitudes, such as we might expect since **Y** is apparently modelled after **X**. Astrolabe **Y** bears astrolabic markings for latitude 37° (and apparently no other latitude). It is worthwhile to look at the latitudes given in the two medieval geographical tables mentioned above, albeit in the light of Ptolemy's geographical coordinates[11] and the Islamic tradition of mathematical geography.[12] From these it seems that the astrolabic markings on **Y** were intended for Sicily.[13] Why an astrolabe apparently made in Urbino should have such markings (and no others) is a mystery.

The historical importance of the two astrolabes **X** and **Y** is not restricted to their serving to explain the design of the astrolabe featured in the Urbino intarsia, for at least two reasons.

The first reason is that the inspiration for the design of astrolabe **X** is clearly the astrolabes of the prolific Abu Bakr ibn Yusuf of Marrakesh ca. 1200 – see Fig. 8 for an example.[14] Both the throne on **X** and the star-pointers were simply copied from his designs. The rete of **X** with its distinctive two sets of circular arcs is, on the other hand, an original, ingenious and aesthetically-pleasing modification of Abu

---

7 What is required is the equivalent of Kennedy & Kennedy, *Islamic Geographical Coordinates*, in which the data from some 125 medieval Islamic sources is arranged in four different ways: alphabetically by place-names, by source, and by increasing longitudes and latitudes.

8 Published together in North, *Horoscopes and History*, pp. 186–95.

9 The Italian geographical table has 41°50' for Rome. The English table has 42° for Pisa. Neither has an entry for Urbino.

10 The Italian table has 43°30' for Pisa. The English table has 44° for Bologna.

11 On Ptolemy see the article by Gerald Toomer in *Dict. Sci. Biog.*, especially pp. 198–200 on the *Geography*.

12 See note 7 above.

13 Ptolemy's latitude for Palermo was 37°. The 15th-century Italian geographical table has 37° for Syracuse and 37°30' for Palermo.

14 On Abu Bakr and his works see Mayer, *Islamic Astrolabists*, pp. 32–3. To the Strasbourg astrolabe (#124) published in detail in Sarrus, 'Astrolabe construit au Maroc', and known to Gunther, can now be added the Toulouse (#1090), Rabat (#2209) and Madrid (#1069) astrolabes, together with another later Maghribi astrolabe in the Science Museum, London (#1057). The Madrid piece has been altered by a European and the main inscription removed – see García Franco, *Catalogo critico*, pp. 277–82 (no. 19), and *Madrid MAN 1992 Exhibition Catalogue*, p. 56.

Fig. 8: The front of the Toulouse astrolabe of Abu Bakr ibn Yusuf (#1090).
[Courtesy of the Musée Paul Dupuy, Toulouse.]

Bakr's distinctive rete-design. Did this influence from Morocco perhaps come *via* Sicily? On the other hand, it is possible that Abu Bakr's distinctive rete-design, so different from early Eastern Islamic patterns and early Andalusi patterns,[15] was not his own inspiration, but rather was inspired by a very early Italian design. To support this hypothesis we can point to a remarkable instrument clearly of early Italian origin (#4556),[16] with precisely the design of rete, complicated and cumbersome, that might have influenced Abu Bakr – see Fig. 9. On the other hand, we can cite

15 Typified by, for example, two astrolabes from 10th-century Baghdad and 11th-century al-Andalus – see Gunther, *Astrolabes*, I, pl. LII (no. 99) and pl. LX (no. 118).

16 This piece, now in a private collection, is published in detail in *London Christie's 29.9.1994 Catalogue*, pp. 34–9 (lot 136); the front is illustrated in King, 'Earliest European Astrolabe', fig. 21. See also note 28 below.

Fig. 9: The front of a remarkable early medieval Italian astrolabe (#4556).
[Courtesy of Christie's, London.]

another Italian astrolabe (#4507),[17] of uncertain date but displaying both medieval and Renaissance features, whose rete was certainly inspired by the distinctive rete of an Islamic astrolabe from 11th-century Toledo (#116).[18]

The second reason is that the design was appropriated in a substantially modified form in 15th-century Vienna – whence we have an undated astrolabe probably

[17] This instrument is preserved in the Museo Nazionale del Bargello in Florence. It has been deemed an early modern forgery; if it is, then it was copied from an original of exceptional historical interest. It is described with some misgivings regarding its authenticity in *Florence MNB 1989 Exhibition Catalogue*, pp. 444–6, and *Florence MNB 1991 Catalogue*, pp. 48–54 (no. 22); rete illustrated in King, 'Astronomical Instruments', fig. 19.

[18] See the detailed description in Woepcke, 'Arabisches Astrolabium', and the summary in Gunther, *Astrolabes*, I, pp. 251–2 (no. 116).

Fig. 10: The front of the astrolabe presented by Regiomontanus to Cardinal Bessarion in 1462 (#640). [Courtesy of Christie's, London.]

made by Georg Peuerbach himself.[19] And in a simplified form it was used shortly thereafter by Johannes Regiomontanus in the astrolabe that he presented in 1462 to his patron the Cardinal Bessarion in Rome (#640) – see Fig. 10.[20] (It is rather ironical that the design should return to Italy so soon after its export to the German-speaking world.) The prolific Georg Hartmann of Nuremberg used the rete-design of Regiomontanus in the early years of the 16th century.[21] But by this time the

---

19 On this see *Nuremberg GNM 1992–93 Exhibition Catalogue*, pp. 582–5 (no. 1.75).

20 On this see King & Turner, 'Bessarion's Astrolabe', and also *Nuremberg GNM 1992–93 Exhibition Catalogue*, pp. 586–9 (no. 1.76).

21 See Gunther, *Astrolabes*, II, pp. 436–40, especially pl. CVIII opposite p. 437.

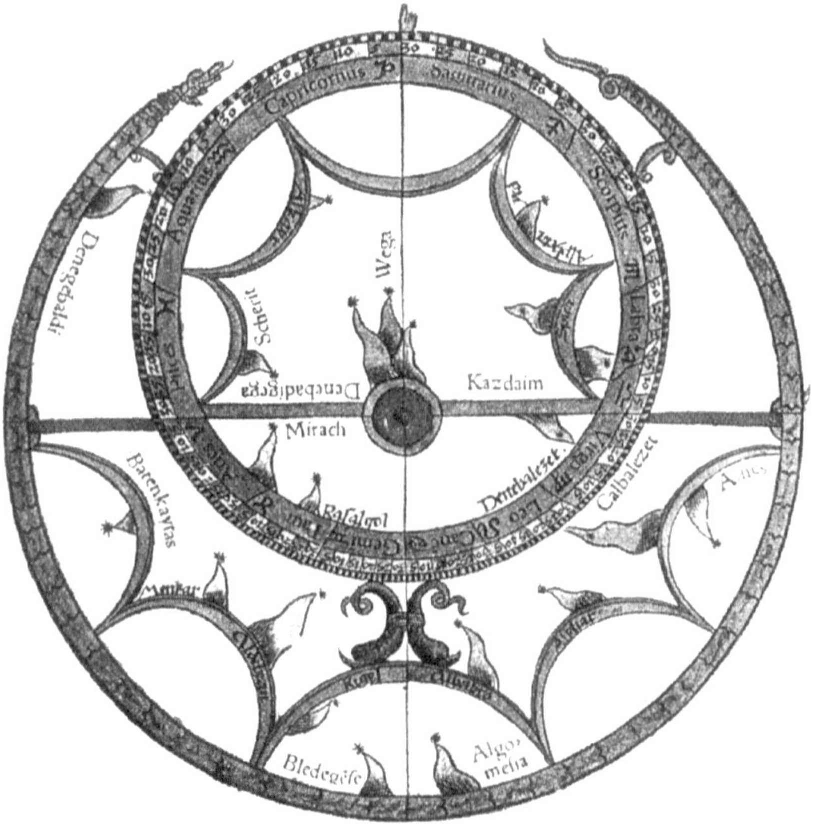

Fig. 11: The rete of the astrolabe illustrated in the treatise of Johannes Stöffler, 1524.
[From Gunther, *Astrolabes*, II, p. 437.]

original Italian design had been substantially modified. It is therefore heartening to find it again in its original form in the *Elucidatio* of Johannes Stöffler published in 1524 – see Fig. 11.[22]

But there is more. The history of the astrolabe in medieval Italy is obscure indeed.[23] There are disappointingly few surviving astrolabes, yet there are more than one would have thought: some two dozen pieces can safely be associated with

[22] See, for example, *ibid.*, II, pp. 427–31 (no. 253).

[23] Note that R.T. Gunther refrained from writing an introduction to his section on Italian astrolabes (*Astrolabes*, pp. 316–38).

Italy from before ca. 1500. Some of these have been known since the time of R.T. Gunther.[24] A substantial number in collections mainly in Milan have been catalogued by Dr Tullio Tomba of Milan.[25] Others have come to light during cataloguing of instruments in various collections such as Nuremberg.[26] Yet others have been identified more recently. Two medieval pieces are of exceptional historical interest, the one featuring a replacement Byzantine plate and additional markings in Hebrew,[27] and the other fitted with a rete which bears no relation to any other known piece.[28] Another, dated 1420, is of extraordinary historical interest, the maker having used ecliptic coordinates instead of equatorial ones to construct his rete.[29] Four medieval astrolabes with Hebrew inscriptions appear to be of Italian origin, perhaps from Bologna ca. 1400.[30] The end of the medieval tradition is to be witnessed in the productions of Falconi *ca.* 1500.[31] The dearth of surviving examples of medieval Italian astrolabes does explain why one remarkable Renaissance astrolabe (#171)

---

[24] Gunther listed 24 'Italian astrolabes', numbered from 166 to 189. His #166 is possibly Italian, not least because the rete-design bears no relation to the Western Islamic / Iberian tradition. It was purchased in 1899 in Venice, but the latitudes serve the climates and there are no clearly Italian features. #167 can hardly be Italian since the plate serves 47° and 48°. #168 is French. #169 is a universal Italian astrolabe with a double myrtle ecliptic (and astrolabic markings for only 24°, the second climate, rather than a set of plates for each climate). Gunther labelled it 'Sicilian' because of an incorrect reading of the latitude. #170 is an Italian 'astrological' astrolabe, probably 15th century. #171 is a spectacular Italian Renaissance astrolabe – see note 33 below. #172 is not an astrolabe. #173 is probably German, and three other pieces from the same workshop are now known – see King & Turner, 'Bessarion's Astrolabe', p. 192, note 69. #174 is certainly Italian, probably from ca. 1500. #186 is an Italian astrolabe from ca. 1450. ##175–85 and 187–9 are Renaissance Italian astrolabes, but are not relevant to the present study.

[25] Some ten medieval Northern Italian astrolabes, mainly with quatrefoil decoration on the rete, display French, and thus ultimately Spanish, influence. They are discussed in the various valuable studies of Tullio Tomba listed in the bibliography.

[26] Three previously unknown medieval Italian astrolabes came to light when the instruments in the Germanisches Nationalmuseum were catalogued some ten years ago (#558, #547 and #548 – see *Nuremberg GNM 1992–93 Exhibition Catalogue*, II, pp. 574–81).

[27] #4509, now in a private collection, merits a detailed study.

[28] On #4556 see note 16 above.

[29] This astrolabe in a private collection in Germany (#4523) was signed by Antonius de Pacent in Lanzano in the year 1420. Not only are the star-positions on the rete problematic (coordinate systems confused) but the plates which are labelled for various latitudes are in fact all engraved for 45°. This piece is now published in Stautz, 'Astrolab aus dem Jahr 1420'.

[30] These are: #158 in the British Museum (see Gunther, *Astrolabes*, II, p. 304 (no. 158); *London BM Catalogue*, p. 113–14 (no. 328) and pl. LIIIa; and also Goldstein, 'Hebrew Astrolabe'); #159 in the Adler Planetarium, Chicago (see Gunther, *Astrolabes*, II, p. 304 (no. 159); Goldstein, 'Hebrew Astrolabe'; and now *Chicago AP Catalogue*, I, pp. 58–60 (no. 7)); #3906, in a private collection (see Bandeira Ferreira, 'Astrolábio hebraico', and also Goldstein, 'Hebrew Astrolabe'). See also #621 in the Deutsches Museum, Munich, a composite medieval European (Italian?) astrolabe with a very early rim, rete and plates, and a back from an astrolabe with Hebrew inscriptions (see *Munich Astrolabe Catalogue*, pp. 161–76 (no. 2), also illustrated in King, 'Earliest European Astrolabe', fig. 13.

[31] These have been studied by Silvio Bedini.

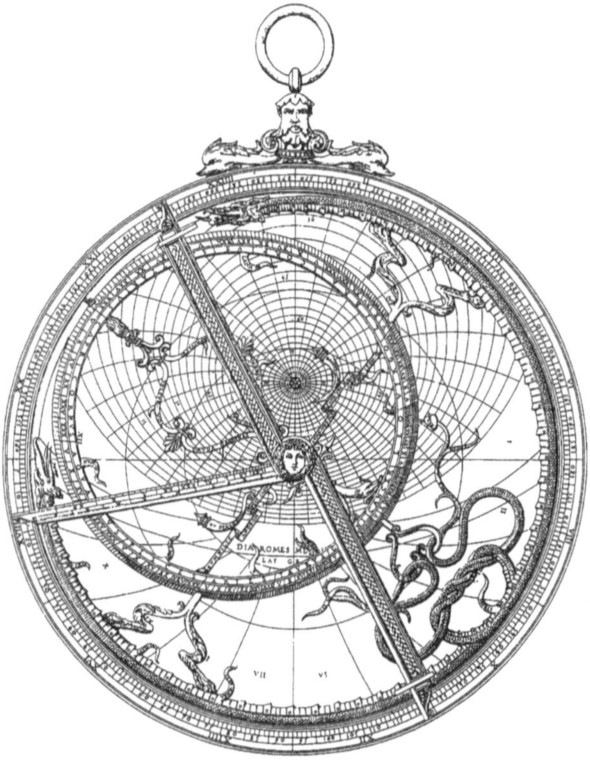

Fig. 12: A superb Renaissance Italian astrolabe (#171). [From Gunther, *Astrolabes*, II, p. 323.]

from Northern Italy ca. 1500 is so full of surprises – see Fig. 12.[32] Also, very few of the available textual sources – in the libraries of Italy alone there must be dozens of medieval manuscripts dealing with the construction and use of the astrolabe – have been studied.[33]

There are problems enough with the surviving instruments, not least of which is the existence of a 10th-century astrolabe from Catalonia that seems to point

---

[32] See Gunther, *Astrolabes*, II, pp. 322–5 (no. 171). This piece, preserved in the Museum für Kunstgewerbe in Hamburg, is regarded by some authorities as a fake, but it is not. See the forthcoming catalogue of the Hamburg astrolabes by Karin Eckhardt.

[33] Several treatises by Andolò di Negro (Genoa, ca. 1260–ca. 1340) are known – see Sarton, *IHS*, III:1, pp. 645–8, especially p. 647. One of these was published by G. Bertolotto in 1892 – see the bibliography.

to a hitherto-unknown Roman tradition of astrolabe-making.[34] Now three Italian astrolabes, **X**, **Y** and the model for **U**, are shown to be defiantly independent of the main Western Islamic → Christian Catalonian and Spanish → French, *etc.* tradition,[35] and indeed clearly derive from yet another tradition by which at least aspects of astrolabe design came to Europe, namely, Morocco → Italy.

Acknowledgements:

I am grateful to François Charette, Reinhard Glasemann and Siegfried Müller, all participants in my instrument seminar at the Johann Wolfgang Goethe University in Frankfurt, for various suggestions of a technical nature. I take this opportunity to thank Franca Principe of the Istituto e Museo di Storia della Scienzia in Florence for the superb photos of the Florence astrolabe; Giles Hudson of the Museum of History of Science at Oxford for copies of photos of the stolen Moulins astrolabe; Nadine Berthelier of the Musée Anne de Beaujeu in Moulins for various items of information on the Moulins astrolabe; and – last but not least – Roberto Mantovani not only for the photos of the Urbino intarsia, but also for his encouragement and patience.

Appendix A

Description of X, a medieval Italian astrolabe

International Instrument Checklist no. 493.

Location: Florence, Museo di Storia della Scienza, inv. no. 1107 (earlier numbering 6785, 1017, 596 and 782).

Material: Gilded brass.

Dimensions: Diameter – 99 mm, thickness – 5 mm.

Photographic documentation: A full set of high-quality photos is available, courtesy of the Museum.

---

[34] #3042, preserved in the Institut du Monde Arabe, Paris – see Stevens *et al.*, eds., *Oldest Latin Astrolabe*, for a variety of opinions on this piece, and King, 'Earliest European Astrolabe', pp. 384–5, on the possible Roman connection. On p. 384 the section beginning 'An astrolabe from ca. 1300 ... ' should have been introduced by the words 'Added in proof'.

[35] Some examples of the middle stage of this transmission – various medieval Christian Catalan and Spanish astrolabes – are discussed in King & Maier, 'Catalan Astrolabe'; King, '14th-Century Spanish Astrolabe'; and *idem*, 'Medieval Catalan Astrolabes' (in Catalan, to appear).

Documentation: Unpublished. Briefly mentioned in *Florence 1954 Catalogue*, p. 62 (no. 1107), and *Florence 1987–88 Exhibition Catalogue*, p. 14 (no. I.6).

The astrolabe has a distinctly medieval appearance. The lettering and numeral forms are carefully engraved in a distinctive hand.

The throne bears some resemblance to two tadpoles facing each other, and each standing on a short vertical front leg, with the tails tapering off along the rim on either side. The suspensory apparatus consists of a shackle with a swivel and ring attached. The scale around the rim of **the mater** is divided and labelled for each 5°, subdivided for each 1°. The mater itself is devoid of astronomical markings. There is a crack across the lower half.

The rete is distinguished by five short circular arcs inside the southern (upper) ecliptic and two pairs of larger arcs separated by a broad Gothic arch on the lower circumferential frame. The equinoctial (horizontal) bar is counter-changed three times on each side, once outside the ecliptic, once at it, and again inside it. There is a short bar joining the central disc to the northern (lower) ecliptic and between the top of the central arch on the circumferential frame and the ecliptic there is a wine-glass-shaped frame. The scale of the ecliptic is divided in 6°-intervals for each sign, with subdivisions for each 2°: the divisions are not labelled. The names of the signs are standard with three exceptions indicated in bold (the '9' is a standard medieval abbreviation for the ending *-us*):

ARIES TAV GEM CAN LEO VIR
LIBRA **SCORPI9 SAGITARI9** CAPRICOR AQVARI9 **PISCIS**

The star-pointers are distinctive, being shaped mainly like hooks. They serve 23 stars, named as in following list (with unusual forms in bold). Regulus (COR LEO) is at *ca.* Leo 20°, and both the selection of stars and the star-positions are related to the star-catalogue compiled in Paris by John of London in 1246 (Kunitzsch Typ VI).[36] The names given in the principal source of that star-catalogue used by Paul Kunitzsch and his numbering and modern identifications are also shown.

---

36 Stautz, *Mathematisch-astronomische Darstellungen*, p. 115 and fig. 5.2a.

The stars on the Italian astrolabe #493, with the corresponding entries in Kunitzsch's Typ VI and identifications

Star-name	Kunitzsch VI	Derivation	Identification
1 **BATVL**	3 *batenkaitoz*	*batn qaytus*	ζ Ceti
2 MENCE	7 *menkar*	*minkhar*	α Ceti
3 **ALAV**	8 *algemb*	*al-janb*	α Per
4 **ALDBA**	10 *aldebaran*	*al-dabarân*	α Tau
5 **BEDEL**	13 *bedalgeuze*	*yad al-jawzâ'*	α Ori
6 ALABOR	14 *alhaabor*	*al-'abûr*	α CMa
7 ALGOME	16 *algomeisa*	*al-ghumaysâ'*	α CMi
8 **ALRI**	19 —	*al-rijl*	μ UMa
9 **ALAB**	20 *alfart*	*al-fard*	α Hya
10 COR LEO	21 *cor leonis*	*cor leonis*	α Leo
11 **ALGORE**	24 *algorab*	*al-ghurâb*	γ Crv
12 ALCHI	25 *alchimech*	*al-simâk (al-a'zal)*	α Vir
13 ALRAMECH	26 *alramech*	*(al-simâk) al-râmih*	α Boo
14 COR SCOR	29 *cor scorpionis*	*cor scorpionis*	α Sco
15 **VEDH**	28 *yed*	*yad*	δ Oph
16 **RAZT**	31 *razcaben*	*ra's al-tinnîn*	γ Dra
17 **BAZALA**	30 *razalangue*	*ra's al-hawwâ'*	α Oph
18 [V/WEGA]	32 *alwega*	*(al-nasr) al-wâqi'*	α Lyr
19 **ADDE**	36 *aldigege*	*al-dajâja*	α Cyg
20 DELFI	34 *delfin*	*dulfin*	γ Del
21 **DENELA**	35 *deneb algedi*	*dhanab al-jady*	δ Cap
22 ALFERAS	37 *alferaz*	*al-faras*	ε Peg
23 DENEB	40 *denebkaytoz*	*dhanab qaytus*	β Cet

There is **one original plate**, the other two being later (see below). The former is marked with altitude circles for each 6° and there are no azimuth circles. The arguments for the altitude-circles are labelled 6–24 on the equinoctial circle and 30–90 on the meridian. The latitudes served are:

1a: 41° and 1b: 42°.

There are construction marks around the circle for the summer solstice on both sides. On the first side there are additional dots inside this circle, a feature not attested elsewhere.

The back bears four altitude scales divided and labelled for each 5° and subdivided for each 1°. The solar scale is contiguous with these and is labelled for each 5° of each sign. The names of the signs are:

ARIES TAVRVS GEMINI CANCER LEO VIRGO
LIBRA SCORPIO **SAGITARI**
CAPRICORNI AQVARIVS **PISCTS**

The 'T' in PISCIS is curious. There is a concentric calendrical scale divided and labelled for each 5 days of each month, subdivided in 1-day intervals, and the names of the months are:

IANVARI9 **FEBROARI9** MARTI9 APRILIS
MAY9 IVNI9 IVLI9 AVGVST9
SEPTEMBER OCTOBER NOVEMBER DECEMBER

The forms are Latin, with MAYUS an orthographical variant for MAIUS, and with O for the first U in FEBROARIUS. Most unusual is the fact that the four seasons are labelled:

VER ESTAS AVTVRN9 YEMS

names which could surely provide information on the provenance.[37]

The equinox is at March 12. In addition there is a circular scale within the other scales, with divisions labelled 1–28 starting at the equinox, and in the lower half of the central space there is a double shadow-square with scales divided and labelled for each 4 units and subdivided for each single unit. The scales are marked:

---

[37] Merlo, *Nomi delle stagione e dei mesi*, pp. 34 and 41–2 (*ver*); 29–31 (*aestate*); 64–7 (*autumnus*); 20 (*hiems*). Apart from the first form *ver*, the others are not specifically mentioned by Merlo.

X

INVMBRA DIRECTA and **INVMBRA** VERSA

The form INVMBRA is unusual.

The alidade is counterchanged at the centre and the vanes each have a V-shaped cut-out at the middle of the top and a circular hole in the middle. Around the central circular disc there are eight stylized floral patterns. The pin is simple and the wedge is short and cylindrical in form.

Additional or substitute parts: The second plate bears altitude circles for each 6° and azimuth circles for each 10°. The two sides serve latitudes:

2a: 43° and 2b: 44°.

The forms of the numerals are Renaissance Italian. On both sides the azimuth base line is visible and on the second there are marks around the circle for Cancer. The third plate has markings only on one side, namely altitude circles for each 5° and azimuth circles for each 10°. The underlying latitude is not stated but is approximately 42°. There are construction marks all around the circle for Cancer. On the back there are a series of lines and a small circle, all without astronomical significance.

Appendix B

Description of Y, a Renaissance Italian astrolabe dated 1462

International Instrument Checklist no. 4506.

Location: Unknown. Formerly in the Musée Départemental (inv. no. 813), now the Musée Anne de Beaujeu, Moulins (Allier), France. Exhibited at the Bibliothèque Nationale, Paris, in 1973. Stolen from the Musée Départemental, Moulins (Allier), France, on August 26, 1977.

Material: Brass.

Dimensions: Diameter – 13 cm.

Photographic documentation: Photocopies of photos of the front and back are available, courtesy of the Museum for the History of Science, Oxford, which houses photos made by Studio Fournier (Camille Gonnard), 16, rue du 4 septembre, Moulins (Allier), but no longer accessible. The Musée Anne de Beaujeu has a good quality photo of the front and back.

Documentation: Unpublished. Briefly mentioned in *Paris BN 1973 Catalogue*, p. 75 (no. 176), with illustrations of the front and back on p. 57.

The astrolabe has a distinct Renaissance appearance. The lettering and numeral forms are in an elegant Italic script (the old form for the '5' is still used and the '2' is still shaped like a 'Z'). The back of the throne bears the inscription 'KP'.[38] Below the shadow squares on the back is an inscription '1462 VRBINI K'.

The throne has a raised circular part in the middle, from the base of which two long leaf-shaped parts fall elegantly onto the rim, and a shorter one falls vertically in the middle. The suspensory apparatus is missing. On the back of the throne is the inscription mentioned above. The scale around the rim of the mater is divided and labelled for each 5° and there are subdivisions for each 1°. There is a small hole on the outer side of the scale for each 15°, and perhaps the hours are numbered on the outer edge. The mater itself is engraved – see below.

The rete is elegantly decorated with five shallow circular arcs inside the southern (upper) part of the ecliptic ring and five more rounded arcs on the lower half of the circumferential ring. This ring has a dragon's head at the left extremity and a tail of sorts at the other end. Of the latter the central one is extended upwards to join a short horizontal collar and 'wine-cup'-shaped frame with its two bifurcations attached to the middle of the northern (lower) ecliptic ring. This ring is divided into six unlabelled 5°-intervals for each sign. The names of the signs are:

ARIES **THAVR9** GEMI CANCER LEO VIRGO
LIBRA SCORPIO **SAGITTAR9**
CARRICOR9 AQVAR9 PISCES

The engraver has allowed himself some licence with regard to the endings of the names of the 9th–11th signs, the '9' standing for *-ius*, *-nius*, and *-ius*, respectively. The 'R' for a 'P' in CARRICORNUS is apparently a slip of the engraver.[39] There is a second, smaller dragon's head serving as pointer at the winter solstice on the ecliptic scale.

The 22 star-pointers are hook- or flame-shaped and are named as follows. The star-positions are taken from a medieval star-catalogue (Kunitzsch Typ VIII) that is in part derived from the one discussed above (Typ VI).[40]

---

38 Since the two letters are separated I deem it unlikely that the reading 'JC P' was intended.

39 'R' for 'B' or 'P' (apparently deliberate) is attested (albeit sporadically) on two medieval Catalan astrolabes – see King & Maier, 'Catalan Astrolabe', pp. 688 and 690.

40 See Kunitzsch, *Typen von Sternverzeichnissen*, pp. 51–8 (Typ VIII), and Stautz, *Mathematisch-astronomische Darstellungen*, p. 115 and fig. 5.2b on p. 280.

The stars on the 1462 astrolabe #4506, with thc corresponding entries in Kunitzsch's Typ VIII and identifications

Star-name	**Kunitzsch VIII**	**Derivation**	**Identification**
1 FINI	5 *finis fluxus* (sic)	*finis fluxus* (sic)	θ Eri
2 **AREN**	7 *algenib*	*al-janb*	α Per
3 ALDEBERAM	9 *aldebaran*	*al-dabarân* α Tau	
4 RIGEL	11 *rigil*	*(al-)rijl*	β Ori
5 ALLABOR	13 *alhabor*	*al-'abûr*	α CMa
6 ALGOMEZA	15 *algomeyza*	*al-ghumaysâ'*	α Cmi
7 ALFARAC	19 *alfart*	*al-fard for al-faras*	α Hya
8 COR L	20 *cor leonis*	*cor leonis*	α Leo
9 DVB	23 *dubhe*	*(al-)dubb*	α UMa
10 CORVVS	22 *coruus*	—	
11 **ALGER**	25 *algorab*	*al-ghurâb*	γ Crv
12 ALCHIMEC	26 *alchimec*	*al-simâk*	α Vir
13 ALRAMIC	28 *alramech*	*al-râmih*	α Boo
14 FITA	29 *alfeca*	*al-fakka*	α CrB
15 **LAD**	31 *yed*	*yad*	δ Oph
16 COR S	32 *cor scorpii*	*cor scorpii*	α Sco
17 ALHAVERI	33 *alhaue*	*al-hawwâ'*	α Oph
18 VVEGA	35 *wega*	*(al-)wâqi'*	α Lyr
19 ALTAIR	36 *altair*	*al-tâ'ir*	α Aql
20 DENEP	44 *denebalgedi*	*dhanab al-jady*	δ Cap
21 **SERA**	45 *sceach (var.:* ***sera****th)*	*sâq*	δ Aqr
22 ALFOR	46 *alferaz*	*al-faras*	β Peg

Note the confusion in no. 21 caused by the similarity of the Gothic letters for c, e, r and t between the already corrupt and curious *sceach* and its variant *serath*, ultimately derived from Arabic *sâq*.

Before and after the names there is either a single dot or a triad of dots. No information is available on the back of the rete.

The mater is engraved with altitude circles for each 5° and azimuth circles for each 10°, as well as with curves for the astrological houses. Both sets of circles are labelled for each 10°, also the seasonal hour curves are numbered. The curves for the houses are labelled in Roman numerals. The latitude served is (according to the Paris catalogue) 37°. There appear to be no plates (at least, none are mentioned in the same catalogue).

The back bears two altitude scales with labelled divisions for each 5° and subdivisions for each 1°. The lower scales are divided in the same way but are unlabelled. Both the 5°- and 1°-degree markings double as a solar scale. The names of the signs are written exactly as on the rete (see above). There is an excentric calendrical scale divided for each 5° and subdivided for each 1°, but the divisions are not labelled. The names of the months are (? represents a letter obscured by the alidade on the only available photo):

IANVAR9 FEBRVAR9 MA??VS
APPRIL9 MAIVS IVNIVS
IVLIVS AVGVST9 SEP**TE**???
OCTVB˜ NOVEMB˜ DECEMB˜

The ‘V’ in MAIVS is an inverted ‘A’, and the ‘TE’ in SEPTE[MB](ER) is engraved as a ligature. The equivalent endings ‘9’ and ‘(I)VS’ are not used consistently. Of linguistic interest are the double P in APPRILUS and the form OCTVB(ER). The equinox is at ca. 12.5 (obscured by the alidade). In the upper central space there is a double universal horary quadrant and below this there is a double shadow square with scales divided for each 3 units and subdivided for each 1 unit, but unlabelled. The inscriptions on the latter are:

VNBRA RECTA and V. **AVERSA**

both VNBRA and AVERSA being unusual forms. Inside the shadow squares and extending over the horary quadrants is a circular scale with divisions numbered 1–28 starting at the vernal equinox. Below the shadow squares is the inscription mentioned above.

The **alidade** is counter-changed at the centre and slightly tapered towards the ends, and one of the two vanes is missing, The head of the pin has a diametrical

X

groove across it, perhaps for screwing the pin into the back of the 'washer' on the front, for there appears to be no wedge.

Appendix C

Identification of the stars on U using those on X and Y

The following list displays the stars on **X** and **Y**. It also displays the stars on **U** which are a subgroup of those on **Y**. These are indicated by asterisks, since no names are given. One questionable connection is boxed. The original Arabic or Latin (A/L) of the names on **X** and **Y** are given.

X	**Y**	**U**	**origin**
BATVL			*A batn qaytus*
	FINI	*	*L finis flux (sic)*
MENCE			*A minkhar*
ALAV	AREN	*	*A al-janb (!)*
ALDBA	ALDEBERAM	*	*A al-dabarân*
BEDEL			*A yad al-(jawzâ')*
	RIGEL	*	*A rijl (al-jawzâ')*
ALABOR	ALLABOR	*	*A al-'abûr*
ALGOME	ALGOMEZA	*	*A al-ghumaysâ'*
ALRI			*A al-rijl*
	ALFARAC	*	*A al-faras for al-fard*
ALAB			*A al-janb*
COR LEO	COR L	*	*L cor leonis*
	DVB (!!)		*A al-dubb*
	CORVVS	*	*L corvus*
ALGORE	ALGER		*A al-ghurâb*
ALCHI	ALCHIMEC	*	*A al-simâk (al-a'zal)*
ALRAMECH	ALRAMIC	*	*A (al-simâk) al-râmih*
	FITA		*A al-fakka*
VEDH	LAD	*	*A yad*
COR SCOR	COR S	*	*L cor scorpionis*
RAZT			*A ra's t(innîn)*
BAZALA	ALHAVERI	*	*A al-hawwâ'*
[VEGA]	[VVEGA]	*	*A (al-nasr) al-wâqi'*

X	Y	U	origin
ADDE			*A al-dajâja*
	ALTAIR	*?	*A (al-nasr) al-ta'ir*
DELFI		*?	*A al-dulfin*
DENELA	DENEP	*	*A dhanab (al-jady)*
	SERA		*A sâq (!)*
ALFERAS	ALFOR	*	*A al-faras*
DENEB			*A dhanab (qaytus)*

Bibliography and bibliographical abbreviations

AIOS: Fuat Sezgin *et al.*, ed., *Arabische Instrumente in orientalistischen Studien*, 6 vols., Frankfurt: Institut für Geschichte der Arabisch-Islamischen Wissenschaften, 1990–91.

Andalo di Negro: see Bertolotto.

Bertolotto, G., *Il trattato sull'astrolabio di Andalo di Negro riprodotto dall'edizione ferrarese del 1475*, Genoa: Società Ligure di Storia Patria, 1892.

Chicago AP Catalogue, I: *Historic Scientific Instruments of the Adler Planetarium & Astronomy Museum*, (general ed., Bruce Chandler, ed., Sara Schechner Genuth), vol. I: *Western Astrolabes*, by Roderick and Marjorie Webster, Chicago, Ill.: The Adler Planetarium & Astronomy Museum, 1998.

Christie's: see London Christie's.

Dict. Sci. Biogr.: *Dictionary of Scientific Biography*, 16 vols., New York, N.Y.: Charles Scribner's Sons, 1970–80.

Edgerton, *Linear Perspective*: Samuel Edgerton, *The Renaissance Rediscovery of Linear Perspective*, New York, N.Y., etc.: Harper & Row, 1975.

Enc. Islam, 2nd edn.: *Encyclopaedia of Islam*, new edn., 8 vols. to date, Leiden: E.J. Brill, 1960 to present.

Florence MNB 1989 Exhibition Catalogue: *Museo Nazionale del Bargello – Arti del Medio Evo e del Rinascimento – Omaggio ai Carrand 1889–1989*, Florence: Museo Nazionale del Bargello, 1989. [Catalogue of an exhibition held from 20.3.1989 to 25.6.1989.]

Florence MNB 1991 Catalogue: Mara Miniati, ed., *Strumenti scientifici della Collezione Carrand*, Florence: Museo Nazionale del Bargello, 1991.

Florence MSS 1954 Catalogue: *Catalogo degli strumenti del Museo di Storia della Scienza*, Florence: Leo S. Olschki, 1954.

Florence MSS 1987–88 Exhibition Catalogue: Mara Miniati, ed., *L'Età di Galileo – Il secolo d'oro della scienza in Toscana*, Florence: Istituto e Museo di Storia della Scienza, 1987. [A catalogue of an exhibition held at the Museum from 1.7.1987 to 9.1.1988.]

Frankfurt IGN Festband: Anton von Gotstedter, ed., *Ad radices – Festband zum fünfzigjährigen Bestehen des Instituts für Geschichte der Naturwissenschaften Frankfurt am Main*, Stuttgart: Franz Steiner, 1994. [A volume of studies including several on instruments.]

García Franco, Salvador, *Catálogo crítico de los astrolabios existentes en España*, Madrid: Instituto Histórico de Marina, 1945.

Goldstein, Bernard R., 'The Hebrew Astrolabe in the Adler Planetarium', *Journal of Near Eastern Studies* 35 (1976), pp. 251–60, reprinted in idem, *Theory and Observation in Ancient and Medieval Astronomy*, London: Variorum, 1985, XVIII.

Gunther, Robert T., *The Astrolabes of the World* ... , 2 vols., Oxford: University Press, 1932, reprinted in 1 vol., London: The Holland Press, 1976.

Hartner, 'Astrolabe', A-B: Willy Hartner, 'The Principle and Use of the Astrolabe' (A), a chapter in Pope, ed., *Survey of Persian Art*, reprinted without the original plates in *Willy Hartner – Oriens-Occidens – Ausgewählte Schriften zur Wissenschafts- und Kulturgeschichte-Festschrift zum 60. Geburtstag*, Hildesheim: Georg Olms, 1968, pp. 287–311, and again with them in *Astrolabica* (Paris: Société Internationale de l'Astrolabe), 1 (1978), as well as *idem*, article 'Asturlab' (B) in *Enc. Islam*, 2nd edn., reprinted in *idem*, *Oriens-Occidens*, 1968, pp. 312–18.

Kennedy & Kennedy, *Islamic Geographical Coordinates*: Edward Stewart and Mary Helen Kennedy, *Geographical Coordinates of Localities from Islamic Sources*, Frankfurt am Main: Institut für Geschichte der Arabisch-Islamischen Wissenschaften, 1987.

King, 'Astronomical Instruments': David A. King, 'Astronomical Instruments between East and West', in Harry Kühnel, ed., *Kommunikation zwischen Orient und Okzident*, Vienna: Österreichische Akademie der Wissenschaften, 1994, pp. 143–98.

—, 'Earliest European Astrolabe': *idem*, 'The Earliest Known European Astrolabe in the Light of Other Early Astrolabes', in Stevens *et al.*, eds., *The Oldest Latin Astrolabe*, pp. 359–404.

—, '14th-Century Spanish Astrolabe: *idem*, 'An Astrolabe from 14th-Century Christian Spain with Inscriptions in Latin, Hebrew and Arabic – A Unique Testimonial to an Intercultural Encounter', *Suhayl – Journal for the History of the Exact and Natural Sciences in Islamic Civilisation* (Barcelona) 3 (2002/03), pp. 9–156, with a new version in *idem*, *In Synchrony with the Heavens*, vol. 2, XV. [For a summary see *London Christie's 15.4.1999 Catalogue*, pp. 98–107 (lot 52).]

—, 'Geography of Medieval Instruments': *idem*, 'Bringing Astronomical Instruments back to Earth: The Geographical Data on Early Astrolabes (to ca. 1100)', in *North Festschrift*, pp. 3–53, with a new version in *idem*, *In Synchrony with the Heavens*, vol. 2, XVI.

—, 'Medieval Catalan Astrolabes': *idem*, 'Astrolabis de la Catalunya medieval', in *La ciència en la història dels Països Catalans*, I: Dels àrabs al Renaixement, Barcelona: Institut d'Estudis Catalans, 2004, pp. 161–204.

—, 'Strumentazione': *idem*, 'Strumentazione astronomica nel mondo medievale Islamico', in Gerard L'E. Turner, ed., *Gli strumenti*, 2nd edn., Turin: Giulio Einaudi, 1991, pp. 154–89 and pp. 581–5 (bibliography).

—, In Synchrony with the Heavens, vol. 2: *idem*, *In Synchrony with the Heavens – Studies in Astronomical Timekeeping and Instrumentation in Islamic Civilization*, vol. 2: *Instruments of Mass Calculation. Studies X–XVIII*, Leiden & Boston: Brill, 2005.

King & Maier, 'Catalan Astrolabe': *idem* and Kurt Maier, 'The Medieval Catalan Astrolabe in the Society of Antiquaries, London', in Josep Casulleras and Julio Samsó, eds., *From Baghdad to Barcelona. Studies in the Islamic Exact Sciences in Honour of Professor Juan Vernet*, 2 vols., Barcelona: Instituto 'Millás Vallicrosa' de Historia de la Ciencia Árabe, 1966, II, pp. 673–718.

King & Turner, 'Bessarion's Astrolabe': *idem* and Gerard L'E. Turner, 'The Astrolabe Presented by Regiomontanus to Cardinal Bessarion in 1462', *Nuncius – Annali di Storia della Scienza* (Florence) 9 (1994), pp. 165–206.

Kunitzsch, *Arabische Sternnamen in Europa*: Paul Kunitzsch, *Arabische Sternnamen in Europa*, Wiesbaden: Otto Harrassowitz, 1959.

—, *Sternnomenklatur der Araber*: *idem*, *Untersuchungen zur Sternnomenklatur der Araber*, Wiesbaden: Otto Harrassowitz, 1961.

—, *Studies*: *idem*, *The Arabs and the Stars*, Northampton: Variorum, 1989.

—, *Typen von Sternverzeichnissen*: *idem*, *Typen von Sternverzeichnissen in astronomischen Handschriften des zehnten bis vierzehnten Jahrhunderts*, Wiesbaden: Otto Harrassowitz, 1966.

London Christie's 29.9.1994 Catalogue, London: Christie's, 1994.

London Christie's 15.4.1999 Catalogue: *Exceptional Scientific and Engineering Works of Art, Instruments and Models, 15 April 1999*, London: Christie's, 1999.

Madrid MAN 1992 Exhibition Catalogue: Juan Vernet, Julio Samsó, *et al.*, *El legado cintífico andalusí*, Madrid: Ministerio de Cultura, 1992. [Catalogue of an exhibition held during April–June, 1992 at the Museo Arqueológico Nacional.].

Maier, 'Astrolab aus Córdoba': Kurt Maier, 'Ein islamisches Astrolab aus Córdoba mit späteren altkatalanischen Inschriften', in *Saltzer Festschrift*, pp. 119–33.

—, 'Romanische Monatsnamen', A-B: *idem*, 'Bemerkungen zu romanischen Monatsnamen auf mittelalterlichen Astrolabien', in *Frankfurt IGN Festband*, pp. 237–54 (A); and *idem*, 'Zeugen des Mehrsprachigkeit: mittelalterliche romanische Monatsnamen auf islamischen astronomischen Instrumenten', in Jens Lüdtke, ed., *Romania arabica. Festschrift für Reinhold Kontzi zum 70. Geburtstag*, Tübingen: Gunther Narr, 1996, pp. 251–70 (B).

Maier: see also King & Maier.

Mayer, Leo A., *Islamic Astrolabists and Their Works*, Geneva: Albert Kundig, 1956.

Merlo, Clemente, *I nomi romanzi delle stagioni e dei mesi studiati particolarmente nei dialetti ladini, italiani, franco-provenzali e provenzali*, Turin: Ermanno Loescher, 1904.

Munich Astrolabe Catalogue: Burkhard Stautz, *Die Astrolabiensammlungen des Deutschen Museums und des Bayerischen Nationalmuseums*, Munich: Oldenbourg, 1999.

North, 'Astrolabe': John D. North, 'The Astrolabe', *Scientific American* 230 (1974), pp. 96–106, reprinted in *idem*, *Stars, Minds and Fate – Essays in Ancient and Medieval Cosmology*, London & Ronceverte, W.V.: The Hambledon Press, 1989, no. 14.

—, *Horoscopes and History*: *idem*, *Horoscopes and History*, (Warburg Institute Surveys and Texts, vol. 12), London: The Warburg Institute, 1986.

North Festschrift: Arjo Vanderjagt and Lodi Nauta, eds., *Between Demonstration and Imagination: Essays in the History of Science and Philosophy Presented to John D. North*, Leiden: E.J. Brill, 1999.

Nuremberg GNM 1992–93 Exhibition Catalogue: Gerhard Bott, ed., Focus *Behaim-Globus*, 2 vols., Nuremberg: Germanisches Nationalmuseum, 1992. [Catalogue of an exhibition held at the Museum during 2.12.1992–28.2.1993.]

Paris BN 1973 Exhibition Catalogue: *Nicolas Copernic ou La révolution astronomique*, Paris: Bibliothèque Nationale, 1973

Poulle, Emmanuel, *Instruments astronomiques*: *idem*, *Les instruments astronomiques du moyen âge*, Paris: Alain Brieux, 1983 (first published in *Le Ruban Rouge* 32 (March, 1967), pp. 18–29).

—, 'Peut-on dater les astrolabes?': *idem*, 'Peut-on dater les astrolabes médiévaux?', *Revue de l'histoire des sciences et de leurs applications* (Paris) 9 (1956), pp. 301–22.

Saltzer Festschrift: Peter Eisenhardt, Frank Linhard and Kaisar Petanides, eds., *Der Weg der Wahrheit – Aufsätze zur Einheit der Wissenschaftsgeschichte – Festgabe zum 60. Geburtstag von Walter G. Saltzer*, Hildesheim: Olms, 1999.

Sarrus, Frédéric, 'Description d'un astrolabe construit à Maroc en l'an 1208', *Mémoires de la Société du Muséum d'Histoire Naturelle de Strasbourg* 4:2–3 (1850), pp. 1–32 and 6 pls. (separatum dated 1852), reprinted in *AIOS*, I, pp. 193–230.

Sarton, *IHS*: George Sarton, *An Introduction to the History of Science*, 5 parts in 3 vols., Baltimore, Md.: Wilkins & Wilkins, for the Carnegie Institution of Washington, 1927–48, reprinted Huntington, N.Y.: Robert E. Krieger, 1975.

Schweinfurt 1993 Exhibition Catalogue: Uwe Müller, ed., *450 Jahre Copernicus 'De revolutionibus' – Astronomische und mathematische Bücher aus Schweinfurter Bibliotheken*, Schweinfurt: Stadtarchiv (Veröffentlichung Nr. 9), 1993. [Catalogue of an exhibition held at the Bibliothek Otto Schäfer during 21.11.1993–19.6.1994.]

Stautz, Burkhard, 'Ein Astrolab aus dem Jahr 1420', *NTM: Internationale Zeitschrift für Geschichte und Ethik der Naturwissenschaften, Technik und Medizin* 5 (1997), pp. 142–59.

—, *Untersuchungen von mathematisch-astronomischen Darstellungen auf mittelalterlichen Instrumenten islamischer und europäischer Herkunft*, Bassum: Verlag für Geschichte der Naturwissenschaften und der Technik, 1997.

Stautz: see also *Munich Astrolabe Catalogue.*

Stevens *et al.*, eds., *Oldest Latin Astrolabe*: Wesley Stevens, Guy Beaujouan, and Anthony J. Turner, eds., *The Oldest Latin Astrolabe*, *Physis* (Rome) 32:2–3 (1995) [published 1996].

Tomba, 'Astrolabio italiano': Tullio Tomba, 'Un astrolabio del XIV secolo di probabile origine italiana', *Physis* 12 (1970), pp. 82–7.

—, ' Due astrolabi latini', A-B: *idem*, ' Due astrolabi latini del XIV secolo conservati a Milano', *Physis* 8 (1966), pp. 295–306 (A), and *idem*, 'Nuove osservazioni sui due astrolabi latini del XIV secolo conservati a Milano', *Physis* 10 (1968), pp. 119–22 (B).

—, 'Tre astrolabi latini': *idem*, 'Tre astrolabi latini, del XIV secolo, conservati in Italia', *Rassegna di Studi e di Notizie* (Milan), 20 (Anno XXII [= 1996]) (the offprint is unpaginated) (E).

Toulouse MPD Catalogue: Raymond D'Hollander, *L'astrolabe – les astrolabes du Musée Paul Dupuy*, Toulouse: Musée Paul Dupuy, and Paris: Association Française de Topographie, 1993.

Trento, Paolo, *L'astrolabio – storia, funzioni, costruzione*, Rome: Biblioteca del Vascello, 1989.

Washington NMAH Catalogue: Sharon Gibbs with George Saliba, *Planispheric Astrolabes from the National Museum of American History*, Washington, D.C.: Smithsonian Institution Press, 1984.

Woepcke, 'Arabisches Astrolabium': Woepcke, Franz, 'Über ein in der königlichen Bibliothek zu Berlin befindliches arabisches Astrolabium', *Abhandlungen der Königlichen Akademie der Wissenschaften zu Berlin* (1855), pp. 1–31 (separatum), reprinted in Fuat Sezgin, ed., *Franz Woepcke – Études sur les mathématiques arabo-islamiques*, 2 vols., Frankfurt am Main: Institut für Geschichte der Arabisch-Islamischen Wissenschaften, II, 1986, pp. 131–65.

XI

THE ASTROLABE PRESENTED BY REGIOMONTANUS TO CARDINAL BESSARION IN 1462

DAVID A. KING

GERARD L'E. TURNER

1. On astrolabes and the state of the art

The astrolabe was the pre-eminent astronomical instrument of the Middle Ages and the Renaissance. It is a model of the universe that one can hold in one's hands, representing the heavens with respect to the horizon of a given locality.[1] One takes an altitude

[1] The best introductions to the astrolabe are North, «Astrolabe», and Hartner, «Astrolabe». See also Michel, *Astrolabe*, Trento; *Astrolabio*, and London NMM *Astrolabe*.

measurement of the sun, or a star, with the sighting device on the back and sets the position of the sun, or the pointer for that star on the rete (the celestial part) on top of the appropriate altitude circle on the plate (the terrestrial part). The combination portrays the configuration of the heavens at the time the sighting was made. All manner of astronomical problems, particularly timekeeping, can be solved by means of the astrolabe.

Today, over one thousand Islamic and European astrolabes are preserved, the majority not properly documented.[2] Astrolabes are historical sources. Like manuscripts they demand attention to their basic design (equivalent to the text and its literary and geographic milieu) as well as to details such as inscriptions and special markings (equivalent to colophon, copyist, script, and marginalia). We still lack a reliable corpus of data on the surviving instruments of the European Middle Ages and early Renaissance, but at least the earliest ones (up to *c.* 1550) will find a place in a new critical catalogue currently being prepared at Frankfurt.[3]

When the first (Price, 1959) and second (G.L'E. Turner, 1989) descriptions of the Bessarion astrolabe were published,[4] no other fifteenth-century German astrolabes had ever been studied in detail. Indeed, very few were known to exist, and some that were known were not recognized as German.[5] Now Bessarion's astrolabe can be compared with several other contemporaneous pieces.[6]

---

[2] The basic work on individual astrolabes remains the monumental study listed in the Bibliography as GUNTHER, *Astrolabes*. However, it treats only one quarter of the astrolabes now known, many only summarily. The errors in Gunther's work have not been corrected in the main, and the section on early German astrolabes is particularly weak. Specifically, many of the most historically important astrolabes – including virtually all of those mentioned in this study – were not available to Gunther. Some reliable catalogues of important collections or exhibitions have appeared in recent years: Milan MAA *Catalogue* and PA *Catalogue* (Tullio Tomba), Rockford TM *Catalogue* (A. J. Turner), Washington NMAH *Catalogue* (Sharon Gibbs and George Saliba). A German translation of a catalogue of the important Nuremberg collection (by D. A. King) is in Nuremberg 1992-1993 *Catalogue*, I, pp. 101-114, and II, pp. 568-602, 640-643. Some of the largest and most important collections, including the Museum of the History of Science, Oxford, the Science Museum, London, the National Maritime Museum, Greenwich, the Deutsches Museum, Munich, and the Museo di Storia della Scienza, Florence, are without catalogues (or without an adequate catalogue in the case of the last).

[3] See KING, «Report», on this new catalogue, and idem, «Astronomical Instruments».

[4] PRICE, «First Renaissance Instrument»: G. L'E. TURNER, *The Regiomontanus Astrolabe*.

[5] In this study, «German» refers to the German-speaking world in order to include both Vienna and Nuremberg under one rubric. Even though the Bessarion astrolabe is based on an Italian rete design, it is still a «German» astrolabe.

[6] In what follows we refer to these instruments by the numbers used in PRICE, *Checklist*, with any additions (4001 and above) designated by D. A. King for his forthcoming *Catalogue*. These numbers are written thus: IC 123.

2. Regiomontanus and Cardinal Bessarion

The academic exchanges between the German astronomer, Regiomontanus, and Bessarion, the Greek bishop turned Catholic cardinal, are well documented and need only be summarized here.[7] Paul L. Rose, in his book *The Italian Renaissance of Mathematics*, wrote:

> Regiomontanus was the first European mathematician to undertake a complete restoration of mathematics and astronomy and the large measure of success which he achieved must in part be attributed to his connections with Italian humanists and classicists, most of all his close friendship with Cardinal Bessarion. Through the cardinal, Regiomontanus was brought to Italy from Vienna, set to work on his important critique of the *Almagest*, introduced to the leading humanist circles where he was taught Greek, and given a free rein to peruse both literary and mathematical manuscripts. Though he began his career north of the Alps, and spent only a few years in Italy, Regiomontanus' work was intimately related to the flowering of an Italian humanist culture sympathetic to mathematics.

Johannes Müller was born on 6 June 1436 at Königsberg in Franconia. He latinized his name as «Joannes de Regio monte», from which comes Regiomontanus. From 1447 to 1450 he studied astronomy and logic at the University of Leipzig, and was attracted to the University of Vienna in 1450 by the reputation of the humanist and astronomer, Georg Peuerbach (1423-1461). The latter had arrived at the university in 1446, and from 1454 he lectured on the Latin poets and also continued the Viennese astronomical tradition founded by John of Gmunden (d. 1442). Peuerbach is known to have visited Italy, but the most decisive turn of events occurred in May, 1460, when Bessarion arrived in Vienna. Bessarion (1403-1472), originally one of the Greek Orthodox representatives at the Council of Florence, and renowned for his scholarship, had remained in the West, and had been made a cardinal and appointed papal legate to the Holy Roman Empire. The Cardinal had planned to prepare a new translation of Ptolemy's *Almagest*, but once in Vienna he handed the project over to Peuerbach. Since Peuerbach did not command enough Greek for a new translation, he started to prepare an epitome from an earlier Latin translation, which was

---

[7] See, for example, Zinner, *Regiomontanus*, especially chapter 5; Rose, *Italian Mathematics*, chapter 4, Regiomontanus in Italy, pp. 90-117; and the article on Regiomontanus by Edward Rosen in *Dict. Sci. Biog.*, XI, 348-352.

completed by Regiomontanus after the untimely death of Peuerbach in April 1461. By providing easier access to Ptolemy's complex masterpiece, the Peuerbach-Regiomontanus *Epitome* contributed to current scientific research by giving astronomers an understanding of Ptolemy that they had not before had access to. Finally printed in Venice in 1496, it was read by Copernicus, and helped him to turn his attention to problems with the Ptolemaic system of planetary motions. Copernicus was to overthrow this system with his own heliocentric system in his book of 1543: *De Revolutionibus*.

In the September of 1461, Bessarion returned to Rome, taking Regiomontanus with him. Regiomontanus stayed in Italy until 1467, when he moved to Hungary to take up a professorship of mathematics at the newly-established University of Pressburg (modern Bratislava).[8] In 1471 he moved to Nuremberg, because of the availability of instruments and ease of communication with learned men. The city was on the great trade routes of central Europe.

Regiomontanus set up an observatory and installed a printing press in his own house to publish scientific texts, which were normally avoided by printers because of diagrams and symbols they did not understand. Johann Gutenberg (1397-1468) invented printing by moveable type at Mainz by 1448. Regiomontanus, the greatest astronomer of the fifteenth century, was the first publisher of astronomical and mathematical literature with texts relatively free of scribal and typographical error. The point was that sound texts in large numbers could only be obtained quickly and cheaply by printing, but to ensure accuracy an expert astronomer would have to be involved in the printing process. He also used the humanist «antiqua» type instead of the German «gothic» in his printing. His first book was Peuerbach's *Theoricae novae planetarum* (*c.* 1474), and the second his own *Ephemerides* (1474) for the years 1475 to 1506, a copy of which was taken by Columbus on his fourth voyage of 1504.

In 1475, Regiomontanus responded to a papal summons to participate in the plans for calendar reform by returning to Rome. It was there that he died, most likely a victim of the plague, in July 1476. His earlier benefactor, Bessarion, had already died in 1472.

Among the astronomical and mathematical works of Regiomontanus that he dedicated to the Cardinal in Rome are the *Epytoma Almagesti* started by Peuerbach and completed by himself

---

[8] ZINNER, *Regiomontanus*.

(1462), *De compositione et usu cuiusdam meteoroscopi* (1462), and *De triangulis* (1463). It is hard to imagine that the «Ioannis» in the astrolabe's inscription refers to any other than Regiomontanus. In about the year 1465, Regiomontanus wrote a treatise for Cardinal Bessarion on an armillary sphere, which he called *Meteoroscopium armillare* (printed in Nuremberg in 1514).[9] This shows that the astrolabe was not the only instrument that the two of them might have discussed.

Regiomontanus dedicated several of his known works to Bessarion. What more appropriate gift for the Cardinal than a model of the universe that he could hold in his hand? In fact, the instrument is fully in keeping with contemporary German instrument making, and incorporates a novel feature, the moveable grid on the back, about which Regiomontanus had learned from his Viennese teachers. One other contemporary German astrolabe incorporates the same feature (Section 5 below: 2a, IC 256). Other astrolabes of this school incorporate a lunar volvelle on the back (Section 5 below: 1a, IC 549; 1b, IC 452), another feature that was novel at that time.

In brief, the Bessarion astrolabe is important for the reasons that Price and Turner indicated, namely, because it is unique amongst European instruments from the thirteenth to the sixteenth century in possibly bearing the name of a distinguished astronomer. The instrument fits squarely in the scene of fifteenth-century German astrolabe making, being one of eleven known instruments from the same school. Although there is no proof that Regiomontanus himself made the instruments, there is written evidence from 1527 that he devised several astrolabes and that they all had a certain feature, the choice of latitudes for the plates, and that feature is shared by the instrument he dedicated to Bessarion.[10] But there are good reasons – the elegance of the piece and the quality of its construction, the accuracy of the astronomical markings, the use of antiqua (specifically roman) type face throughout, the full use of Latin star

---

9 ZINNER, *Instrumente*, pp. 481-482.

10 See Section 5, note 40. The only published queries about the authenticity were written by «trade-journal» writers who knew nothing about German instrument making in the fifteenth century, and who were not familiar with any other astrolabes of the Regiomontanus type. See ROY (Paris), and ANONYMOUS (London). In passing, we may mention that the doubts raised even by Derek Price on the so-called «Peuerbach» and «Regiomontanus» brass diptych sundials, of which 7 are known, are ill-founded. Some are dated, and the range is from 1451 to 1463. See PRICE, «Review», p. 132. In WARD, «Sundials», some evidence for their authenticity is provided, but more is available. This becomes apparent when the instruments are studied in detail and compared one with another.

names, and the standardized «universal» span of latitudes for the plates – to retain the description of this astrolabe by a leading historian. Price called it simply: «the first scientific instrument of the Renaissance»; certainly no other instrument qualifies for this noble title.

3. Cardinal Bessarion's astrolabe: description and provenance

i. *Description*

The whole instrument is made in brass. The dimensions (all in millimetres) are:

Mater, outer diameter 115; thickness at rim 3.8
Plates, diameter 108.2; thickness 0.3.
Rete, diameter 108.3; thickness 1.0
Disc on the reverse; diameter 91.5; thickness 0.6
Rule, length 115; thickness 1.0
Alidade, length 116; thickness 1.7
Pin, length 12.5; diameter of shaft 4.0 (may not be original)
Wedge, now lost, replacement fitted

The mater is constructed from a plate, about 0.5mm thick, formed from a casting filed flat. One region shows fissures formed in casting. This plate is held by four rivets to a rim (width 12mm) on the back. From the front a faint inscribed circle can be seen just 12mm from the outer edge, which shows the inner limit of the rim on the back. This rim, which is 1mm thick, is so made that one side of the throne is contiguous with it. The other side of the throne is integral with a narrow (3.5mm) rim on the front of the mater. The two portions of the rim are tightly joined, presumably by solder. The thin line of the join can be seen right through the instrument. At the throne the join is more noticeable because the thicker, front part has cracked across having been stressed against the thin part behind, which has not cracked.

On the reverse, the rim is only some 0.8mm proud of the central disc. The engraving of the rim into its degree divisions has been done after assembly because one can easily see how the engraver's burin slipped forward on the cutting stroke and came to a stop in the disc leaving a nick. On the front there are very few nicks, presumably because the rim is some 2mm proud of the plate, leaving greater depth for the forward cut to be arrested. Those nicks that do occur on this side are much further towards the centre than on the reverse side.

The rim on the face side is engraved in one-degree intervals and numbered 0° to 360°. On the reverse the rim is engraved in degrees, numbered 90°-0°-90°-0°-90°. Inside the degree scale is the Zodiac, each sign divided to 30°, and inside this is a calendar scale, the months divided into their days. Finally, the innermost scale is in two parts. The upper semicircle is divided in degrees, 0°-90°-0°. The lower forms a shadow square centred on the centre of the astrolabe, both quadrants divided 0-12-0. This is used in height/distance measurements, as well as to find shadow lengths.

Into the mater fit three plates, each with a locating lug at the top to prevent rotation. Each plate is engraved with circles of altitude (almucantars) and azimuth at 5°-degree intervals. Punched dots have been employed to lay out the co-ordinate lines. The latitude for which the plate is engraved is marked on each side.

1.	LATIT̄	39 GŘ	/	LAT̄	54
2.	LATIT̄	48 GŘ	/	LATIT̄	45 GŘ
3.	LAT̄ ROMAE	42 Ǧ	/	LATIT̄	51 GŘ

The three plates cover Europe from 39° to 54° in 3° intervals.

Other engraved circles indicate the tropics, equator, and solar depression at twilight. There are also lines of unequal hours, numbered 1 to 12. The plate engraved for 48° is also engraved with the boundaries of the astrological houses.

The position of the Vernal Equinox – first point of Aries – is indicated on the rim on the reverse, and it is opposite a position 10.5 days into March.

Both the Zodiac and the calendar scales have been labelled in an unusual way, but only to save space. On most contemporary German astrolabes one finds that each sign and each month occupy an annular space above the degree and day divisions. But on this astrolabe the names of the signs are abbreviated to their first two letters, and these are placed where normally 30° would be engraved. The calendrical scale is marked in an analogous manner, where the first two letters occupy the space normally filled by 28, 30, or 31. In the case of February, the final 3-day space is too small to accommodate the letters FE, so they have been placed in the earlier 5-day space.

The rete is delicately made from cast brass. It has 30 star

pointers, 18 inside the ecliptic circle, and 12 outside. Each end of the Capricorn band terminates in carefully worked decoration, while adjacent to the first point of Capricorn is a trilobed motif. The pointer for *Cor Leonis* is decorated by an eight-pointed star at its root. The degrees on the ecliptic circle have been marked out in 5° intervals on the otherwise plain back of the rete. These are layout lines for transference to the front where the edge is bevelled. The one-degree intervals are then added, presumably by eye. The alignment is by no means perfect. The lettering on the rete has been executed with great care.

It is astonishing that the above description of the rete can serve in every detail for the rete on a different astrolabe (see Section 5, 1b; IC 452). The floral chasing is to the *identical pattern*, and the root of *Cor Leonis* is the same. The layout lines on the back and the letter forms on the front follow exactly as here. The minor difference lies in the naming of some stars, which are all in Latin on Bessarion's astrolabe, and partly derived from Arabic on the second instrument. There can be no question that both of the astrolabes were made by the same hand.

In the recess on the reverse of the mater fits a rotating disc, engraved with an angel above and the dedication below the *organum Ptolemei*, which runs across the middle. This is discussed fully below. The back of this disc is plain, and shows casting flaws and file marks. There are two small circles with a centre prick that have been made using scribing compasses. A scribed arc passes through 10 prick points, an end pair being divided by four prick points into five divisions. Perhaps this was an aid to dividing the ecliptic circle on the rete. The dedication reads:

SVB DIVI BESSARIONIS DE CARDINE DICTI
PRAESIDIO ROMAE SVRGO IOANNIS OPVS:- 1462

[Under the protection of the divine Bessarion called Cardinal
I arise in Rome as the work of Johannes: 1462]

The rule is divided in celestial declinations, and the arms are engraved: MERIDIES SEPTEMTRIO.[11] The plain, cast alidade has

---

[11] Sometimes the comment is made that the M in Septemtrio is incorrect, the use of N being classical Latin (see note 24 below). One can regularly find either find either spelling on astrolabes, and any Latin dictionary will give, for example, «septemtriones (septentriones)» for the seven stars in the plough, or Great Bear.

lost the vanes from its hinged sights. The pin may not be original; the wedge has been lost.

The throne has a shackle and ring. It is decorated on the front with a motif comprising three roses. On the reverse there is one rose above, and below, a pair of six-pointed stars.

Around the outer edge of the mater are twenty-four divisions representing the hours in a day/night period. Each section has been engraved with the hours, so the numbering runs from 1 to 12 twice. The style of the numbers is similar to those on the other parts of the instrument, but using later (modern) forms of 4, 5, and 7.

Two metallurgical examinations have been made on the brass of this astrolabe. By modern standards the brass used to construct this astrolabe is a very dirty *alpha* brass alloy. The results, and the problems of metal analysis for historical dating purposes, are considered below.

ii. *The Names of the 30 Star Pointers on the Astrolabe*

As engraved	*Full name*	*Modern name*	*Popular name*
VM AND	Umbilicus Andromedae	β Andromedae	*Mirach*
CAVDA CETI	Cauda Ceti	β Ceti	*Diphda*
VENT CET	Venter Ceti	ζ Ceti	*Bata kaitos*
NAR CET	Naris Ceti	α Ceti	*Menkar*
CAP GOR	Caput Gorgonis	β Persei	*Algol*
OCVL TAV	Oculus Tauri	α Tauri	*Aldebaran*
HIR AG	Hircus Aurigae	α Aurigae	*Capella*
PES DEX OR	Pes dexter Orionis	β Orionis	*Rigel*
CAN MA	Canis Major	α Canis Majoris	*Sirius*
CAN MI	Canis Minor	α Canis Minoris	*Procyon*
YDRA	Hydra	α Hydrae	*Alphard*
COR LE	Cor Leonis	α Leonis	*Regulus*
VAS	Vas Aquarium*	α Crateris	
CAV LE	Cauda Leonis	β Leonis	*Denebola*
CORVVS	Corvus	γ Corvi	*Algorab*
SPICA	Spica Virginis	α Virginis	*Spica*
CAV VR	Cauda Ursae	η Ursae Majoris	*Alkaid*
BOET	Boetes	α Boötis	*Arcturus/Alramech*
CORONA	Corona	α Coronae Borealis	*Alphecca*

As engraved	*Full name*	*Modern name*	*Popular name*
MAN SER	*Manus Serpentarii*	δ Ophiuchi	*Yed*
HER	Hercules	α Herculis	*Rasalgethi*
CAP SER	Caput Serpentarii	α Ophiuchi	*Rasalhague*
CA DR	Caput Draconis	γ Draconis	*Eltanin*
VVL CA	Vultur Cadens	α Lyrae	*Vega*
VVL VOL	Vultur Volans	α Aquilae	*Altair*
DELPH	Cauda Delphini	ε Delphini	*Deneb Adolfin*
GAL	Gallina	α Cygni	*Deneb*
MV EQ	Musita Equi	ε Pegasi	*Enif*
CRVS PEG	Crus Pegasi	β Pegasi	*Scheat*
HV EQ	Humerus Equi	α Pegasi	*Markab*

* Variously named Vas, Patera, Crater = bowl for libations.

The identifications of the engraved star names with the prominent stars have been made on the basis of the usage of the time. As may be expected, the star pointers being thin metal can be fairly easily bent away from their intended positions. If measurements are made of right ascension and declination of the positions of the tips of the star pointers, one finds, however, very little discrepancy after allowing for precession. This shows the quality of the craftsmanship that went into making this astrolabe and also the care with which it has been kept for 527 years. It might be thought that a confusion might arise when a star is named, for example, merely as Hercules, as there are three stars that could have been chosen. But the right ascension shows that α *Herculis*, a variable brightness star, is the correct one. The engraver has named β *Orionis* as *Pes dexter Orionis* instead of the customary *Pes sinister Orionis*, and the co-ordinates confirm that β *Orionis* is the star intended. The confusion may have arisen through the name given to α *Orionis*, *Dexter humerus Orionis* (Betelgeuse), which does not feature on this astrolabe.

iii. *The Provenance*

A short mention of Bessarion's astrolabe was published in *The Times* of London on 9 February, 1959, headed: «15th century Astrolabe at Maritime Museum». The report began:

Commander M. H. Hardcastle has lent a small brass astrolabe, made in 1462 by Johannes Muller of Konigsberg, to the National Maritime Museum, Greenwich, which is now putting it on view. Muller, more generally known by his Latin name, Regiomontanus, stands at the beginning of the renaissance of science, and this astrolabe was the first scientific instrument made by him. He made it for Cardinal Bessarion, who had taken the young scholar to Rome and given him access to his manuscripts of classical Greek astronomy. Regiomontanus afterwards returned to Nuremberg and there built an observatory, a printing press for scientific books, and a workshop for scientific instruments. It was these activities that, in the words of the musueum's announcement, «marked the re-birth of astronomy and indeed of all modern science».

The National Maritime Museum's inventory (No. A.56/58-207) adds little, except that the first point of Aries was datable as 10.75 March, and the latitudes on the three plates were given.

Clearly, such an interesting astrolabe would attract the attention of the late Professor Derek Price, whose very detailed monograph on a fourteenth-century manuscript dealing with an equatorium had been published in 1955 by the Cambridge University Press.[12] In the first issue of the new Italian journal, *Physis*, which appeared in 1959, Price published a short paper briefly illustrating and describing the Regiomontanus astrolabe.[13] Here Price gave no more than the bare bones of a description, though he did add a paragraph on provenance, which he presumably obtained from the owner. The paragraph reads as follows:

> The provenance of this historic specimen has not been without interest. It remained in Rome apparently until March 1848 when it was presented by some person unknown to Dr. Somerville, husband of Mary Somerville (*Connection of the Physical Sciences*, London, 1834) who was then wintering in Italy with his family, as was their custom at this time. It was then given by the Somervilles to their close friend, Sir John F. W. Herschel, the astronomer, and it has continued in the possession of that family ever since, being now the property of Commander M. H. Hardcastle, R. N. He has been kind enough to allow the present author to have it on indefinite loan for study and exhibition in Europe and America.

In passing, one must remark that this passage is not entirely accurate. The Somervilles lived a peripatetic life in Italy from 1840. Apart from visits to England in 1844 and 1848, they saw out their lives abroad. Both William Somerville (1771-1860) and his wife, Mary Somerville (1780-1872) died in Italy.[14] And the instrument did leave the Herschel family.

Sir John Herschel was a renowned astronomer and mathematician, and was a friend of Dr. Somerville for a long time: he was a supporter for Somerville's election to Fellowship of the

---

12 PRICE, *Equatorie*.

13 PRICE, «The First Renaissance Instrument».

14 PATTERSON, *Life*; PATTERSON, *Mary Somerville*.

Royal Society in 1817. Mrs. Somerville dedicated *Physical Geography*, published in 1848, to her scientific advisor for over twenty years, Sir John Herschel. If in 1848 the Somervilles were journeying to England, it would not be surprising that they should take a present of an astrolabe for an old friend. But where did the astrolabe come from? Possibly the Rome observatory. The old, conservative Pope Gregory XVI died in 1846, and was succeded by the younger Pius IX. Since he encouraged modernization, it may be that in clearing out the mess of ages in some institution, the ancient brass astronomical instrument was thought by some Roman astronomer to be a worthy gift to his world-famous English colleague.

Sir John Herschel died on 11 May, 1871. The Will of Herschel records that Edward Hardcastle, a Manchester merchant, and Herschel's widow were the two executors. The Will says nothing more specific than «my scientific books and instruments». It would seem that the astrolabe passed to the Hardcastle family through this connection. Commander Hardcastle lent the astrolabe at the end of 1956 to Derek Price who was about to embark on a lecture tour of the United States. In October 1958 it was placed on extended loan with the National Maritime Museum, and was withdrawn twenty-seven years later. On 28 September, 1989, it was sold at Christie's South Kensington, lot 170.

4. Remarks on features of the Bessarion Astrolabe

i. *The engraved letters and numbers*

All the letters are majuscules (upper-case, commonly called capitals), and they are engraved with skilled control. Their form comes from the Italian scribal hand, and it was this that was the model for much of the new printing industry. Typefounders followed the desires of their extremely conservative customers and produced type faces to look as nearly like manuscript as possible. Thus for German readers the type was «gothic», while for Italian and French readers the type was «antiqua».[15] It is usual to divide «antiqua» and

---

[15] Steinberg, *Printing*.

talk of roman and italic.[16] Thus it was natural that the first book to be printed in Italy, Cicero's *De oratore*, imitated the round hand of the Renaissance scribes. It was thought that for the spread of humanism the use of «littera antiqua» was the most appropriate. This influence was soon seen in the instrument trade. The forms on the Bessarion astrolabe are similar to those on, to take one example, the Viennese compendium dated 1491 in the British Museum (No. 94 6-15 1) bearing the name of the Benedictine monk, Hans Dorn (before 1440-1509). Even so, it should be noted that the form of the 4 is still gothic.

It is obvious that the design of the Regiomontanus astrolabe and the several others like it became the inspiration for the somewhat similar astrolabes produced in the next century by Georg Hartmann (1489-1564), who was vicar of the Church of St. Sebaldus in Nuremberg from 1518 to 1544. Here he wrote on sundials, and began to manufacture dials, astrolabes, and other mathematical instruments. His astrolabes copied those of the previous century, down to the shape of the rete and of the throne, repeating the rose motif. There is one signal difference: letters and numbers are punched and not engraved. This allowed for faster production in batches.[17]

The astrolabe is engraved with arabic numbers that, apart from 4, 5, and 7, are in their modern form. Around the outer edge there are numbers all in the modern form. In the old style, 4 is a loop with the two ends below; 5 is a vertical stroke, like 1, with a small curve to the left that is open at the top; the 7 is an inverted V. The modern forms of these numbers were in use in Italy by about 1400.[18] The usage varied among trades, the lapidaries being amongst the first at the beginning of the fifteenth century. Medallists were employing some of the new forms by 1417, and printers, a trade that started mid-century, were using all modern by about 1467, and painters likewise in about 1472. To the north the change-over was slower; German printed books were all modern by 1488, but in the Low

---

[16] MORISON, *Printing*. Pointing out that the terminology is inexact, Morison writes: «the Germans have the best of it; for our «roman», their equivalent is «antiqua», for «italic», «kursiv» (p. 17).

[17] Batch production was employed by Hartmann of Nuremberg. G. L'E. Turner has found an identifying number or letter scratched on all the parts of several of his astrolabes. This implies a parts shaper, a mathematical line engraver, and a letter/number puncher.

[18] See, for example, HILL, *Arabic Numerals*; MENNINGER, *Zahlwort und Ziffer*, pp. 213-265 («Die Wanderung der indischen Zahlzeichen nach Westen»); FLEGG, *Numbers*.

Countries the modern form had not penetrated all trades until after 1500. In England one has to look to 1550 before the modern forms of all the numbers are in general use.

It seems, therefore, that the use of the older forms of 4, 5, and 7 on the astrolabe could be accounted for by the designer coming from Vienna, or the craftsman who made the engravings being of the older tradition. The numerals used to mark the hours around the edge of the instrument could have been added shortly after it was completed, or sometime within perhaps fifty years of completion. The manuscripts of Regiomontanus show that he gradually changed from the old to the new forms of 4, 5, and 7. Even so, his printed *Kalender* of 1474 employed the old forms, most probably because his readers were used to them.

ii. *The* Organum Ptolemei *on the reverse*

In a manuscript of a series of treatises by earlier Viennese astronomers copied by Regiomontanus in his own hand *c.* 1456, the grid known as the *organum Ptolemei* and its use for timekeeping by the sun at any terrestrial latitude are described.[19] Regiomontanus made no claim to have invented this «instrumentum universale ad inveniendum horas in quocunque climate fueris fabricare», and he was well aware that the underlying theory was due to Ptolemy. In addition, the instrument is described in several late-fifteenth-century German texts (in Latin), and features in writings of Apian in the early sixteenth century. In Paris there was published in 1550 a more complete grid based on the same projection and also serving the stars: this is now known as the Rojas projection.[20]

The only studies of this mathematical grid in the modern literature are by E. Zinner (on the *organum Ptolemei*), published in 1956, and F. R. Maddison (on the de Rojas projection), published in 1966.[21] The former dealt with the early history of the projection in Germany, which is overlooked in the latter. Maddison concluded his account of the de Rojas projection by recording the existence of a few instruments of earlier date than the publication of de Rojas's book in 1550, «which nevertheless use the same projection described

---

[19] ZINNER, *Instrumente*, p. 132.

[20] JUAN DE ROJAS, *Commentariorum in astrolabium*... (Paris, 1550; 2nd edn 1551). See also GUNTHER, *Astrolabes*, pp. 353-354; London NMM *Astrolabe*, pp. 32, 40.

[21] ZINNER, *Instrumente*, pp. 131-134; MADDISON, «Rojas Projection».

by him». One of these was the Bessarion astrolabe (at that time displayed at the National Maritime Museum, Greenwich); another, in Cracow, attributed to Hans Dorn (before 1440-1509); another at Florence, also attributed to Dorn, although the lettering appears dissimilar to that on the only known signed Hans Dorn instrument (dated 1491) at the British Museum.

We can now point to a couple of mid-fifteenth-century Regiomontanus type astrolabes with the *organum Ptolemei* on the back, one in a private collection at Paris (see Section 5, 2a; IC 256) and the other Bessarion's. The fact that de Rojas first described this projection in 1550 and that it is commonly called the «de Rojas projection», is one more example of an unfortunate misnomer, and underlines how little we knew about contemporaneous German astronomy until the work of Ernst Zinner.

The *organum Ptolemei* was developed as a device for reckoning time from solar altitude *accurately in equinoctial hours*. It was an alternative to the universal horary quadrant popular on astrolabes both before and after the fifteenth century. This was an inferior device for reckoning time from solar altitude, giving approximate values in seasonal hours. Both devices were popular because they worked for all latitudes, but the quadrant was easier to construct and was far more widely used.

On most examples of the grid, including those on the Bessarion astrolabe and the other mid-fifteenth-century German example, and the astrolabe attributed to Hans Dorn, the grid rotates within an altitude-cum-latitude scale. As such it functions as an observational instrument; the alidade is aligned towards the sun, and the grid is aligned parallel to the celestial equator for a specific latitude – one then simply reads off the time from the grid. In this fashion the instrument is described by Regiomontanus in his treatise. If the grid is fixed with respect to the surrounding scale, then it functions as a calculating device. One must first measure the altitude of the sun by means of the alidade, note it, and then feed the value into the grid and read off the time.

The disc on Bessarion's astrolabe could not be engraved on both sides because of abrasion when rotated. The other similar astrolabe also has a rotating disc that is blank on the inside. Consequently, the grid, angel, and dedication scroll all had to be fitted on to the one side.

iii. *The Dedication*

The attribution of the astrolabe to Regiomontanus rests on the inscription in Latin on the back plate of the instrument. This gives the date, 1462, the name Ioannes (Johannes) as that of the maker, the place of dedication as Rome, and refers to Cardinal Bessarion as the maker's patron.

The study of inscriptions on medieval and early Renaissance scientific instruments is in its infancy; there are very few experts on medieval scientific Latin, and none who have turned their attention to inscriptions on scientific instruments. The dedication on the Bessarion astrolabe is one of several on medieval and Renaissance instruments in which the instrument «speaks» in the first person.[22] The verse form of the dedication is described as «elegiac couplets» by Dr. Scholderer in a Note on Price's paper of 1959.[23] Writing for librarians and bibliophiles, he said:

> All the inscriptions on the instrument are in majuscules of much the same style as those which Regiomontanus cut for his smaller type when he came to print at Nuremberg in 1474, and indeed the splendid craftsmanship of the piece is obviously akin to the skill and accuracy so impressively displayed in the pages of his printed ephemerides.

Some objections have been raised about the quality of the Latin. After the sale, the *Antiques Trade Gazette* quoted Dr. J. V. Field, at that time employed at the Science Museum, London, as having found mistakes in the Latin. Her view was reported thus:

> ... she is unhappy about the association with Regiomontanus because there are mistakes in the Latin on the rete and on the label and one does not expect mistakes from Regiomontanus. «As a humanist he was a leading proponent of the use of correct classical Latin».[24]

This last sentence is true, but irrelevant since there are no demonstrable mistakes; more to the point is the fact that humanists enjoyed learned puns. An even more ill-considered opinion was printed in a Parisian paper,[25] where the dedication is described as:

---

[22] See KING, «Medieval Picard Astrolabe», p. 60-61, especially note 24.

[23] V. SCHOLDERER, «A Note on the Astrolabe of Regiomontanus», *The Library*, 5th series, 15 (1960), p. 209.

[24] Anonymous (London). For rete, read rule. See note 11 above.

[25] ROY (Paris).

«en latin de cuisine». Professor Noel Swerdlow of the University of Chicago, who has worked a great deal on the writings of Regiomontanus, sees no problem with the Latin dedication.[26] Dr. Nancy K. Zumwalt, of the University of California, has pointed out that the inscription is neat and clever:[27]

> The writer of this dedicatory inscription was learned in classical Latin. The inscription shows skilful command of periodic style. It also contains two witty astronomical puns.

These puns can only be understood by someone familiar with astronomy. The reference to Bessarion as «called *de cardine*» could relate to the *cardines*, a concept of prime importance in astronomy and astrology, and would then constitute a pun on the title Cardinal, the original form of which was, indeed, *de cardine*, and possibly a further pun based on the use of *cardo* to mean the East (Bessarion came to Rome from the East).[28] Zumwalt saw a pun in the word *surgo*, «I arise», or «I come forth», the association being with an astronomical rising (usually described by the verbs: *ascendere*, to mount, ascend, go up; *orire*, to arise, spring from). In any case, the choice of words was to some extent forced by the verse form of the inscription.

iv. *The Angel*

Who is going to dare to explain why any craftsman decided to choose a particular motif with which to ornament his handiwork? On the astrolabe dedicated to Cardinal Bessarion there are floral motifs on the rete, roses and six-pointed stars in rosettes forming the throne, a *trompe l'oeil* scroll that carries the dedication, and an angel. This figure was part of the original construction as becomes obvious through the hands, which were cut before the straight line border that they overlap.

---

[26] Private communication to D. A. King.

[27] In a report made available to Christie's of London.

[28] The «cardinal» points of the compass are North, South, East, West. The «cardinal» signs in astrology are those that mark the seasons: Aries, Cancer, Libra, Capricorn. Similarly, the astrologer's four main Houses of Heaven that relate to one's surroundings are called the «angles» or «cardines». See EADE, *Astrology*. Both cardines and cardinal derive from the Latin *cardo*, a hinge. This embraces the idea of swinging, a point around which anything turns, a pole, the ends of the Earth's axis; *Cassell's New Latin-English... Dictionary*, 5th edn (London, 1968).

Those who are unthinking claim the rose as English Tudor, and the star as that of David. It is quite otherwise. In ancient Rome, roses were obligatory at ceremonies. In heraldry the rose forms the cadency mark of a seventh son. In early and medieval Christianity it symbolised the Virgin Mary and Paradise. The rosette was used in ancient Egyptian ornament, and in Assyria, the Middle East, and in Greek and Romanesque art. «The rosette is one of the most widely disseminated of all decorative motifs».[29] The interior of a rosette may have a rose or floral form, with symmetries of 4, 6, 8, 10, or 12-fold; 5-fold symmetry is not obligatory. This is why the rosettes in the throne with 6-fold symmetry need not be associated with David: it is just a pattern. The rose motif is found on the other Regiomontanus-type astrolabes (see below), and on those made by Georg Hartmann.

The angel is a typical example of those found in many parts of Europe in the mid-fifteenth century. Not that angels are not found at other times; it is the very long wings, the plain tunic with simple collar, the hair and the clear «young» face that are characteristic of the time. In fact, the brass angel could easily have flown out of the superb *A Choir of Angels* painted on wood by Simon Marmion from Flanders, who was active in 1449, and who died in 1489. Any visitor to Tuscany cannot avoid all the angels in the many lunettes and panels done in glazed white terracotta on a blue ground. Of those by Luca della Robbia (1400-1482), one may point to the four angels in the lunette of the *Resurrezione* in the Duomo (1443) at Florence, and also to the *Madonna della Cintola* at Arezzo (*c.* 1490) by Andrea della Robbia (1435-1525).[30]

There is evidence that some other astronomical instruments of the fifteenth century feature angels. Johannes of Gmunden (*c.* 1380-1442), the predecessor of Peuerbach in Vienna, described, before 1435, a bronze plate with a representation of the seven celestial spheres with the Trinity, the Virgin Mary, and various angels. An unsigned surviving instrument from the fifteenth century features precisely these.[31] Then there is the celestial globe supported by two angels that was placed by King Matthew of Hungary in his

---

[29] Osborne, *Decorative Arts*, p. 673.

[30] Simon, Marmion, National Gallery, London. Duomo plate 20, Arezzo plate 58 in *Della Robbia: A Family of Artists*, Milan: Scala, 1992. A Bohemian chasuble embroidered in silk and gold thread, dated 1487, shows the Madonna with four flyng angels of the same style; Moravian Gallery, Brno, Czech Republic, inv. no. 27411.

[31] Zinner, *Regiomontanus*, p. 323.

library at Buda after he became King of Bohemia on 3 May 1469. The instrument was fixed to show the heavens at the time of his coronation.[32] Insofar that angels are heavenly beings, the presence of one of them on an astronomical instrument dedicated to a cardinal could be seen as entirely appropriate. One need not labour the point: always popular in Christian imagery, they proliferated during the fourteenth and fifteenth centuries, usually sexless in appearance reflecting the Hellenic ideal of the beautiful youth.[33]

v. *The Metal*

A metallurgical examination has been made on the brass of this astrolabe by light and scanning electron microscopy, and X-ray dispersive analysis.[34] The brass was found to be of very poor quality. The mater appears to be a single phase *alpha* brass, which contains many small grain boundary cracks that appear to have been produced at the time of manufacture. These cracks are attributable to embrittlement of the grain boundaries of the metal during cooling or subsequent mechanical handling. The cracks contain many inclusion particles based on zinc sulphide, together with lesser amounts of iron and some antimony. The disc on the reverse (with engraved angel) is also *alpha* brass, but with a higher zinc content than the mater. The mater and the disc are made from different batches of *alpha* brass. The disc has undergone mechanical deformation, and it may be assumed that the original metal was mechanically worked into a thin plate from which the disc was cut.

By modern standards the brass used to construct this astrolabe is a very dirty alloy, and the zinc sulphide would cause the observed embrittlement. Metallic zinc was not known in the West until the sixteenth century, and for the production of brass, zinc ores (for example, carbonate or sulphide) were used. Such a technique would not have been employed in western Europe in modern times (after 1800).

A completely separate analysis was made using a different technique at a different laboratory. The results in both cases are consistent in their conclusions.

The second analysis was part of a pilot experiment made to

---

[32] ZINNER, *Regiomontanus*, p. 160.

[33] OSBORNE, *Decorative Arts*, pp. 45-46.

[34] Matcon Report No. 1603, August 1989.

compare Bessarion's astrolabe with a few other German astrolabes. These were by Hartmann of Nuremberg, made some 60 years later. The work was done by Dr. R. Brownsword, Department of Applied Physical Sciences, Coventry University, Coventry.[35] It would, of course, be worthwhile to perform a similar analysis on all ten of the known Regiomontanus-type astrolabes. Comparison with contemporaneous brass of known origin might reveal differences in the composition of pieces from, say, Nuremberg as opposed to Vienna. Dr. Brownsword's technique does not involve transporting an instrument to his laboratory, but requires a very small sample of about 20 milligrams, which is not consumed by the X-ray fluorescence analysis. In this programme of chemical analysis there are nine elements that constitute the key: the big four, all under the metal-founder's control: copper, zinc, tin, lead; the small five, not under the founder's control: antimony, arsenic, nickel, iron, silver. The profile of these nine metallic elements has been proved to be useful in dating and geographically locating certain classes of brass specimens.[36]

Table 1 - Chemical analysis of Bessarion's astrolabe dated 1462

Metal	mater	rete	45/48° plate	disc	alidade	rule
Cu	73.64	68.71	72.44	73.16	78.15	76.65
Zn	23.22	28.24	25.58	25.74	19.45	21.44
Sn	0.98	0.91	0.26	0.15	0.59	0.15
Pb	1.04	0.48	0.39	0.17	0.63	1.38
Sb	0.13	0.14	0.20	0.13	0.21	0.08
As	0.37	0.51	0.39	0.10	0.26	0.14
Ni	0.28	0.29	0.33	0.08	0.49	0.29
Fe	0.25	0.65	0.35	0.46	0.15	0.00
Ag	0.07	0.07	0.07	0.01	0.07	0.02

Percentage of nine elements present in six different parts.

[35] Dr. Brownsword's report is in the hands of the authors. The alloy of a Hartmann astrolabe dated 1524 is generally similar to the alloy of the Bessarion, but the Hartmann is higher in lead content.

[36] Brownsword, «Metalwork».

The nature of the material used by craftsmen is one source of historical information about scientific instruments that has not hitherto been tapped. A thorough and soundly based programme of research and analysis is desirable, in particular, on mathematical instruments before about 1600. There are many astrolabes, quadrants, armillary spheres, etc., where, although they are genuine enough, it is as yet impossible firmly to fix a date for them more closely than within one hundred or even two hundred years, and next to impossible to allocate them to a region of manufacture. Without a programme that builds up a data bank of a very considerable number of analyses, such problems can hardly be solved. The ratio of the percentage of copper to zinc in the alloy forming the brass in a single sample has been erroneously regarded as an adequate basis for establishing, or undermining, authenticity; indeed, as a *surer* basis than connoisseurship.

Professor A. M. Pollard, then of the Research Laboratory for Archaeology and the History of Art, Oxford, chemically analyzed by X-ray probe many scientific instruments at the National Maritime Museum, Greenwich, and elsewhere, while his colleague, Miss C. Mortimer, examined some instruments in the Museum of the History of Science, Oxford.[37] Amongst Pollard's findings was that chemical analysis cannot be used to date a piece of brass:

> ... all that can be done is to compare its composition with those of known genuine pieces... It is also possible that perfectly genuine instruments will have analyses outside the expected range for a particular country and date...

Mortimer's conclusion reads:

> Although some compositional patterns are emerging in certain workshops and countries, it still seems unlikely that judgements on authenticity can be made purely on the basis of major element chemical analysis.

Another approach is more invasive (it involves polishing and etching a small area) but more telling: this is metallographic examination using a special optical microscope. This kind of examination reveals the microstructure of the metal, and shows whether it is cast, beaten, rolled, annealed, etc. One such

---

37 POLLARD, «Authenticity of Brass Objects»; MORTIMER, «X-Ray Fluorescence».

examination has been made by Dr. Robert B. Gordon on a Hartmann astrolabe dated 1537.[38]

5. Remarks on «Regiomontanus-Type» Astrolabes

Bessarion's astrolabe has, for over three decades, presented a problem in that it seemed to stand alone. It is only since 1989 that a significant group of German astrolabes of the fifteenth century has been discovered, and systematic study of them begun. So far, ten closely related instruments form a context in which the astrolabe of Bessarion holds a key position. These eleven astrolabes will be referred to as «Regiomontanus-type».[39]

There is evidence that Regiomontanus did devise astrolabes. Georg Hartmann, the leading instrument maker in Nuremberg in the first half of the sixteenth century, mentions, in his treatise on astrolabe construction written in his own hand in 1527, a feature of all the astrolabes of Regiomontanus that he had seen, namely that the plates served the latitudes 39°, 42°, 45°, 48°, 51°, and 54°.[40] This is a feature of the Bessarion astrolabe, and one other, amongst the Regiomontanus-type instruments. There is other documentary evidence that Regiomontanus himself devised astrolabes, namely:

a) About 1470 he designed a large astrolabe with three plates and with «Italian lettering», i.e. antiqua, now lost, for King Matthew Corvinus of Hungary. In 1512 this was passed on to Kurfürst Friedrich of Saxony.[41]

b) In the autumn of 1478, the second year after Regiomontanus' death, King Matthew of Hungary sent Hans Dorn to Nuremberg to purchase his books and instruments. Dorn, although he remained

---

[38] Gordon, «Metallography»; Gordon, «Technology».

[39] It is a sad reflection on the state of the field that these astrolabes were not properly documented decades ago. A similar situation prevails for fifteenth-century French and also Italian astrolabes, let alone for earlier astrolabes from these three, and other, regions. Detailed descriptions of all of the astrolabes mentioned in this study are in King, *Catalogue*.

[40] See Zinner, *Regiomontanus*, p. 222, and Zinner, *Instrumente*, p. 480. The relevant passage in MS Vienna Staatsbibliothek 12768, fol. 2*r*, translates roughly as follows: «Johannes de Monte Regio made [hat gemacht] astrolabes in brass. He usually [gewonlich?; reading uncertain] took three plates and put the pole elevation on all sides. On the first he made 39° pole elevation, on the other 42°, the third 45°, the fourth 48°, the fifth 51°, and the sixth 54°. The[se] pole elevations he used [illegible] on all astrolabes made by him that I have seen».

[41] Zinner, *Regiomontanus*, p. 160, quoting MS Jena Universitätsbibliothek Msk. Appendix 22; and Zinner, *Instrumente*, p. 480.

several months in Nuremberg, failed to achieve his mission.[42] In April 1479, Bernhard Walther purchased some of Regiomontanus' brass instruments. A catalogue of 1512 includes mention of two astrolabes and three incomplete ones.[43] In 1480, according to Walther, the executors of Regiomontanus' legacy had several instruments made by Regiomontanus deliberately destroyed and their metal sold to the locksmiths. It is not clear which of the instruments now in the Germanisches Nationalmuseum are to be associated with these two complete and three incomplete astrolabes.[44] A prospectus of books that Regiomontanus hoped to print in Nuremberg, published in about 1473 or 1474,[45] ends thus:

> In the workshop an astrarium is constantly under construction, an achievement certainly worthy to be viewed as a wonder; some astronomical instruments for observations of the heavens are also being made, and likewise other instruments for common daily use, to recite the names of which would be tedious.

An astrolabe often associated with Regiomontanus in the modern literature is most probably only a piece that was owned by him if not merely associated with him by later tradition.[46] It is an astrolabe plate with gothic inscriptions serving a latitude of about 46 1/2°, suitable for some locality in the Italian Alps, and it is dated 1468. It is difficult to imagine Regiomontanus designing such a «medieval» instrument six years after his stay in Italy. Also, if the piece was in Nuremberg in the early sixteenth century Hartmann would surely have seen it, and had he thought that Regiomontanus had made it he would not have written his sweeping remark about all of Regiomontanus' astrolabes which he had seen.

Earlier (fourteenth century?) German astrolabes are a varied

---

42 ZINNER, *Regiomontanus*, p. 245, and ZINNER, *Instrumente*, p. 293.

43 ZINNER, *Regiomontanus*, pp. 221, 246.

44 For a discussion of the provenance of these instruments, see Nuremberg 1992-1993, *Catalogue*, part 1.

45 Information kindly supplied by Professor Noel Swerdlow, University of Chicago.

46 IC 550: Nuremberg, Germanisches Nationalmuseum, inv. no. WI 5. See GUNTHER, *Astrolabes*, pp. 280-281 (no. 137), with a lithograph of the front captioned «Astrolabe of Regiomontanus» (in the text this piece is confused with a twelfth-century Syrian astrolabe with Zodiac figures depicted on the rete); ZINNER, *Regiomontanus*, pp. 217, 222 (on Regiomontanus' scientific estate), and figs. 72-73 on plate 40 (minuscule reproductions of the front and back); ZINNER, *Instrumente*, p. 480 (brief description, in which the latitude is miscalculated to 50°); Nuremberg 1976-1977 *Catalogue*, pp. 90-91 (no. 47); Linz 1990 *Catalogue*, p. 33 (no. 30: «Astrolabium des Regiomontanus»), with an illustration of the back only; Nuremberg 1992-1993 *Catalogue*, item 1.76 (first detailed description by D. A. King).

group, including a quatrefoil astrolabe owned by Ludolf de Scicte of Einbeck,[47] and an astrolabe displaying French influence.[48] An illustration of an astrolabe in a fourteenth-century treatise, perhaps by Heinrich of Langenstein, shows that rete types were used in that century bearing no obvious relation to any surviving German instruments.[49] An interesting example of transmission southwards across the Alps is afforded by an astrolabe signed by Bernadinus Aurifex in 1565,[50] which has the same unusual rete design as the one featured in the fourteenth-century German manuscript.

German astrolabes of the fifteenth century can be divided into four main groups: instruments of the Regiomontanus-type (eleven pieces), with retes based on an Italian model and precursors of the better-known Hartmann astrolabes (Type A, ten pieces), or more complicated, and precursors of the Hans Dorn-type (Type B, one piece); simplified astrolabes associated with Hans Bamman of Schmalkalden (six pieces), dated between 1484 and 1493;[51] some instruments usually associated with Hans Dorn (four pieces);[52] the

---

[47] IC 2027: Cracow, Jagellonian University Museum, inv. no. 4043. See ZINNER, *Regiomontanus*, plate 42, figs. 78, 79; ZINNER, *Instrumente*, pp. 535-536.

[48] IC 610: Cologne, Städtisches Museum, inv. no. KSM-L 181. On this piece, see DIECKHOFF, «Arsenius-Astrolabium», pp. 23-24.

[49] See ZINNER, *Instrumente*, p. 142, for details but no illustration.

[50] IC 2058: Fano, Museo Civico, inv. no. 103. See PANICALI, *Orologi*, pp. 38, 39, 46 for the front, back, and signature.

[51] On some of these, see GUNTHER, *Astrolabes*, pp. 423-424 (IC 248 and IC 249), and ZINNER, *Instrumente*, p. 244.

[52] On Dorn and the instruments associated with him (except for IC 492), see ZINNER, *Regiomontanus*, pp. 46, 245, and plate 41 (IC 642); ZINNER, *Instrumente*, pp. 292-297. Also A. J. TURNER, *Instruments*, pp. 35-39, figs. 25, 26 (IC 642); Washington NGA 1992 *Catalogue*, pp. 221-224 (by F. R. MADDISON). The *only one* of his instruments that is signed is a compendium in the British Museum, London (IC 250); the attribution of the others to him still needs to be confirmed. Astrolabic instruments associated with Dorn, which are quite different in conception (except for the rete design) and execution to the «Regiomontanus-type», are:

IC 4519: dated 1480; celestial globe, with sundial on horizontal frame and *organum Ptolemei* on astrolabic astrological plate attached to the top: Cracow, Jagellonian University Museum, 4039-37/V; Ameisenova, *Bylika Globe* (detailed description by art historian).

IC 492: dated 1483; astrolabe, fitted with *organum Ptolemei* and eclipse calculator (in fact part of an albion): Florence, Museo di Storia della Scienza, inv. no. 1096; Florence, 1954 *Catalogue*, pp. 57-59 (no. 1096) (identified as German); Florence, 1991 *Catalogue*, p. 8, I. 14; GOLDSTEIN, «Transversal Scale», pp. 103-104, figs. 1, 2 (front and detail); G. L'E. TURNER, *Strumenti*, p. 73 (front).

IC 642: dated 1486; astrolabe: Cracow, Jagellonian University Museum, 4040-38/V;

IC 250: dated 1491; signed «HANNS DORN»; compendium with astrological astrolabic plate: London British Museum, inv. no. 94 6-15 1; London BM *Catalogue*, p. 74, no. 207, plate 28 (back); G. L'E. TURNER, *Strumenti*, p. 120 (both sides).

Two astrolabic plates for a single latitude in Florence and Salzburg are related to work attributed to Dorn, if not actually by him.

astrolabic plate dated 1468 and serving the latitude of a locality in the Italian Alps.

i. *The «Regiomontanus-Type» Astrolabes*

Regiomontanus-type astrolabes can be classified as follows:

Groups 1-4: Astrolabes with Type A retes

Group 1: Astrolabes with a lunar volvelle on the back

a) Nuremberg - dated 1457, gothic script;[53]

b) Paris - undated, gothic script as on 1a on the plates and back, and antiqua as on 2b on the rete.[54]

Group 2: Astrolabes with an *organum Ptolemei* on the back

a) Paris - undated (with some early-sixteenth-century additions on the back of the movable plate), gothic script resembling that on 1a and 1b, some antiqua majuscules as on 1b and 2b;[55]

b) London - dated 1462, signed by Ioannes (Regiomontanus), dedicated to Cardinal Bessarion, antiqua script as on the rete of 1b.

Group 3: Standard astrolabes

a) Darmstadt - dated 1466, gothic script, some antiqua capitals;[56]

b) Munich - undated, gothic script;[57]

c) Strasbourg - dated 1481, gothic script as on 3d;[58]

---

[53] IC 549: Nuremberg, Germanisches Nationalmuseum, inv. no. WI 129. See ZINNER, *Regiomontanus*, p. 222, and plate 40, figs. 70-71 (front and back); ZINNER, *Instrumente*, pp. 150, 465, and plate 50.2 (back only); PILZ, *Astronomie*, p. 54; Nuremberg 1976-1977 *Catalogue*, p. 71 (no. 15); Linz 1990 *Catalogue*, vol. 2, p. 32 (no. 29), (illustration of the back only); Nuremberg 1992-1993 *Catalogue*, item 1.75 (detailed description).

[54] IC 452: Currently in a private collection; formerly in the Greppin Collection in Brussels, sold at Galerie Köller in Zurich on 17.11.1975. See Brussels 1957 *Catalogue*, p. 9 (no. A13); and GREPPIN *Catalogue*, p. 32 (no. 3122: «probablement par Georg Hartman [*sic*], vers 1460»), with an illustration of the back only on plate 13. We are grateful to the present owner for allowing us to study this instrument.

[55] IC 256: Currently in a private collection, formerly in the Nachet Collection. See NACHET *Catalogue*, p. 44 (no. 126), illustration of the front on plate 14, facing p. 42; GUNTHER, *Astrolabes*, p. 434 (IC 256), and also KING, «Report», p. 7, fig. 5 (illustration of throne and inscription on mater). The inscriptions recorded by Gunther postdate the construction of the instrument by about 75 years. We are grateful to the present owner for allowing us to study this instrument.

[56] IC 592: Darmstadt, Hessisches Landesmuseum, inv. no Kg. 25.15. Unpublished.

[57] IC 620: Munich, Deutsches Museum, inv. no. 70678. See ZINNER, *Instrumente*, plate 41, fig. 1 (front only: «Gotisches Astrolab»).

[58] IC 530: Strasbourg, Musée Historique, inv. no. IL 74. See POULLE, «Fabrication des astrolabes», p. 118 (illustrations of front and back); ZINNER, *Instrumente*, pp. 150, 421 (s.v. Arnold Kunig); Strasbourg, 1987 *Catalogue*, p. 1 (illustration of detail of plate), front cover (colour illustration of back), and back cover (colour illustration of front).

d) Cracow – dated Schmalkalden, 1482, gothic script as on 3c.[59]

Group 4: Astrolabes with unusual plates and three different styles of inscription

a) Paris – undated;[60]
b) New York – undated.[61]

Group 5: Astrolabe with Type B rete

a) Cracow – undated rete and plates, engraved in the same gothic hand as 1a, fitted to a fourteenth-century Italian mater.[62]

Table 2 - Size, number of stars on rete, and date of vernal equinox on the Regiomontanus-type astrolabes.

No. location	IC no.	size	stars	equinox	date
1a Nuremberg GNM	549	129/6	30	10 1/3	1457
1b Paris PC [Greppin]	452	137/7	30	11	-
2a Paris PC	256	168/9	27	11	-
2b London PC	640	115/4	30	10 1/2	1462
3a Darmstadt HLM	592	144/7	27	10 1/2	1466
3b Munich DM	620	132/6	27	10 1/3	-
3c Strasbourg MH	530	170/4	27	13	1481
3d Cracow MN *	4501	152/6	27	12	1482
4a Paris CNAM	516	132/5	27	10 1/4	-
4b New York CUL	545	133/7	27	10	-
5a Cracow JUM	644	**	34	**	-

* anomalous throne
** mater not original.

No: number in Groups 1-5 above
IC number: International Checklist number
Size: diameter and thickness in millimetres
Stars: number of stars on the rete
Equinox: date in March of vernal equinox as read from calendrical scale.

---

[59] IC 4501: Cracow, Muzeum Narodove, Czartoryski Collection, inv. no. XIII/753. Unpublished.

[60] IC 516: Paris, Conservatoire National des Arts et Métiers, inv. no. 3882/1. Unpublished (see the unpublished 1851 handlist available in the Museum, p. 40, no. F-b 20; brief mention).

[61] IC 545: New York, Columbia University Special Collections, inv. no. 27-256. Unpublished. See New York IM *Handlist*, p. 59 (no. 13) (brief mention).

[62] IC 644: Cracow, Jagellonian University Museum, inv. no. 4042-40/V. See ZINNER, *Regiomontanus*, figs. 74-75 (front and back); ZINNER, *Instrumente*, p. 296; BURCZYK-MARONA, «Astrolabium» (detailed description with illustrations of all parts, including all four quarters of the rete).

The inspiration for the distinctive style of the retes on these astrolabes came originally from Italy. Two Italian pieces provide the evidence: a fourteenth-century (?) astrolabe (Fig. 10),[63] and an astrolabe based on the same rete design made in Urbino and dated 1462.[64] The latter is signed «KP», initials more likely to have represented a German (or Greek) name than an Italian one. One of these pieces, or another with identical details, including the shape of the individual star-pointers, and also the rete pattern, is depicted on the intarsia in the study of Archduke Ferdinand of Urbino.[65]

The most intricate of the Type A retes is that on the 1457 piece. The Type B rete, whose sole surviving example is somewhat larger than any of the Type A ones, is also to be found later on the even larger astrolabes associated with Hans Dorn, dated 1483 and 1486.

The astrolabes of Type A provided some of the inspiration for Hartmann's *Volkastrolabien*, made in Nuremberg in the first half of the sixteenth century.[66] Some two dozen of his works survive, and with one exception, they all resemble at first sight the Regiomontanus Type A instruments. Hartmann simply used the same retes that had been popular for more than half a century. The rete form may have come to Hartmann by way of his earlier contemporary Johannes Stöffler (1452-1531), whose retes are still closer to the Italian originals mentioned above.[67] Hartmann also used rosettes on the thrones of his instruments, again imitating the works of his predecessors. In fact he even copied the stars used on the earlier instruments, sometimes taking over 20, and at other times 30 stars. These instruments are sufficiently well-documented in the modern literature that no detailed description is necessary here.[68] However, no detailed comparative study has ever been conducted.

---

[63] IC 493: Florence, Museo di Storia della Scienza, inv. no. 1107. See Florence IMSS 1954 *Catalogue*, p. 62 (no. 1107); Florence IMSS 1991 *Catalogue*, p. 8 (no. I.6).

[64] IC 4506: Moulins (Allier), Musée Départemental – stolen. See Paris BN 1973 *Catalogue*, p. 75 (no. 176), with illustrations of the front and back on p. 57.

[65] Illustrated most recently in CARDINI & FUMAGALLI, *Antiche Università*, p. 37.

[66] On Hartmann, see GUNTHER, *Astrolabes*, pp. 436-440; ZINNER, *Instrumente*, pp. 357-368; A. J. TURNER, *Instruments*, pp. 39-40, 45; KLEMM, *Hartmann*; KING, *Catalogue*, Section 7.2.

[67] See, for example, GUNTHER, *Astrolabes*, p. 428, for an illustration.

[68] For detailed descriptions of individual Hartmann astrolabes, see Rockford TM *Catalogue*, pp. 128-131 (no. 17); Washington NMAH *Catalogue*, pp. 146-150 (no. 262); Ecouen MR *Catalogue*, pp. 106-107 (no. 82); and Nuremberg 1992-1993 *Catalogue*, items 1.69 and 1.70. See also GORDON, «Metallography» and «Technology».

Some other early sixteenth-century German astrolabes continued the tradition of the Regiomontanus and Dorn astrolabes, namely:

a) four undated astrolabes by the same maker with distinctive Y-shaped frames inside the ecliptic and below (one – the only one known to Gunther – was misidentified as Italian and dated c. 1500; a later dating may be more appropriate);[69]

b) two undated astrolabes by the same maker with a vaguely rectangular frame on the rete developed from Type B.[70]

ii. *Some common features of the «Regiomontanus-type» astrolabes*

The eleven associated astrolabes display competent workmanship throughout, with various hands in evidence in the engravings. Only the distinctive gothic of 1a, 1b (plates and mater) and 5a, and the distinctive antiqua of 1b (rete) and 2b, are unmistakable. We are dealing with the productions of a series of individuals, possibly in a single workshop.

The diameters of the ten astrolabes with Type A retes vary between 115mm (2b) and 170mm (3c), with the single Type B rete (5a) fitting into an earlier mater of diameter 252mm. A diameter of 132 or 133mm is found on three instruments (3b, 4a, 4b); otherwise the diameters are all different.

The basic design of the thrones on all but one piece (3d) is identical. Each instrument has a throne consisting of three rosettes, the lower two supported by a trilobed motif at either side and the suspensory apparatus being attached to the upper one. The predominant decoration for these rosettes is a five-petal rose, which also appears occasionally on the pin holding the astrolabe together

---

[69] IC 173: Oxford, Museum of the History of Science, inv. no. LE 2065, diameter 197 mm. See GUNTHER, *Astrolabes*, pp. 327-328, and plate 73 (front and back).

IC 453: Point Lookout, N. Y., Collection of Leonard Linton, formerly Brussels, Greppin Collection (no. 3124), sold at Galerie Köller in Zurich on 17.11.1975, diameter 197 mm. See GREPPIN, *Catalogue*, p. 33 (no. 3124: «français vers 1500»), with illustrations of the front and back on p. 14; LINTON, *Catalogue*, p. 133 (no. 186), colour plate of front and back following.

IC 4505: with leather case, probably original, dated 1542: Munich, Deutsches Museum: 1991-223, 1 (astrolabe) + 2 (case). Purchased in 1991 from private owner in Jena, diameter 197 mm. Unpublished.

IC 4543: Rome, Osservatorio Astronomico, diameter not known.

[70] IC 491: Florence, Museo di Storia della Scienza, inv. no. 1110, diameter 134 mm. See Florence IMSS 1954 *Catalogue*, p. 62 (no. 1110); Florence IMMS 1991, *Catalogue*, p. 8 (no. I.7); POULLE, «Instruments astronomiques», pp. 20-21 (illustration of front).

IC 4544: Stuttgart, Württembergisches Landesmuseum (loan of the Fürstlich Fürstenbergische Sammlungen): L1985-200, diameter 187 mm. Unpublished.

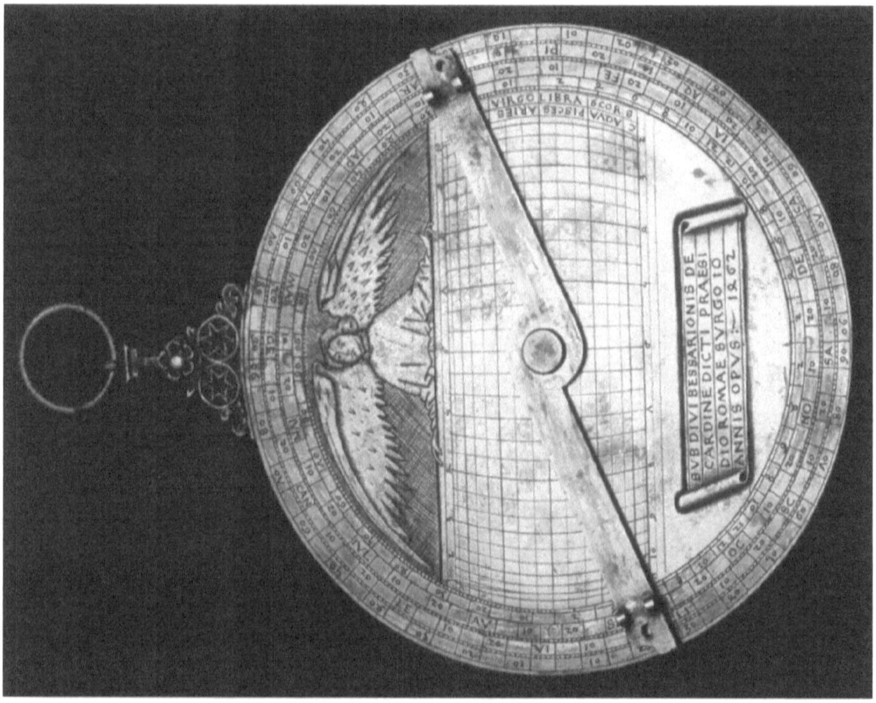

Fig. 2 - IC 640; 2b Bessarion's astrolabe, back.

Fig. 1 - IC 640; 2b Bessarion's astrolabe, front.

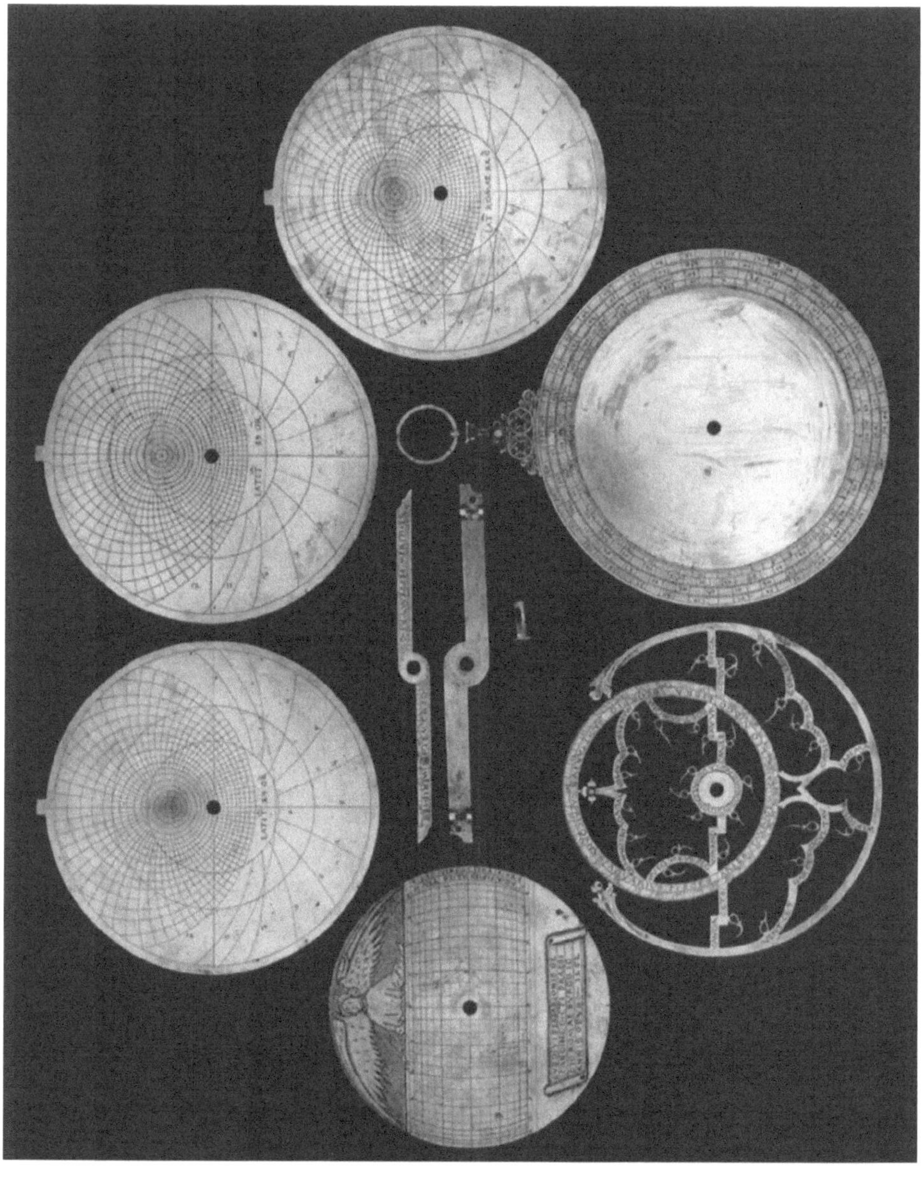

Fig. 3 - IC 640; 2b Bessarion's astrolabe, constituent parts: mater, plates, rete, disc with dedication, rule, alidade, pin.

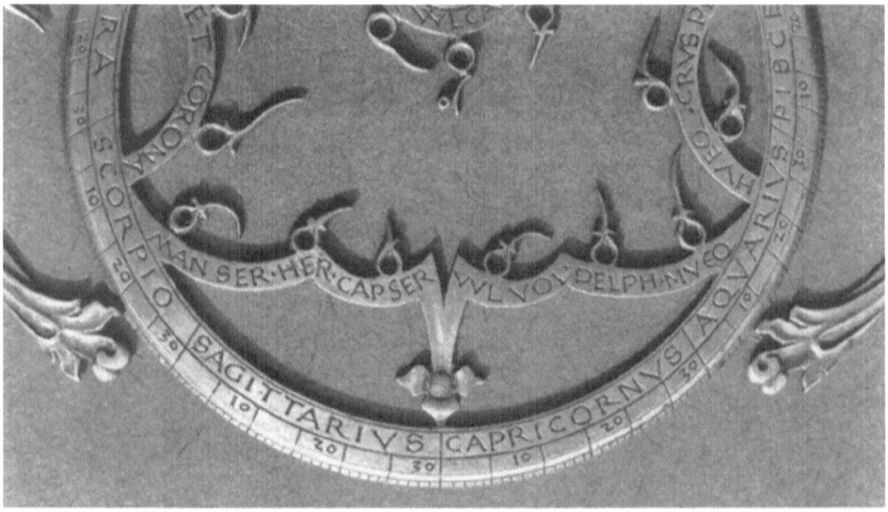

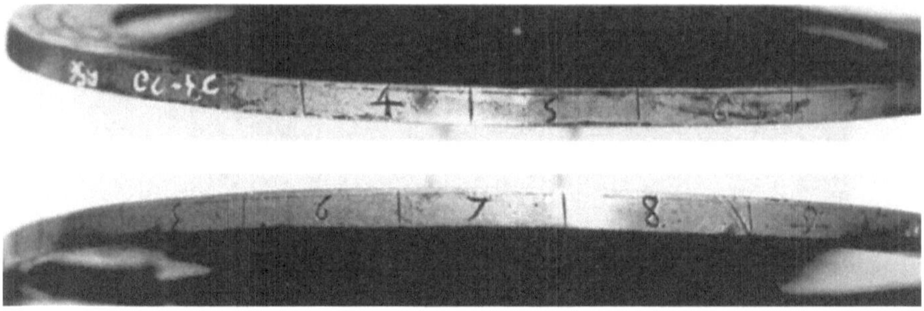

Fig. 4 - IC 640; 2b Bessarion's astrolabe, letter forms on the rete.

Fig. 5 - IC 640; 2b Bessarion's astrolabe, letter and number forms on disc.

Fig. 6 - IC 640; 2b Bessarion's astrolabe, number forms around outer edge of rim.

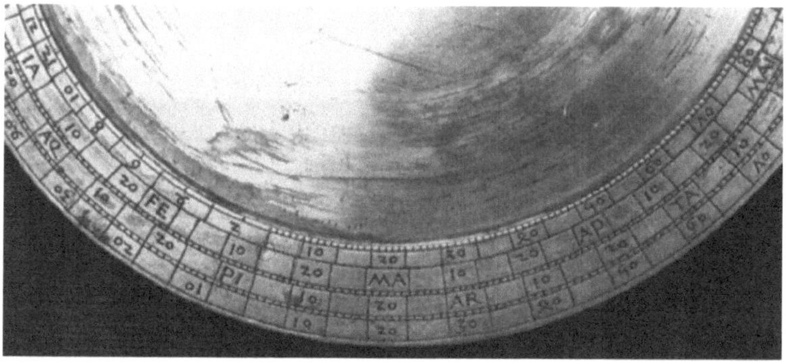

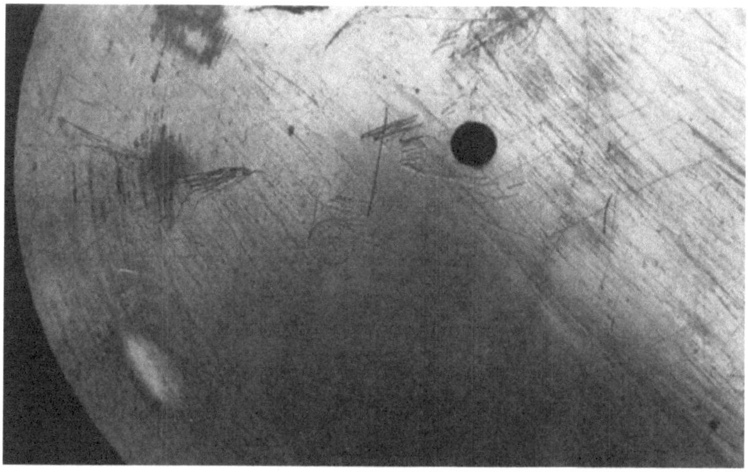

Fig. 7 - IC 640; 2b Bessarion's astrolabe, back of the astrolabe showing the burin jabs produced after scribing degree lines.

Fig. 8 - IC 640; 2b Bessarion's astrolabe, back of rete, showing 5-degree layout lines on the ecliptic circle.

Fig. 9 - IC 640; 2b Bessarion's astrolabe, back of disc, showing a scribed arc with degree dots, scratches, and doodles.

Fig. 10 - IC 493; Florence IMSS inv. no. 1107; fourteenth-century (?) astrolabe rete.

Fig. 11 - IC 549; 1a Rete; note beasts' heads and appendage to side of star-pointer for Wega, marking centre of ecliptic.

Fig. 12 - IC 549; 1a Front of throne, with roundels containing the arms of Vienna.

Fig. 13 - IC 549; 1a Back of astrolabe, showing the throne with two roundels bearing blank shields, and date 1457. Note calendar scale with elongated S.

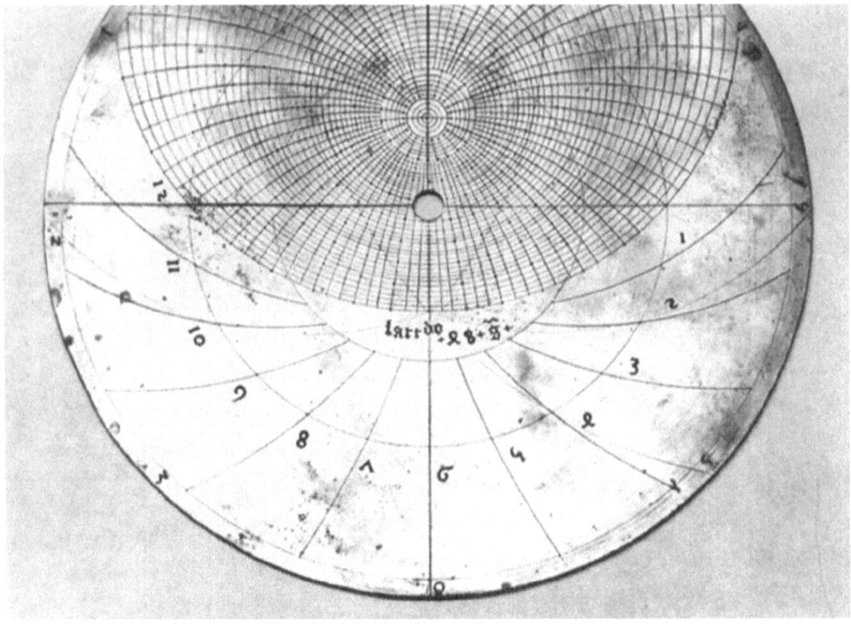

Fig. 14 - IC 452; 1b Rete, showing identical features and antiqua lettering by the same hand as on 2b, Bessarion's astrolabe.

Fig. 15 - IC 452; 1b Part of plate for latitude 48°, showing gothic script.

Fig. 16 - IC 452; 1b Front of throne, with roundels containing the Sun and the Moon.

Fig. 17 - IC 452; 1b Back of throne, showing roundels with shields.

Fig. 18 - IC 256; 2a Back of astrolabe, showing *organum Ptolemei*, and throne with magnetic compass.

Fig. 19 - IC 256; 2a Detail of back of astrolabe with disc removed showing later engraving with date 1535. Note gothic lettering with elongated S.

Fig. 20 - IC 592; 3a Details: front of throne; the date, 1466; plate for latitude 48°.

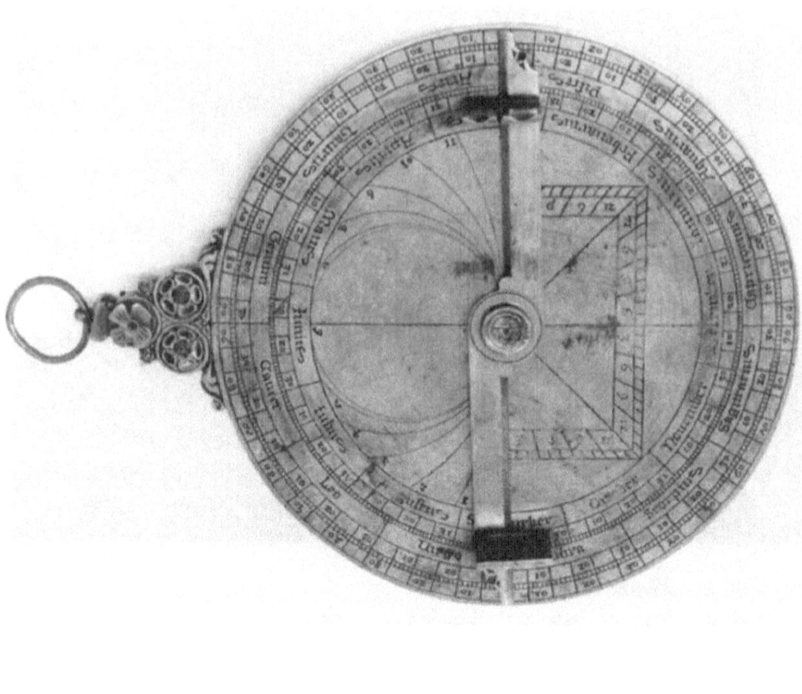

Fig. 21 - IC 620; 3b Back of astrolabe.

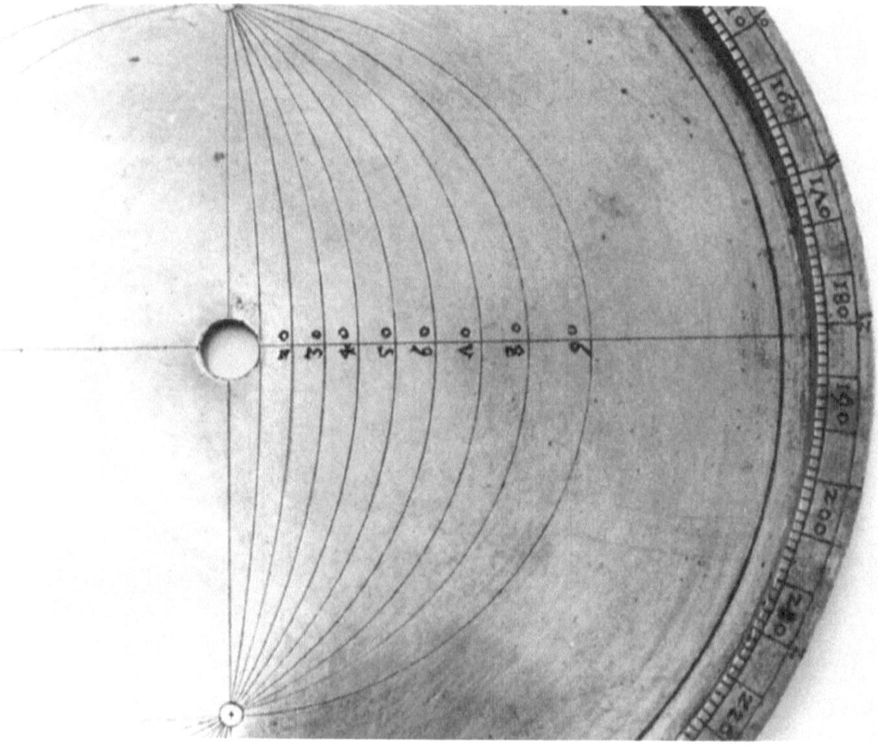

Fig. 22 - IC 620; 3b Front of astrolabe, showing part of mater inscribed with tablet of horizons.

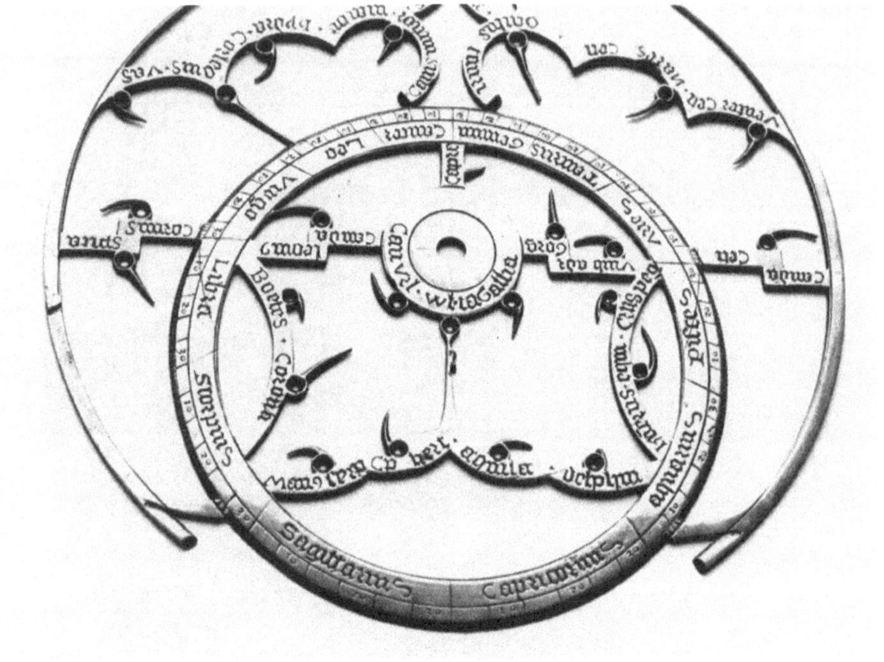

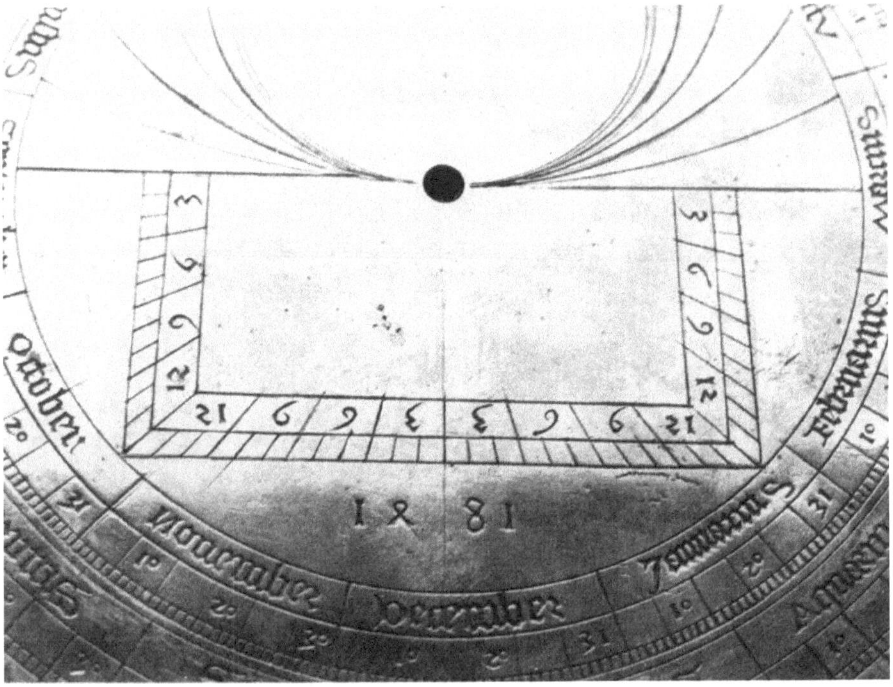

Fig. 23 - IC 530; 3c Part of rete.

Fig. 24 - IC 530; 3c Back of astrolabe, showing date 1481.

Fig. 25 - IC 4501; 3d Front of astrolabe, showing throne with variant decoration including one rosette.

Fig. 26 - IC 4501; 3d Back of astrolabe, showing date 1482.

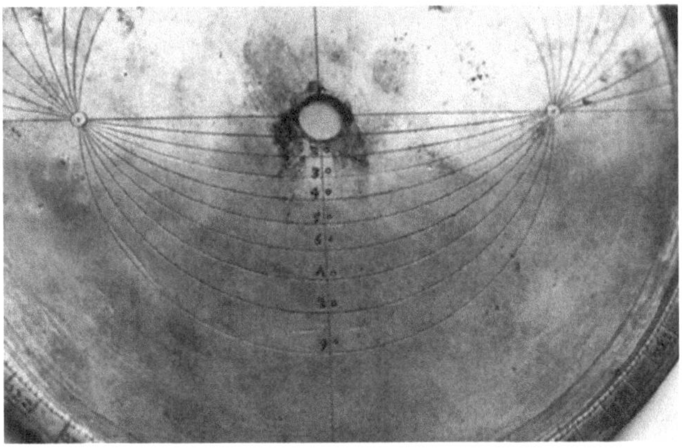

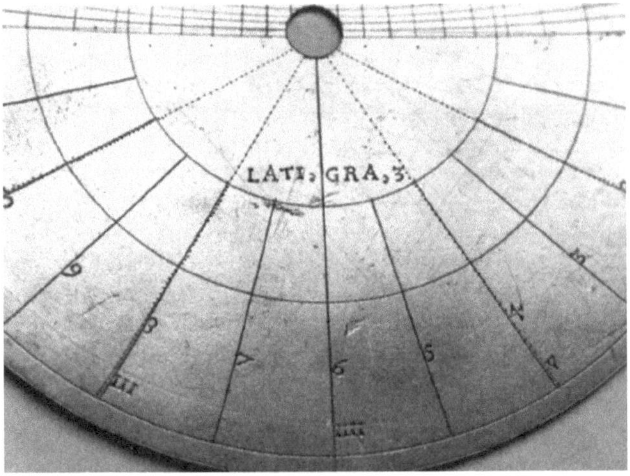

Fig. 27 - IC 516; 4a Detail of back of astrolabe with throne.

Fig. 28 - IC 516; 4a Detail of mater, showing tablet of horizons.

Fig. 29 - IC 516; 4a Detail of lettering on plate for 3°, lettters and numbers punched.

Fig. 30 - IC 644; 5a Astrolabe with rete of Type B.

(see below). A star (2b, 3a) or representations of the sun and moon (1b) sometimes replace the rosette. On three pieces rosettes are replaced by roundels with coats of arms (arms of Vienna on 1a, arms of von Wallenstein on 1b,[71] empty on 3a), and on one (2a) a magnetic compass.[72] The anomalous throne on 3d consists of a single rosette supported by a lattice with foliate motifs.

The basic design of the retes is very similar, with one larger and more developed than the rest (Type B, 5a) and one of the remainder (1a) at an intermediary stage of development. There is a roughly rectangular frame inside the southern ecliptic, formed by two arcs standing on the sides of the ecliptic and four circular arcs «parallel» to the equinoctial bar. A trilobed motif features at the middle of the upper side on the two astrolabes 1b and 2b (Bessarion's). The rete of 1a is more sophisticated than the rest of Type A, featuring elegant details such as a vertical bar shaped with a beast's head above the central disc. On the Type A retes there is a roughly Y-shaped frame below the ecliptic, formed by a series of circular frames, sometimes with a central part modified from a half-quatrefoil (1a and 2b). On several of the retes (1b, 2b, 3a, 4a, 4d) there is a small circular appendage attached to the pointer for Vega to indicate the centre of the ecliptic. On others (1a, 1b) there is a rectangular frame outside the ecliptic circle at the winter solstice. Type B has six arcs of circles along the top of the «rectangular» frame inside the ecliptic and a vertical solstitial bar spanning the ecliptic which is counter-changed three times above the central circular disc. There is a prominent trefoil design supporting the frame below the ecliptic.

The equinoctial bars on the Type A retes are all counter-changed according to the same pattern, here stylized:

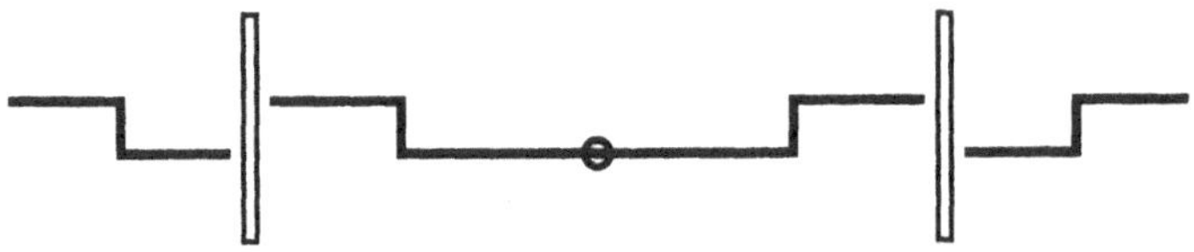

That on Type B is the same but upside down.

---

[71] The engraving reads AEW, apparently for Albrecht Václav Eusebius von Wallenstein (1583-1634), Bohemian soldier and statesman, commanding general of the armies of the Holy Roman Emperor, Ferdinand II, during the Thirty Years' War. See, for example, *Encyclopedia Britannica*, XII, pp. 470-471.

[72] See KING, «Report», p. 7, fig. 5, for an illustration.

The basic design of the star-pointers is identical. A wavy pointer is attached to a circular base, often with one or two «collars», and occasionally to a base made up of two circles. The base of the pointer for Regulus is distinguished by an asterism around the circular base on 1b, 2b, and 3a. The reason is simply that a crown is intended; the name Regulus means «little king».

The letter forms are often gothic, but in two cases, clearly involving a single engraver, antiqua is used. The gothic inscriptions are characterized by a prolonged final S on its side (Figs. 13, 17, 19, 21); and a «1» which is higher than the «0». The «2» is the typical German form resembling a «Z» (Figs. 13, 21). The elongated S was also used on the simplified Schmalkalden astrolabes and the larger instruments attributed to Hans Dorn. Some of the inscriptions are stamped rather than engraved. For example, on 4a all the lettering and numbering on three of the plates is punched. The letters are antiqua majuscules, and the numbers are all modern. The letters resemble those used by Hartmann, so it may be the case that three plates were added or substituted (or even modified) in the sixteenth century in this older astrolabe. Regiomontanus also gradually changed from the old gothic numerals to the new forms in his manuscript works.[73] His contacts with Italy also led him to abandon the gothic script for the antiqua. On the astrolabes of Hartmann, the inscriptions are stamped not engraved, and are in antiqua; however, in his manuscripts Hartmann used a gothic script.

On the Strasbourg instrument (3c) the engraving is at first sight different; for example, a different form of the «2» is used and some capital N's are reversed. The astronomical markings are achieved with less precision; in particular the calendrical scale is inaccurate, with a date for the vernal equinox that is centuries out-of-date (March 13 instead of March 10 1/2). The instruments 4a and 4b require special consideration because of anomalous plates and varied hands.

The star-names on all the instruments are not identical, but are closely related. They are of particular historical interest because they reflect an important trend of the time: the abandonment of Latinized Arabic names and the adoption of Latin names. This trend

---

[73] ZINNER, *Regiomontanus*, p. 49; G. L'E. TURNER, *The Regiomontanus Astrolabe*, p. 11. On the development of medieval and Renaissance numerals, see HILL, *Arabic Numerals*, an important work based, however, mainly on manuscripts. An investigation of the development of numeral forms as attested by instruments is a task for the future.

was not irreversible: on later instruments we witness the gradual but persistent reversion to Arabic names. It is as a result of this late trend that we still use some Arabic star-names to this day. The choice of stars and the forms of the Arabic names used for them on the Type A retes indicate the influence of the star-table of John of Gmunden (d. Vienna, 1442).[74] In particular, *alhes* for α *Crateris* derives from a corruption of Arabic *al-ka's* for «drinking glass», and *alkair* for α *Aquilae* from a Germanized rendering of *alcair*, derived from *altair* (Arabic *al-ṭā'ir*, «flying (eagle)») through confusion of c and t in gothic script.

Preliminary investigations of the positions of the star-pointers by Burkhard Stautz of Frankfurt indicate that the positions correspond to contemporaneous star catalogues. These investigations are still in progress. The same catalogues were used by Hartmann decades later, which explains why the star positions on Hartmann's astrolabes correspond to a date several decades before his time.

We have already commented on the lunar volvelles which feature on the backs of two pieces (1a and 1b), and the *organa Ptolemei* on two others (2a, 2b). On the former pair the shadow scale is found inside the calendrical scale. On the latter pair it is ingeniously divided into four rectilinear parts in the annulus between the scale for the lunar mansions and the outer scales. With the exception of 1a, 1b, 2a, and 2b, all instruments have a double universal horary quadrant and a double shadow square on the back.

Two pieces (3b, 4a) have a set of horizons for each 10° engraved on the mater. The latitudes used for the plates are not identical, but there are favoured patterns. The increments of 3° between 30° and 54°, which are a common feature, are intended to make the astrolabes «universal». The specific latitudes then correspond roughly to the boundaries and middles of the climates of Antiquity (with extensions to serve Northern Europe); the continued explicit and implicit use of the climates throughout the Middle Ages and the early Renaissance is only now being recognized.[75] The latitudes served by the plates on the eleven astrolabes are:

---

[74] See Kunitzsch, *Arabs and Stars*, chapter XXII, and also Kunitzsch, *Sternverzeichnisse*, pp. 111-116 («Typ XVII»). On John of Gmunden see, for example, Zinner, *Instrumente*, pp. 322-323; *Dict. Sci. Biog.* VII, pp. 117-122.

[75] On climates, see Harley & Woodward, *History of Cartography*, I, especially pp. 182-183; Nuremberg, 1992-1993 *Catalogue*, I, p. 107; King «Astronomical Secrets».

Table 3 - The latitudes engraved on the plates

1a:	32°	37°	48°	51°	(one plate missing)			
1b:	32°	35°	37°	42°	45°	48°	51°	
2a:	30°	36°	39°	42°	45°	54°	(one plate missing)	
2b:	39°	42°(R)	45°	48°	51°	54°		
3a:	36°	39°	42°	45°	48°	51°	54°	57°
3b:	39°	42°	45°	48°	51°	54°		
3c:	42°	45°	48°	51°				
3d:	42°	45°	48°	51°				
4a:	0°	3°	6°	9°	12°	51°	54°	90°
4b:	39°	42°	51°	54°	*			
5a:	48°	49°	50°	51°	52°	67°		

R = ROMAE; * two sets of markings not yet explained.

Only the latitudes used on the Bessarion astrolabe (2b) and the undated Munich piece (3b) correspond precisely to those on Regiomontanus' astrolabes which Hartmann had seen (see above). The latitudes used by Regiomontanus for Vienna in his written works were 47° 46′ (from John of Gmunden) and later 48° 0′, 48° 15′, 48° 22′, and 48° 27′; for Nuremberg he used 49° 0';[76] it was typical of medieval and Renaissance instrument-making to use integral degrees. On several instruments (including 1a, 1b, 3a, 3b, 4a) there are markings on the lugs of the plates, either a reminder of the latitude or marks to distinguish one plate from another. The latitudes used on 4a are unique in surviving European instruments; they were clearly chosen for pedagogic purposes.

Those original alidades that survive are elegantly fashioned. Some (2a, 3b, 3c, 3d) have folding sights. One (1a) has ends decorated with beasts' heads (also found on the retes of 1a and 5a). None bears any astronomical markings. Several pieces have a radial rule for use on the front; two (1a, 2b) have a diametral rule. (The radial rule without markings on 3c and the diametral rule without markings on 2a are replacements). The northern and southern declinations are indicated by Septentrionalis (3a), Latitv Septen (3b), Septemtrionalis (1a, 1b), or Septemtrio (2b), and Meridionalis (1a), Lati Meri (3b), or Meridies (2b) respectively.

Two instruments aside from the Bessarion astrolabe bear marks of ownership. On 2a are the date 1535 and two names (Fig. 19)

---

[76] ZINNER, *Regiomontanus*, pp. 39, 60, and plate 30, fig. 55.

engraved on the blank back of the mater over which the *organum Ptolemei* rotates; on the back of the latter are inscribed the date 1537 and initials. On the second (3d), the initials SWIII (= SWM?, or the third instrument produced by SW?) occur inside the shadow square.

One instrument aside from the Bessarion astrolabe bears mention of a location. On Bessarion's astrolabe (2b) Rome is mentioned, most probably as the place where the astrolabe was presented. The 1482 piece with the anomalous throne (3d) bears the name Schmalkalden (near Suhl in Thuringia) by the date. We can be fairly certain that of the eleven astrolabes only this one was made there. One could speculate that the maker was visiting Hans Bamman, several of whose astrolabes, all contemporaneous with this piece, are marked with the name of this town.

iii. *Concluding remarks*

We are dealing here with a newly-identified corpus of historical material. Only recently (since 1989) have most of the items been identified in the course of the Frankfurt cataloguing project. Still more recently (since 1991) have they been inspected by us consistently, paying particular attention to the engraving and construction marks. But we have not been able to compare any two of these instruments with each other except by using photographs. We lack reliable information on the workshop(s) where they were made. It is possible that the location of the workshop was in Vienna. We know that Peuerbach, Regiomontanus, and Dorn devised or made astrolabes. In the earliest dated piece (1a from 1457) Peuerbach may have played a role. One could conjecture, and this is pure speculation, that Peuerbach started the piece in two different hands (1b) and Regiomontanus, still in Vienna and still favouring the old Arabic star-names, finished it. Dorn is the only one of the three who was mainly in Vienna throughout the 1450s and 1460s, and he may have had a role in the later productions. But the instruments attributed to him (of which he signed only one) are quite different in character from the ones we have described here.

Inevitably the corpus raises as many questions as its discovery solves, but these questions are now on a different plane. Two of these astrolabes, the 1457 piece and the Bessarion astrolabe, were exhibited together at the Germanisches Nationalmuseum in

Nuremberg in 1992-1993,[77] and the Bessarion astrolabe in the Bibliothek Otto Schäfer in Schweinfurt in 1993-1994.[78] If sufficient interest in this remarkable corpus of instruments can be aroused, it would be worthwhile to exhibit all of them together. Ideally, the entire corpus of German astrolabes up to about 1550, a significant date when new trends are detectable, should be studied in detail, along with the available texts.

Acknowledgements

The authors are grateful to the directors and staff of all those museums that have allowed instruments to be examined and photographed. They are also indebted to a number of private collectors for similar facilities. They wish especially to thank Dr. W. F. Ryan and the late Jon Darius for their comments on the text.

BIBLIOGRAPHY

AMEISENOWA, *Bylika Globe*
Zofia Ameisenowa, *The Globe of Martin Bylika of Olkusz and Celestial Maps in the East and the West*, translated by Andrzej Potocki, Wrocław, Cracow, Warsaw: Polska Akademia Nauk, 1959.

ANONYMOUS (London)
«Under the protection of the divine Bessarion called Cardinal I arise in Rome... 1462», *Antiques Trade Gazette*, 28 October, 1989, p. 27.

BROWNSWORD, «Metalwork»
R. BROWNSWORD, «X-ray Fluorescence Analysis on Non-ferrous Archaeological Metalwork: Part 2. Archaeological Applications», *Analytical Proceedings*, 22 (1985), pp. 73-75.

Brussels 1957 *Catalogue*
[H. MICHEL, *et al.*], *Exposition: Art, Histoire et Sciences. Catalogue*, Brussels: Musées Royaux d'Art et d'Histoire, 1957.

BURCZYK-MARONA, «Astrolabium»
D. BURCZYK-MARONA, «Zegar słoneczny i astrolabium profesora Jana

[77] See Nuremberg 1992-1993 *Catalogue*, items 1.75 (the Nuremberg GNM astrolabe of 1457) and 1.76 (the Bessarion astrolabe).

[78] Schweinfurt, 1993-1994, *Catalogue*, item 179.

Brozka w zbiorach Muzeum Uniwersytetu Jagiellońskiego», *Opuscula Musealia (Zeszytu Naukowe Uniwersytetu Jagiellońskiego)*, [Warsaw and Cracow], 3 (1988).

CARDINI & FUMAGALLI, *Antiche Università*
Franco Cardini and M. T. Fumagalli Beonio-Brocchieri, *Antiche Università d'Europa: Storia e personaggi degli Atenei nel Medio Evo*, Milano: Editoriale Mondadori, 1991.

Dict. Sci. Biog.
Dictionary of Scientific Biography, 16 vols, New York: Charles Scribner's Sons, 1970-1980.

DIECKHOFF, «Arsenius-Astrolabium»
R. DIECKHOFF, «Cosmographia planisphaeria: Ein 'Arsenius'-Astrolabium des späten 16. Jahrhunderts in Kölnischen Stadtmuseum», *Kölner Museums-Bulletin: Berichte und Forschungen aus den Museen der Stadt Köln*, 2 (1990), pp. 23-44.

EADE, *Astrology*
J. C. EADE, *The Forgotten Sky: A Guide to Astrology in English Literature*, Oxford: Clarendon Press, 1984.

Ecouen MR *Catalogue*
ADOLPHE CHAPIRO, CHANTAL MESLIN-PERRIER, and ANTHONY TURNER, *Musée national de la Renaissance, Château d'Ecouen: Catalogue de l'horlogerie et des instruments de précision du début du XVIe au milieu du XVIIe siècle*, Paris: Editions de la Réunion des musées nationaux, 1989.

Encyclopaedia Britannica
The New Encyclopaedia Britannica, 15th edn, Chicago: University of Chicago Press, 1986.

FLEGG, *Numbers*
GRAHAM FLEGG, *Numbers through the Ages*, Basingstoke: Macmillan, 1989.

Florence IMSS 1954 *Catalogue*
M. L. BONELLI, ed., *Catalogo degli strumenti del Museo di Storia della Scienza*, Firenze: Olscki, 1954.

Florence MSS 1991 *Catalogue*
M. MINIATI, ed., *Museo di Storia della Scienza: Catalogo*, Firenze: Istituto e Museo di Storia della Scienza, 1991.

GOLDSTEIN, «Transversal Scale»
BERNARD R. GOLDSTEIN, «Levi ben Gerson: On Instrumental Errors and the Transversal Scale», *Journal of the History of Astronomy*, 8 (1977), pp. 102-112.

GORDON, «Technology»
R. B. GORDON, «Sixteenth-Century Metalworking Technology used in

the Manufacture of two German Astrolabes», *Annals of Science*, 44 (1987), pp. 71-84.

GORDON, «Metallography»
R. B. GORDON, «Metallography of Brass in a 16th-century Astrolabe», *Journal of the Historical Metallurgy Society*, 20 (1986), pp. 93-96.

GREPPIN, *Catalogue*
La Collection Greppin, Instruments scientifiques, Zurich: Galerie Koller, 1975. [Catalogue of auction, 17 Nov. 1975].

GUNTHER, *Astrolabes*
R. T. GUNTHER, *Astrolabes of the World based upon the Series of Instruments in the Lewis Evans Collection... with Notes on Astrolabes in... other Public and Private Collections*, 2 vols [continuously paginated], Oxford: privately printed by the University Press, 1932; reprinted bound as one, London: The Holland Press, 1976.

HARLEY & WOODWARD, *History of Cartography*
J. B. HARLEY & DAVID WOODWARD, *The History of Cartography Volume I: Cartography in Prehistoric, Ancient, and Medieval Europe and the Mediterranean*, Chicago & London: Chicago University Press, 1987.

HARTNER, «Astrolabe»
WILLY HARTNER, «The Principle and Use of the Astrolabe», chapter in Arthur U. Pope, *A Survey of Persian Art*, III, London: Oxford University Press, 1939; reprinted with original plates in *Astrolabica 1*, Paris: Société Internationale de l'Astrolabe, 1978, and the article «Asturlāb» in *Encyclopedia of Islam*, I, pp. 722-728. Both reprinted in W. HARTNER, *Oriens-Occidens*, Hildesheim: Georg Olms, 1968, pp. 287-318.

HILL, *Arabic Numerals*
G. F. HILL, *The Development of Arabic Numerals in Europe Exhibited in Sixty-Four Tables*, Oxford: Clarendon Press, 1915.

KING, «Report»
DAVID A. KING, «Medieval Astronomical Instruments: A Catalogue in Progress», *Bulletin of the Scientific Instrument Society*, no. 31 (1991), pp. 3-7.

KING, *Catalogue*
DAVID A. KING, *A Catalogue of Medieval Astronomical Instruments: Astrolabes, Quadrants, and Sundials*, parts 1-5, Eastern (Islamic and Indian), parts 6-10 European: in preparation.

KING, «Medieval Picard Astrolabe»
DAVID A. KING, «Rewriting History through Instruments: The Secrets of a Medieval Astrolabe from Picardy», in R. G. W. ANDERSON, J. A. BENNET & W. F. RYAN, eds, *Making Instruments Count: Studies in Honour of Gerard Turner*, Aldershot: Variorum, 1993.

KING, «Secrets»
DAVID A. KING, «Some Medieval Astronomical Instruments and their Secrets», in Renato G. Mazzolini, ed., *Non-verbal Communication in Science prior to 1900*, (Biblioteca di «Nuncius», Studi e testi), Firenze: Leo S. Olschki, 1993.

KING, «Astronomical Secrets»
in HARRY KÜHNEL, ed., *Kommunikation zwischen Orient und Okzident, Sitzungsberichte der Österreichischen Akademie der Wissenschaften, Phil.-Hist. Klasse, Veröffentlichungen des Instituts für Realienkunde des Mittelalters und der frühen Neuzeit*, Bd. 17, Vienna: Verlag der Österreichischen Akademie der Wissenschaften, 1994.

KLEMM, *Hartmann*
HANS GUNTHER KLEMM, *Georg Hartmann aus Eggolsheim (1489-1564): Leben und Werk eines fränkischen Mathematikers und Ingenieurs* (Wissenschaftliche und künstlerische Beiträge, Ehrenbürg-Gymnasium Forchheim, Heft 8), Forchheim: Gürtler Druck, 1990. [Revision of catalogue of exhibition at Eggolsheim, 1989, for 500th anniversary of Hartmann's birth.].

KUNITZSCH, *Sternverzeichnisse*
PAUL KUNITZSCH, *Typen von Sternverzeichnissen in astronomischen Handschriften des zehnten bis viersehnten Jahrhunderts*, Wiesbaden: Otto Harrassowitz, 1966.

KUNITZSCH, *Arabs and Stars*
PAUL KUNITZSCH, *The Arabs and the Stars: Texts and Traditions on the Fixed Stars and their Influence in Medieval Europe*, Variorum Collected Studies Series CS307, Northampton: Variorum, 1989. [Includes «The Star Catalogue commonly appended to the Alphonsine Tables», *Journal for the History of Astronomy*, 17 (1986), pp. 89-98].

LINTON *Catalogue*
Scientific Instruments, Rare Books: 1 - Leonard Linton Collection, Paris: Alain Brieux, 1980. [Catalogue of the Linton Collection offered for sale at Nouveau Drouot, Paris, 9-10 October 1980].

Linz 1990 *Catalogue*
WILFRIED SEIPEL, ed., *Mensch und Kosmos: Die Heraufkunst des modernen naturwissenschaftlichen Weltbildes (Katalog: Oberösterreichische Landesausstellung 1990)*, Linz: Landesmuseum, 1990. [Exhibition in Schloßmuseum, 7 May-4 Nov. 1990].

London BM *Catalogue*
F. A. B. WARD, *A Catalogue of European Scientific Instruments in the Department of Medieval and Later Antiquities of the British Museum*, London: British Museum Publications, 1981. [See Price, «Review»].

London NMM *Astrolabe*
The Planispheric Astrolabe, London: National Maritime Museum, 1976.

MADDISON, «Rojas Projection»
FRANCIS R. MADDISON, «Hugo Helt and the Rojas Astrolabe Projection», Separata da *Rivista da Faculdade de Ciências* [Coimbra], 39 (1966), 61 pp. and one foldout [pp. 195-251 of the orignal].

MENNINGER, *Zahlwort und Ziffer*
KARL MENNINGER, *Zahlwort und Ziffer: Eine Kulturgeschichte der Zahl, Band II: Zahlschrift und Rechnen*, 2nd edn, Göttingen: Vandenhoeck & Ruprecht, 1959; translated as *Number Words and Number Symbols*, Cambridge, Mass.: MIT Press, 1970.

MICHEL, *Astrolabe*
HENRI MICHEL, *Traité de l'Astrolabe*, Paris: Cauthiers-Villars, 1947; reprinted Paris: Librairie Alain Brieux, 1976.

Milan MAA *Catalogue*
TULLIO TOMBA and GIUSEPPE BRUSA, *Museo d'Arti Applicati: Strumenti scientifici-orologi*, Milano: Electa Editrice, 1983.

Milan PA *Catalogue*
TULLIO TOMBA, «Gli astrolabi della Collezione Settala nella Pinacoteca Ambrosiana», *Atti della Fondazione Giorgio Ronchi*, 33, part 2 (1978), pp. 297-314, continued in 33, part 4 (1978), pp. 3-14.

MORISON, *Printing*
STANLEY MORISON, *Four Centuries of Fine Printing*, 4th edn, London: Ernest Benn, 1960.

MORTIMER, «X-ray Fluorescence Analysis»
C. MORTIMER, «X-ray Fluorescence Analysis of Early Scientific Instruments», *Proceedings of 1987 Archaeometry Symposium*, pp. 311-317.

NACHET, *Catalogue*
A. NACHET, *Collections Nachet: Instruments scientifiques et livres anciens*, Paris, 1929; reprinted Paris: Librarie Alain Brieux, 1976.

New York IM *Handlist*
Anonymous, *The Industrial Museum of New York, Exhibit of Early Astronomical and Mathematical Instruments (February 1930)*, New York: Museums of the Peaceful Arts [1930].

NORTH, «Astrolabe»
J. D. NORTH, «The Astrolabe», *Scientific American*, 230, no. 1 (January, 1974), pp. 96-106; reprinted with modifications in J. D. NORTH, *Stars, Minds, and Fate: Essays in Ancient and Medieval Cosmology*, London and Ronceverte, West Virginia: The Hambledon Press, 1989, pp. 211-220.

Nuremberg 1976-1977 Catalogue
L. Hennig, ed., *500 Jahre Regiomontan – 500 Jahre Astronomie*, Nuremberg: Germanisches Nationalmuseum [1976].

Nuremberg 1983 *Catalogue*
J. Willers and K. Holzamer, *Treasures of Astronomy: Arabic and German Instruments of the German National Museum*, Nuremberg: Germanisches Nationalmuseum, 1983 [In English, German, Arabic].

Nuremberg 1992-1993 *Catalogue*
Gerhard Bott and Johannes Willers, eds, *Focus Behaim-Globus*, Nuremberg: Germanisches Nationalmuseum, 1992.

Osborne, *Decorative Arts*
Harold Osborne, ed., *The Oxford Companion to the Decorative Arts*, Oxford: Clarendon Press, 1975.

Panicali, *Orologi*
Roberto Panicali, *Orologi e orologiai del Rinascimento Italiano: La scuola Urbinate*, Urbino: Quattro Venti, 1988. [In Italian, French, English].

Paris BN *Catalogue*
Bibliothèque Nationale: Nicolas Copernic ou La révolution astronomique, Paris: Bibliothèque Nationale, 1973.

Patterson, *Life*
Elizabeth C. Patterson, *Mary Somerville 1780-1872*, Oxford, 1979. [Booklet of 44 pp.].

Pilz, *Astronomie*
Kurt Pilz, *600 Jahre Astronomie in Nürnberg*, Nuremberg: Verlag Hans Carl, 1977.

Pollard, «Authenticity of Brass Objects»
A. M. Pollard, «Authenticity of Brass Objects by Major Element Analysis», unpublished proceedings of the 1984 Archaeometry Symposium.

Poulle, «Fabrication des Astrolabes»
Emmanuel Poulle, «La Fabrication des astrolabes au Moyen Age», *Techniques et Civilisations*, no. 22 (vol. 4, part. 4, 1955), pp. 117-124.

Poulle, «Instruments astronomiques»
Emmanuel Poulle, «Les Instruments astronomiques du Moyen Age», Museum of the History of Science, Oxford, Selected Off-prints no. 7 (1969); reprinted from *Le Ruban rouge*, 32 (March 1967).

Price, *Equatorie*
Derek J. Price, *The Equatorie of the Planetis: Edited from Peterhouse MS 75.1*, Cambridge: Cambridge University Press, 1955.

PRICE, «First Renaissance Instrument»
DEREK J. DE S. PRICE, «The First Scientific Instrument of the Renaissance», *Physis: Rivista di Storia della Scienza*, 1, pt. 1 (1959), pp. 26-30.

PRICE, *Checklist*
SHARON L. GIBBS, JANICE A. HENDERSON, and DEREK de SOLLA PRICE, *A Computerized Checklist of Astrolabes*, [New Haven, Conn.: Yale University], November 1973.

PRICE, «Review»
DEREK J. DE SOLLA PRICE, «Instruments in the British Museum» [review of London BM *Catalogue*], *Journal for the History of Astronomy*, 13 (1982), pp. 129-132.

ROCKFORD TM *Catalogue*
BRUCE CHANDLER, ed., *The Time Museum Catalogue of the Collection Vol. 1: Time Measuring Instruments, Part 1: Astrolabes, Astrolabe Related Instruments*, by A. J. Turner, Rockford, Illinois: The Time Museum, 1985.

ROSE, *Italian Mathematics*
PAUL LAWRENCE ROSE, *The Italian Renaissance of Mathematics: Studies of Humanists and Mathematicians from Petrarch to Galileo* (Travaux d'Humanisme et Renaissance CXLV), Geneva: Libraire Droz, 1975.

ROY (Paris)
JOSEPH ROY (pseudonym) «Sous le signe des étoiles», *L'Express*, no. 1997, 13-19 October, 1989.

Schweinfurt 1993-1994 *Catalogue*
UWE MÜLLER, ed., *450 Jahre Copernicus* De revolutionibus: *Astronomische und mathematische Bücher aus Schweinfurter Bibliotheken*, Schweinfurt: Stadtarchiv (Veröffentlichung Nr. 9), 1993.

STEINBERG, *Printing*
S. H. STEINBERG, *Five Hundred Years of Printing*, 3rd edn, Harmondsworth: Penguin Books, 1974.

Strasbourg 1987 *Catalogue*
FRANCIS DEBAUVAIS, *Catalogue des livres et objects scientifiques exposés à la Bibliothèque Nationale et Universitaire de Strasbourg, Section des Sciences et Techniques, à l'occasion du Congrès de la Société Française de Physique, du 6 au 10 juillet 1987*, Strasbourg, 1987.

Trento, *Astrolabio*
PAOLO TRENTO, *L'astrolabio, storia, funzioni, costruzione*, Rome: Biblioteca del Vascello, 1989.

A. J. TURNER, *Instruments*
ANTHONY J. TURNER, *Early Scientific Instruments: Europe 1400-1800*, London: Sotheby's Publications, 1987.

A. J. Turner, see also Rockford TM *Catalogue.*

G. L'E. Turner, *The Regiomontanus Astrolabe*
Gerard L'E. Turner, *The Regiomontanus Astrolabe dated 1462*, Supplement to *Fine Scientific and Philosophical Instruments*, 28 Setptember, 1989, London: Christie's South Kensington, 1989.

G. L'E. Turner, *Strumenti*
Gerard L'E. Turner, *Gli Strumenti*, vol. 1 of *Storia della Scienza*, general editor Paolo Galluzzi, Torino: Giulio Einaudi editore, 1991.

Ward, «Sundials»
F. A. B. Ward, «The Earliest String-Gnomon Sundials», *Antiquarian Horology*, June, 1976, pp. 800-801.

Washington NMAH *Catalogue*
Sharon Gibbs and George Saliba, *Planispheric Astrolabes from the National Museum of American History*, Washington, D.C.: Smithsonian Institution Press, 1984.

Washington NGA *Catalogue*
Jay A. Levenson, ed., *Circa 1492: Art in the Age of Exploration*, Washington, D.C.: National Gallery of Art, and New Haven: Yale University Press, 1991. [Exhibition at National Gallery of Art, 1992.]

Zinner, *Instrumente*
Ernst Zinner, *Deutsche und niederländische astronomische Instrumente des 11.-18. Jahrhunderts*, Munich: Verlag C.H. Beck, 2nd edn 1967; reprinted 1979.

Zinner, *Regiomontanus*
Ernst Zinner, *Leben und Wirken des Joh. Müller von Königsberg genannt Regiomontanus*, 2nd edn, Osnabrück: Otto Zeller, 1968; translated by Ezra Brown as *Regiomontanus: His Life and Work*, Amsterdam: North-Holland, 1990. [NB. References in the text are to the German edition.]

XII

An ordered list of European astrolabes to *ca.* 1500

This list is arranged chronologically by provenance. It dates from the early 1990s and has barely been modified since then. It was generated as the table of contents of the European astrolabe component of a catalogue of medieval instruments (to *ca.* 1500) in the making, which for various reasons was not completed. The German Research Foundation (DFG) generously funded the basic research for several years. This list corresponds to what was to be Section 6 of the catalogue, and references to individual instruments should be in the form '6.m.n', where 'm.n' is the number assigned in the list. (See my *In Synchrony with the Heavens*, XVIII, for a list of Islamic instruments from Sections 1–5.) Subsequent sections dealing with instruments other than astrolabes to *ca.* 1500 have been suppressed here. I present this list in the hope that it may be useful to future researchers.

Note: The symbol # precedes a unique number assigned to each instrument, a tradition started by Derek de Solla Price in his *Astrolabe Checklist* first circulated in 1973.

Abbreviations:

Antwerp NSM	National Maritime Museum (Het Steen)
Barcelona OF	formerly preserved in the Observatorio Fabra
Barcelona MM	Museu Marítim
Bernkastel-Kues NKS	Nikolaus-Kusanus-Stift
Boston, Mass., MFA	Museum of Fine Arts
Brescia MEC	Museo dell'Età cristiana, now Museo di Santa Giulia
Brugine CISST	Centro Internazionale di Storia dello Spazio
Cambridge CC	Caius College
Cambridge WMHS	Whipple Museum of the History of Science
Chicago, Ill. AP	Adler Planetarium & Astronomy Museum
Cracow CZ	Czartoryski Museum
Cracow JM	Jagiellonian Museum
Cologne SM	Stadtsmuseum
Darmstadt HLM	Hessisches Landesmuseum
Edinburgh RSM	Royal Scottish Museums
Florence MNB	Museo Nazionale Bargello
Florence MSS	Museo di Storia della Scienza, now Museo Galileo
Frankfurt am Main HM	Historisches Museum

Ghent BM	Bijloke Museum
Greenwich NMM	National Maritime Museum
Kassel HLM	Hessisches Landesmuseum
Leiden MB	Museum Boerhaave
Istanbul BTTM	Bilim ve Teknoloji ve Teknik Müzesi (Museum of Science, Technology and Technics)
Liège MVW	Musée de la vie wallonne
London BM	The British Museum
London NG	Nur Gallery = collection of Dr David Khalili
London RGS	Royal Geographical Society
London SA	Society of Antiquaries
London SM	Science Museum
London VA	Victoria and Albert Museum
Luneburg MFL	Museum Fürstenturm Lüneburg
Milan MCAA	Museo Civico d'Arti Applicati
Milan PA	Pinacoteca Ambrosiana
Milan TT	former collection of Dr. Tullio Tomba
Moulins (Allier) MD	former Musée départementale, now Musée Anne de Beaujeu
Munich BNM	Bayerisches Nationalmuseum
Munich DM	Deutsches Museum
New York N.Y. CU	Columbia University, Special Collections
Nuremberg GNM	Germanisches Nationalmuseum
Oldenzaal (NL) HM	Historisch Museum 'Het Palthe-Huis'
Oxford MC	Merton College
Oxford MHS	Museum of the History of Science
Oxford OC	Oriel College
Paris BNF	Bibliothèque nationale de France
Paris CNAM	Centre national des arts et métiers
Paris IMA	Institut du monde arabe
Pavia MSU	Museo per la Storia dell'Università di Pavia
PC	private collection
PLU	present location unknown
Prague NTM	National Technical Museum
Rockford, Ill. TM	formerly Time Museum, now dispersed
Rome OA	Osservatorio astronomico, now Museo Copernicano
Salem, Mass., PM	Peabody Museum
Salzburg CA	Museum Carolino-Augusteum
Salzburg PS	Museum des Stiftes St. Peters
Schweinfurt SM	Stadtsmuseum
Stockholm NM	Nordiska Museum
Turin MPMT	Museo Palazzo Madama de Torino
Strasbourg MH	Musée historique

Utrecht UM	University Museum
Washington, D.C., NMAH	National Museum of American History
Wroclau OBS	Observatory, now in Cracow JM

6. Early European astrolabes (to ca. 1500)

1. Miscellaneous early European astrolabes (I)

1 The earliest European astrolabe, from 10th-century Catalonia (#3042 – Paris IMA)

2 An astrolabe for the seven climates (#161 – London BM)

3 An Italian astrolabe (#3037 – PLU < Milan TT)

4 An unsigned, undated (?) astrolabe (#4504 – PLU)

5 A Spanish (?) astrolabe (for Saragossa and other latitudes, with later additions for Paris) (#191 – Oxford MHS)

6 An English or French astrolabe for latitudes between 24° and 60°, including for [Paris] and [London] (#300 – Oxford MHS)

7 A French astrolabe (#428 – Greenwich NMM)

8 An Italian (?) astrolabe with a zoomorphic rete (#2041 – Oxford MHS)

9 An Italian (?) astrolabe with plates for the climates (#166 – Oxford MHS)

10 An Italian (?) astrolabe for the climates (#167 – London BM)

11 A French astrolabe for the upper climates (#420 – Greenwich NMM)

12 An English astrolabe missing its rete (#303 – Oxford MHS < MC)

13 A mater and plates owned by Nikolaus of Cusa (#589 – Bernkastel-Kues NKS)

14 An astrolabe of uncertain provenance for latitudes between 16° and 52° (#558 – Nuremberg GNM)

15 An astrolabe dated 1374 (#3114 – PLU)

16 A medieval Italian astrolabe with a rete design incorporating a frame shaped like a bishop's mitre (#4556 – PLU < Christie's 1994)

4 *An ordered list of European astrolabes to ca. 1500*

2. Miscellaneous early European astrolabes (II)

1 A Catalan astrolabe with a V–shaped frame on the rete (#416 – Greenwich NMM)

2 An Italian astrolabe with a distinctive rectangular frame on the rete (#493 – Florence MSS)

3 A related astrolabe, signed 'KP' and made in Urbino (?) in 1462 (#4506 – Moulins MD – stolen)

4 An Italian (?) astrolabe (#441 – London SM)

5 An astrolabe with an equatorial ring on the rete (#567 – Cracow JM < Wroclau OBS)

6 An astrolabe with an equatorial ring on the rete, dated 1375 and signed by Petrus Raimondus (#3053 – Boston MFA)

7 A composite astrolabe with Hebrew inscriptions on the back of the mater (#621 – Munich DM)

8 An astrolabe with a latitude grid around the ecliptic on the rete and zodiacal figures on the back (#213 – PLU)

9 A 15th-century astrolabe from the Low Countries with a rete in an *art nouveau* style (#536 – Leiden MB)

10 An unsigned, undated astrolabe with an unusual rete (#3096 – Salem PM)

11 An Italian astrolabe with degenerate quatrefoil markings and a *shakkâziyya* grid (#2044 – Oxford MHS)

12 An Italian astrolabic plate for the second climate and a simplified rete combining north and south projections (#169 – Oxford MHS)

13 *deest*

14 A late Italian astrolabe copied from an earlier type (#4507 – Florence MNB)

15–16 Two related astrolabes, one Italian made to look like a Dutch one, and a Dutch copy

15 A composite Italian astrolabe with a dubious signature by Henricus de Hollandia (#548 – Nuremberg GNM)

16 An unsigned 15th-century (?) Dutch astrolabe in imitation of the style of the Nuremberg astrolabe 'signed' by Henricus de Hollandia (#566 – Prague NTM)

17 Two 'lost' astrolabes (#625 and #626 – PLU, formerly Salzburg PS)

3. Astrolabes with quatrefoil decoration (I)

1 An astrolabe with a rectangular frame inside the ecliptic on the rete and plates for latitudes down to that of [Palestine] (#162 – London SA)

2 A French astrolabe with gear mechanism (#198 – London SM)

3 An English astrolabe dated 1342 and signed by Blakeney (#292 – London BM)

4 The 'Great Sloane astrolabe' (#290 – London BM)

5 An Italian astrolabe with zoomorphic representations on the throne and back (#547 – Nuremberg GNM)

6 An Italian astrolabe (#4515 – PLU < Milan TT)

7 An Italian astrolabe dedicated to (Alfonsus ?) Bonhominus (#4516 – Milan MCAA)

8 An Italian astrolabe (#3203 – Turin MPMT)

9 An Italian astrolabe (#2043 – Oxford MHS)

10 A German astrolabe owned by Ludolf de Scicte of Einbeck *ca.* 1330 (#2027 – Cracow JM)

11 An unsigned rete (#294 – Oxford MHS)

12 An English astrolabe (#200 = #295 – Chicago AP)

13 An Anglo-French (?) astrolabe with an owner's mark in Hebrew script (#293 – London SM)

14 An Italian (?) astrolabe (#2062 – Pavia MSU)

4. Astrolabes with quatrefoil and trefoil decoration (II)

1 A French (?) astrolabe (#4751 – Rome OA – stolen)

2 An astrolabic plate for [Paris] (#190 – London VA)

3 An astrolabe with a replacement quatrefoil rete (#627 – Salzburg CA)

4 An astrolabe with a 'broken' half quatrefoil (#4752 – Cambridge WMHS)

5 A German astrolabe with a half quatrefoil on the solstitial bar of the rete (#461 – Ghent BM)

6 *An ordered list of European astrolabes to ca. 1500*

6 A 14th-century Picard astrolabe with a half–quatrefoil on the solstitial bar, featuring monastic cipher numerals (#202 – PC < London Christie's 1991)

7 A French or English astrolabe owned by John of London (?) (#3058 – Greenwich NMM)

8 A North Italian astrolabic plate with a single trefoil on the rete (#4753 – Milan MCAA)

9 A French or English astrolabe with half-quatrefoils on the supports of the equatorial bar, supposedly dated 1501 (#4517 – PLU)

10 A French (?) astrolabe (#2042 – Oxford MHS)

11 An English astrolabe (#4518 – PC)

12 An English astrolabe (#457 – Liège MVW)

13 A French (?) astrolabe (#546 – PC)

14 An English astrolabe (#4508 – Istanbul BTTM)

15 A 'cosmopolitan' Italian astrolabe with a replacement plate from a Byzantine astrolabe and additional markings in Hebrew (#4509 – PC)

16 An Italian (?) astrolabe (#4754 – Milan PA)

17 'Sir John Findlay's astrolabe' (#238 – PLU)

18 A debased trefoil Gothic astrolabe (#573 – PLU formerly Needham Collection)

19 A 14th-century Spanish quatrefoil astrolabe with scratchings in Hebrew and inscriptions in scholastic Latin and Arabic (#4560 – PLU)

5. Astrolabes with inscriptions in Judaeo–Arabic or Hebrew

1 An astrolabe with a rectangular frame inside the ecliptic on the rete, dating from *ca.* 1300 and with inscriptions in Judaeo-Arabic, including a corrupt poem on the nature of an astrolabe (#3915 – London NG)

2 An astrolabe with Hebrew inscriptions from the mid-14th century (#158 – London BM)

3 A 15th-century astrolabe with Hebrew inscriptions (#3906 – Portugal PC)

4 A 15th-century astrolabe with Hebrew inscriptions on the back (#621 – Munich DM)

5 A 16th-century astrolabe with quatrefoil markings on the rete and Hebrew inscriptions (#159 – Chicago AP)

6 Miscellaneous bits and pieces with Hebrew inscriptions and instruments made by Jewish craftsmen
a) A rete with messy Hebrew inscriptions (#2572 – Washington NMAH)
b) A Maghribi astrolabe with Arabic inscriptions made by a Jewish craftsman (#1122 – PC)
c) The back of a 14th–century 'cosmopolitan' astrolabe (#4520 – PC)
d) An owner's mark in Hebrew script (#293)

6. English astrolabes with a Y-shaped frame on the rete in the tradition of Geoffrey Chaucer

1 An astrolabe for Oxford or London (#301 – Cambridge CC)

2 An Anglo-French astrolabe for Montpellier, Oxford and elsewhere, dated 1326 (#291 – London BM)

3 'The Painswick astrolabe' for localities between Toledo and [Thule] (#299 – Oxford MHS)

4 An astrolabic plate for Oxford fitted with an equatorium (#297 – Oxford MC)

5 An astrolabic plate for Oxford with a universal horary quadrant (#296 – Oxford OC)

6 An astrolabe for localities from Dover to Berwick (#298 – London BM)

7 'Parnel's astrolabe' (# 304 – Washington NMAH)

8 An English astrolabe (#4755 – Oxford MHS)

9 An astrolabe with plates for Paris, Cologne and London (#4521 – Milan TT)

10 An English zoomorphic astrolabe (#2006 – Washington NMAH)

7. Early French and Italian astrolabes with an upper frame on the rete

1 An early Italian astrolabe with a bowl-shaped frame (#4756 – Milan PA)

2 A Franco-Italian astrolabe for the climates (#4512 – PLU < Rockford TM)

3 A Franco-Italian astrolabe (#168 – Oxford MHS)

8 *An ordered list of European astrolabes to ca. 1500*

4 A Franco-Italian (??) astrolabe similar to the previous one (#410 – Oxford MHS)

5 A Franco-Italian astrolabe (#4522 – PLU)

6 An astrolabe similar to the previous one (#609 – Luneburg MFL)

7 An astrolabe with an unfinished rete (#4757 – Rome OA – stolen?)

8 An Italian astrolabe dated 1420 (#4523 – PC)

9 A French astrolabe for Nevers (#4524 – Frankfurt HM)

10 An Italian astrolabe with an upper equatorial frame (#186 – Washington NMAH)

11 A German astrolabic plate with a tulip-shaped frame, dated 1468 and serving a latitude in N. Italy, and said to have been owned by Regiomontanus (#550 – Nuremberg GNM)

12 An unsigned Italian astrolabe / astrolabic plate with a thistle-shaped frame on the rete, dated 1500 (#4758 – Milan PA)

13 An Italian astrological plate for the latitude of [Palestine] (#170 – Oxford MHS)

14 A late Italian astrolabe with a bowl-shaped upper frame on the rete (#576 – Cambridge WMHS)

15 A late Italian astrolabe with an upper frame on the rete (#247 – London SM)

16 An Italian astrolabe with two short upper bars (#479 – Florence MSS)

17 An Italian astrolabe from ca. 1500 (#174 – London RGS)

18 An astrolabe of uncertain provenance (#431 – Greenwich NMM)

19 An astrolabe from the Low Countries or from Germany influenced by the 'Fusoris' tradition but with a prominent bowl-shaped frame on the rete (#4525 – Munich DM)

20 An anomalous late-15th-century German astrolabe exhibiting French influence (#601 – Cologne SM)

8. 15th-century French astrolabes in the tradition of Jean Fusoris

1 The 'Thornoe' astrolabe (#337 – Greenwich NMM)

2 An astrolabe related to the Greenwich Thornoe one (#476 – Barcelona OF)

3 A French (?)/ Anglo-French (?) astrolabe (#163 – Oxford MHS)

4 A 'Fusoris-type' astrolabe without an upper frame on the rete (#598 – Kassel HLM)

5 A simplified 'Fusoris-type' astrolabe (#4526 – PC)

6–20 'Fusoris-type' astrolabes

6 A 'Fusoris-type' astrolabe (#196 – Oxford MHS)

7 A 'Fusoris-type' astrolabe (#194 – Oxford MHS)

8 A 'Fusoris-type' astrolabe, maybe slightly later than the rest (#192 – Oxford MHS)

9 The 'Fusoris-type' astrolabe of Robert Gordon of Straloch (#3082 – Edinburgh RSM)

10 A 'Fusoris-type' astrolabe (#3080 – Paris CNAM)

11 A 'Fusoris-type' astrolabe with rete and plates reworked by an Egyptian or a Syrian in the nineteenth (?) century (#460 – Antwerp NMM)

12 'Sir John Findlay's 'Fusoris-type' astrolabe' (#193 – Chicago AP)

13 A 'Fusoris-type' astrolabe (#199 – Chicago AP)

14 A 'Fusoris-type' astrolabe (#195 – London BM)

15 A 'Fusoris-type' astrolabe (#568 – Cracow JM < Wroclaw OBS)

16 A 'Fusoris-type' astrolabe (#4527 – Stockholm NM)

17 A simplified 'Fusoris-type' astrolabe (#193B = #3070 – PLU ???)

18 A 'Fusoris-type' astrolabe (?) (#3083 – PLU)

19 A 'Fusoris-type' astrolabe (?) (#3084 – Barcelona MM)

20 An astrolabe in the 'Fusoris' tradition (#518 – Paris BNF)

21 Two imitation 'Fusoris' astrolabes
a) Undated (#407 – Cambridge WMHS)
b) An early copy of a 'Fusoris-type' astrolabe dated 1447 or 1487 (#623 = #3208 – Munich BNM)

22 An astrolabe in the 'Fusoris' tradition signed by Hooft (#197 – Cambridge CC)

10 *An ordered list of European astrolabes to ca. 1500*

23 An astrolabe in the 'Fusoris' tradition (#4528 – Brescia MEC)

24 An unsigned imitation 'Fusoris-type' astrolabe (#3081 – PLU)

25 Another imitation 'Fusoris-type' astrolabe (#4529 – Brugine CISST)

26 Three 16th-century Dutch astrolabes, imitations of 'Fusoris-type' astrolabes, signed by Johannes Bos
a) Undated (#539 – Utrecht UM)
b) Dated 1591 (#430 – Greenwich NMM)
c) Undated (#534 – Leiden MB)

9. 15th-century German astrolabes in the tradition of Regiomontanus and Hans Dorn

1–5 Some mid-15th-century German astrolabes with similar ('Regiomontanus-type') retes

1 Two astrolabes by the same maker (?) with a lunar volvelle on the back
a) Dated 1457 (#549 – Nuremberg GNM)
b) Undated (#452 – PC)

2–3 Two astrolabes with an organum Ptolemaei on the back

2 An unsigned astrolabe with later additions on the blank side of the plate of the organum (#256 – PC)

3 An astrolabe signed by Ioannes (Regiomontanus) and dedicated to Cardinal Bessarion in 1462 (#640 – PC, England < London Christie's 1989)

4 Three unsigned astrolabes with similar basic features and inscriptions in Gothic script
a) Dated 1466 (#592 – Darmstadt HLM)
b) Dated 1481 (#530 – Strasbourg MH)
c) Undated (#620 – Munich DM)

5 Two astrolabes with unusual plates and three different scripts
a) An astrolabe with plates for tropical latitudes (#516 – Paris CNAM)
b) An astrolabe with universal plates (#545 – New York CU)

6–8 Some related astrolabes, some signed by Johannes / Hans Bamman of Schmalkalden

6 An astrolabe with a 'Regiomontanus-type' rete dated 1482 (#4501 – Cracow CZ)

7 Some simplified astrolabes
a) Dated 1484 (#633 – Schweinfurt SM)
b) Dated 1487 (#604 – PLU < lost from Schloß Kreuzenstein)
c) A mater dated 1490 (#248 – Oxford MHS)
d) Dated 1491 (#249 – Munich BNM)
e) A mater dated 1491 (#540 – Oldenzaal HM)

8 An astrolabe with an unusual rete signed 'hb' and dated 1493 (#4502 – PLU < Milan TT)

9–13 Some instruments signed by Hans Dorn and some others unsigned but probably also due to him

9 An *organum Ptolemaei* on the back of an astrolabic plate attached to a celestial sphere signed by Hans Dorn in 1480 (#4510 – Cracow JM)

10–12 Three astrolabes representing the generation after Regiomontanus, with a single half quatrefoil and a rectangular frame developed from the 'Regiomontanus-type' rete

10 An astrolabe owned by (?) Ioannes Broscius (#643 or #644 – Cracow JM)

11 An unsigned astrolabe dated 1483 and fitted with an *organum Ptolemaei* and an eclipse calculator (#492 – Florence MSS)

12 An unsigned astrolabe owned by Martin Bylica (#642 – Cracow JL)

13 A compendium with an astrological plate signed by Hans Dorn in 1491 (#250 – London BM)

14 An unsigned, undated astrological plate and an equatorium (#4511 – Milan PA)

15 An astrolabe on paper and vellum by Hanns Herghamer dated 1492 (#403 – Oxford MHS)

INDEX

Note: The entries for 'astrolabe', which are admittedly but inevitably inadequate, have been arranged more or less chronologically under Byzantine, Islamic, and European.